T0297894

The Zodiac

Myths And Legends Of The Stars

Richard Hall

ISBN: Softcover 978-1-4653-9887-1
EBook 978-1-4691-1000-4

Print information available on the last page

Rev. date: 05/24/2019

To order additional copies of this book, contact:
Xlibris
0800-443-678
www.xlibris.co.nz
Orders@ Xlibris.co.nz

Front Cover:
Background Eta Carinae Nebula, European Space Agency; Lower left: Mithrus; Lower right: Cybele, Fuente de la Cibeles- the Cybele Fountain, Madrid, Spain. R. Hall

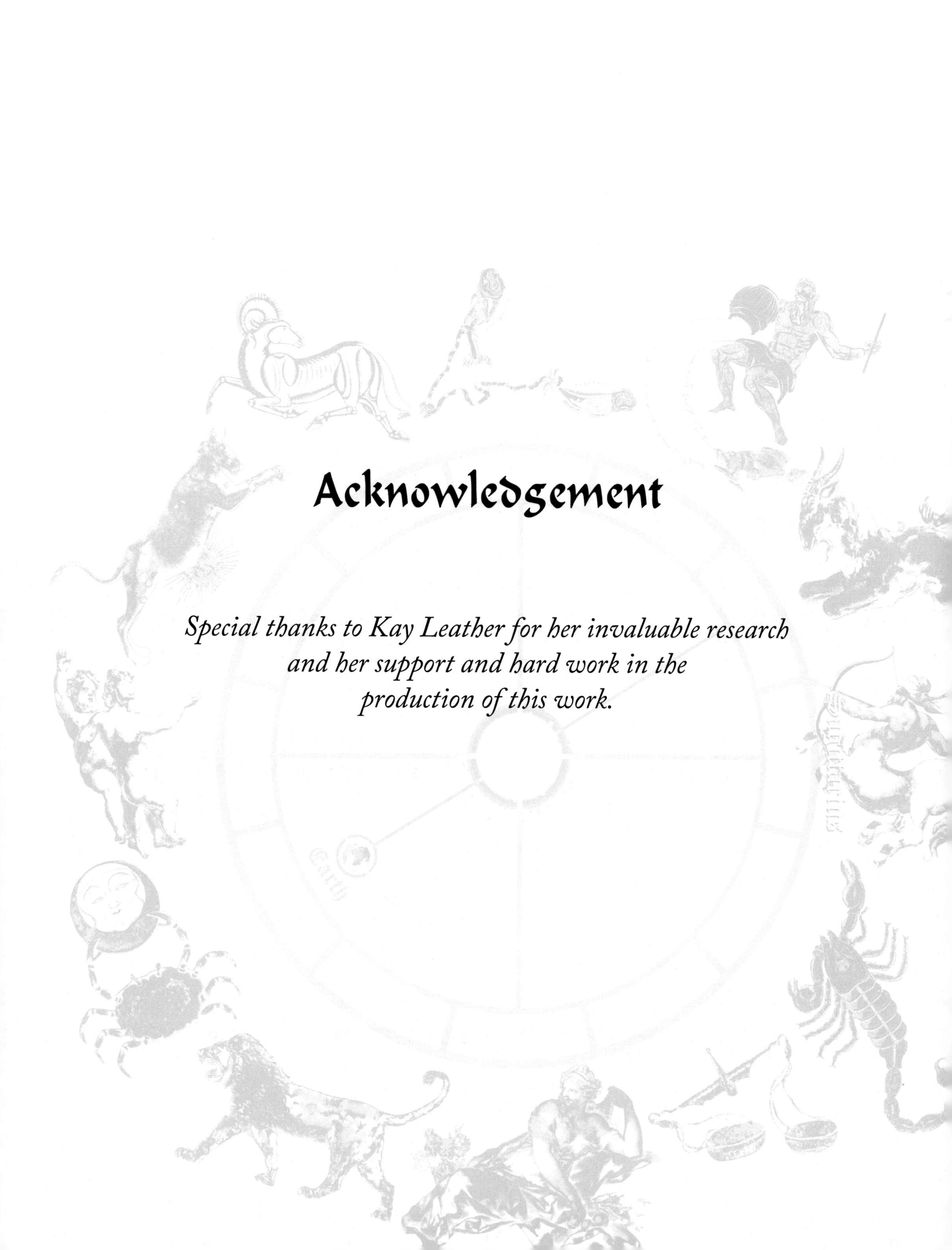

Acknowledgement

Special thanks to Kay Leather for her invaluable research and her support and hard work in the production of this work.

Contents

Introduction

Fig. 1: *Analemma at Stonehenge Aotearoa.* R.Hall

On a hill overlooking the Ruamahunga Valley stands Stonehenge Aotearoa.[1] Near the centre of the stone circle is a five metre tall needle of stone called an obelisk[2]. The obelisk does something very simple – it casts a shadow on the ground. Now, you may think there is nothing special in a shadow. But surprisingly, as our ancestors

1 Stonehenge Aotearoa is a full scale working adaptation of ancient stone circles from around the world. It is located near the town of Carterton in the Wairarapa, Lower North Island of New Zealand.

2 An obelisk is a tall four sided column of stone. In ancient Egypt they were placed near a temple. Dedicated to Re, the Egyptian sun-god, they were believed to be a protective, petrified solar ray.

discovered, there are amazing things that you can discover about the world and beyond using nothing more complicated than a stone or post and the shadow it casts on the ground. At noon on a sunny day the obelisk at Stonehenge Aotearoa casts a shadow down onto a 10 metre long tiled area that has lines and a curious figure of eight pattern, called an analemma (Figure 1). As the point of the shadow falls onto the analemma it tells us the date and the true length of the day (which varies throughout the year). But it also tells us something we cannot see with our eyes – it shows us where the sun is amongst the stars.

You can't see the stars during the daytime but, if you were in space, just above the Earth's atmosphere, you would see our sun against the blackness of space sown with stars. And, over a year, as the Earth orbits around the sun, it would appear from your location as if you were standing still and that the Sun was moving in a great circle around the Earth. The background stars along this path of the sun, form the constellations of the Zodiac. This is the significance of the Zodiac. It encircles the plane of the Solar System and is the path in the sky of the sun, moon and planets. Although you cannot see where the sun is amongst the stars from the surface of the Earth the shadow of the obelisk shows you exactly where it is.

If people know anything about the stars it is their astrological star sign. And, traditionally, your star sign is the constellation the sun was in at the time of your birth. If for example if your star sign is Gemini this means that the sun - the life giving powers of the sun - was in Gemini when you were born.

When visitors at Stonehenge Aotearoa hear that they are not the star sign that they have been led to believe in newspapers and magazines they surge forward and stare at the analemma. I, for example, was born on the 3rd of November which, if I look it up in the newspaper, tells me that I'm a Scorpio. But, the shadow cast by the obelisk tells a different story – the sun was is in Libra when I was born. I'm a Libran! How can this be so? Well, one question leads to another and, to fully explain this question and others about the Zodiac I ended up by writing this book.

The heavens above have always been for people both a spiritual realm and a physical reality. Consequently ancient astronomy is interwoven with spiritualism and religious beliefs. The story of the Zodiac is then, more than just astronomy and astrology. It is also the story of how human beings attempted to understand the complex universe around them and their place and significance within it. It is the story of how our ancestors tried to gain some measure of control over their own destiny both through science and mysticism. And it is the story of the origins of our spiritual beliefs, and many of the traditions and myths we live by.

Where necessary, to aid the understanding of the non-astronomer, I have provided explanations of basic astronomical phenomena. I have also included tables with additional information for those who wish to study the subject in more detail. Unless otherwise stated, when discussing the seasons, solstices and equinoxes, I am referring to those of the Northern Hemisphere where most of the historical drama and traditions relating to the Zodiac arose.

Throughout the book I have used B.C. and A.D. for dating rather than B.C.E. and C.E. because the former are more widely understood by the general public. The term 'ancestors' used on its own applies to all peoples and the term 'early ancestors' refers to people who lived at a time before the rise of civilization, 8,000 B.C. or earlier.

The story of the Zodiac is complex. Myths and meanings come from many different cultures which, over time, and have become intertwined. Consequently, there is no simple pathway to take when telling this story. For this reason I have divided the book into six main parts. Each part focuses on a different theme relating to the zodiac but each part connects and relates to the others. Where possible I have placed the chapters in historical sequence. The six parts are as follows:

Part I: The Cosmic Wheel provides the basic framework around which this grand story of ancient astronomy and star lore is built. We discuss what the Zodiac is and, how and why astronomy and star-lore became a cornerstone to the rise of civilization. This section also provides basic information for the non-astronomer on how it all works - the sky, the seasons, and how they change with the passage of time.

Part II: Caves, Pillars & Gates traces the origin of the zodiac. We discuss early beliefs on the nature of the cosmos and the significance of the zodiacal signs, solstices and equinoxes to our traditions and spiritual beliefs.

Part III: The Great Festivals. In this section we discuss the myths and symbolism built around the equinoxes and solstices and their associated zodiacal signs. Here we find the origins of some of the great stories of antiquity that became incorporated into world religions

Part IV: The Mansions of the Moon, discusses the Lunar Zodiac and its symbolism, from which the Solar Zodiac, the calendars, and our systems of time-keeping emerged. We also look at how people, thousands of years ago, made amazing discoveries about our world and the heavens above using nothing more complicated than stone circles, posts and shadows.

Part V: The Planetary Powers. The central theme in this section is the origin of the horoscope and the historical significance of natal astrology. It begins by looking at ancient cosmologies - early concepts on the nature and origin of the universe - from which astrology and ultimately the sciences emerged. Finally it explores some of the ways in which star lore and astrology, particularly the planetary powers, played a major role in the foundation of Christianity.

Part VI: The Signs and Stars summarizes and brings together information presented earlier in the book on each individual sign. It also provides additional historical, astronomical, astrological and mythological information and data on both the individual signs of the Zodiac and their brighter stars.

The reader should know that I have approached and researched this subject from a science and anthropological perspective. It is not however, anti-astrology or anti-Christian or anti anything else. I have tried to present the stories and beliefs from different religions/cultures with equal weight within the context of historical facts. My own interpretations and conclusions are clearly identified. I am fortunate to live in a country where one can question anything. I say this in the realization that had I written this book a few centuries ago I would probably have been burnt at the stake.

PART 1:

THE COSMIC WHEEL

Long before the rise of civilization, perhaps hundreds of thousands of years ago, our ancestors looked upward at the night sky and wondered, and pondered the question that has been asked down through the ages – 'What are the stars?' These mysterious lanterns of the night stimulated their imagination. The individual stars and the patterns they formed had permanence and became familiar companions of the night. By watching the stars they began to discover connections between the earth and the sky which provided them with important information. This knowledge that they gained from the stars would form a cornerstone to the rise of civilization.

This first part of the book provides the basic framework around which the grand story of ancient astronomy, the zodiac and star lore is built. It also provides basic information for the non-astronomer on how it all works - the sky, the seasons, and how they change with the passage of time.

1. Astronomy, Astrology and Mythology

Do you believe that the sciences would ever have arisen and become great if there had not been before magicians, alchemists, astrologers and wizards who thirsted and hungered after hidden, forbidden powers?

Nietzsche

The Joyful Science, 1886

Long, long ago people gave names to the stars and began ordering them into groups which we call constellations. Many of the names of stars in use today date back more than 2,500 years; some constellations may date back to the very dawn of civilization. These names had meaning and behind each is a story. To those who can read it, the night sky is a picture book of stories from antiquity. Here are the stories of the gods, heroes and mighty events. These tales are a window into the past. They give us insight to the beliefs of our ancestors and provide clues to the origin of both our science and religion.

The most well-known constellations are those of the Zodiac. But what exactly is the Zodiac? To most astronomers/scientists today the Zodiac is just an ancient name given to the constellations that lay along the plane of the Solar System. To the general public the Zodiac is all about astrology, the mystic influence of the stars. Astrology crosses cultural and religious boundaries and is more widely accepted than any other belief system in the world. In India for example, most people wouldn't consider getting married without first consulting an astrologer. Mysticism and pseudo-science is gaining ground in the Western world and a sizeable percentage of the population has a degree of faith in astrology. Astrological star signs and horoscopes can be found in every newspaper and women's magazine.

Today astronomers map the heavens by dividing the entire sky into 88 areas and the stars within each area form a constellation. The traditional Zodiac that most people in the western world are familiar consists of twelve constellations or signs that form a band around the sky. These are:

Aries *the Ram*
Taurus *the Bull*
Gemini *the Twins*
Cancer *the Crab*
Leo *the Lion*
Virgo *the Virgin*
Libra *the Scales*
Scorpio *the Scorpion*
Sagittarius *the Archer*
Capricorn *the Sea-goat*
Aquarius *the Water-carrier*
Pisces *the Fishes*

***Fig 2:** The ZodiacWheel.* R.Hall

The Zodiac Wheel in figure 2 places the twelve constellations, which are represented by a figure and a symbol, in the order in which they are found in the sky. The sun moves through these signs in an anti-clockwise direction. It completes a circuit of the Zodiac, passing through each of the constellations, in a period we call a year. The word year is ancient and is derived from the Persian "yare", which signifies a circle.

All of the Zodiacal constellations were named by ancient astronomer/astrologers. So first I had better explain something that is often confused by the general public, the difference between astronomy and astrology.

Astronomy is the scientific study of the physical universe while astrology is the study of the supernatural influence of the stars and planets. It should be noted that the differentiation between astronomy and astrology is of relatively recent origin, three or four centuries at the most. Astronomy, the oldest of all sciences, was from the earliest of times intertwined with spiritual beliefs. For most of human history astronomy and astrology were part and parcel of the same thing – the study of the heavens. All the great astronomers of antiquity, including Galileo and Kepler, were also astrologers.

Many of the astronomers I know are offended when called an astrologer because, as scientists, they use the scientific method, not the supernatural, to explain phenomena within the universe. However, many astrologers call astrology a science. This is undoubtedly intended to give astrology a seal of established authority, a discipline that has been tested and is based upon facts. But, unlike geology, chemistry or physics, very little of it can stand up to scientific analysis. There are no departments of astrology in our universities.

The above statement is not intended to demean astrology. Because astrology deals with the supernatural, it is more akin to religion and other spiritual beliefs. Indeed, much of it originated from astrologer-priests of ancient religions. Whether you believe in it or not is a matter of faith. A Catholic Priest said to a friend of mine, "If you could prove God existed you wouldn't need Faith". How true. Most people, from my experience, have some form of belief in the supernatural. Very few people claim that they are atheists. And incidentally, several of those astronomer friends of mine who get irritated by mysticism and astrology, turn up to church on a Sunday. I have always found it intriguing how it was acceptable to be an astronomer (a scientist) and also believe in God, but not acceptable to be an astronomer and an astrologer. I guess it all comes down to what the individual classifies as a truth or a myth. That brings us to another question, what is mythology?

A myth is a traditional story involving supernatural or fantastic beings which embodies popular ideas on natural or social phenomena. Accordingly the Bible is a collection of myths. However, today most people use the word myth to label a story as fiction rather than a truth. Is the Bible mythology? Most Christians would say no, non-believers would say yes. Thus mythology has become a term we apply to the beliefs of other people that differ from our own.

Myths are extremely important because they give us insight into the beliefs, values and knowledge of ancient cultures. Before the time of the written word information was stored and passed from generation to generation in stories, poetry and song. Natural phenomena, the forces of nature often became characters in these stories. The adventures and interactions of the characters provided practical information on everything from etiquette and moral values to navigation and seasonal events.

The reality is, whether you believe it is a truth or myth, astrology has played a major role in the shaping of history. It is indeed responsible for the death of princes and fall of kingdoms. The signs of the Zodiac ruled entire nations. In the ancient world each geographical area was ruled by a different sign of the Zodiac. Pisces for example, represented Israel and Taurus ruled Arabia. A celestial event such as a conjunction of planets or the occurrence of an eclipse in a particular Zodiacal constellation was believed to have special significance to the people that lived in the area ruled by that sign.

The Zodiac and its associated star lore form a corner stone of religions around the world, including Christianity. If you doubt this think back to Christmas. I bet some of those Christmas cards you received had a star on them (and yes, it was a star in the Zodiac). The belief was that nothing just happened. For any important event there would be a sign in the sky.

How did these ideas emerge and why have the stars played such an important role in the development of civilization? To answer these questions we must take a journey back in time and explore the development of human thought on the nature of the cosmos.

2. Signs and Symbols in the Sky

And God said, "Let there be lights in the firmament of the heavens to separate the day from the night; and let them be for signs and for the seasons and for days and years, and let them be lights in the firmament of the heavens to give light upon the earth"

Genesis 1-14

We live in a world in which we are surrounded by symbolism. It is a form of hieroglyphic imagery that, if you know what the symbol means, conveys information. This could be as simple as proclaiming ownership or giving instructions, such as a stop sign. Some symbols are far more complex and may contain hidden meanings. Whether or not you know what a symbol means often depends on where you come from or the social group to which you belong. Most New Zealanders', for example, would know the meaning of, and claim ownership, of the 'Silver Fern'. However, someone from another country that was not interested in sport probably wouldn't have a clue what it meant.

There are many symbols with which we are familiar, we see them, but we do not know what they mean. Sometimes this is because the symbol has a great antiquity and its meaning has been lost to most people with the passage of time. The Zodiac is full of ancient symbolism and, the meanings of this symbolism are fundamental to what was known as the 'divine art' of astrology and underpin beliefs that gave rise to major religions.

The most common divine symbol, particularly among the oldest of religions, is the Sun.

"The sun is the most splendid and glorious object in nature. The regularity of its course knows no change. It is "the same yesterday, today, and forever." It is the physical and magnetic source of all life and motion. Its light is a type of eternal truth; its warmth of universal benevolence. It is therefore not strange that man in all ages has selected the sun as the highest and most perfect emblem of God."

Stellar Theology and Masonic Astronomy

Robert Hewitt Brown

1882

Historical records or misinterpretation have often confounded the symbol with the deity. Consequently, an ancient culture may be described as sun-worshipers when in fact the Sun was the symbol of the god, not the god himself (or herself). For example, the Hindu/Persian god Mithra is often described as a sun-god. He was in fact 'the Son of Light', and his emblem was the 'Sun of God'.

The same reasoning should be applied to the Moon, planets and stars. In understanding ancient beliefs they should be seen as symbols and signs of the gods, not the gods themselves.

In ancient astrology the Sun is paramount because it is the symbol of god, or the supreme god. As it made its way through each individual sign of the Zodiac it was said to assume the nature of or, triumph over that particular sign. Thus, when the Sun entered Aquarius it brought the rains. When the Sun entered Capricornus (which 2,500 years ago contained the winter solstice) the Sun triumphed over darkness and, as it emerged from that sign, its warmth and light grew in strength.

When the Sun traveled through a Zodiacal sign it was believed that the power and influence of that sign (which some people today describe as energy) was greatly enhanced and became dominant. This is why the location of the Sun at the time of one's birth determines a person's natal or birth sign.

For thousands of years the Zodiac and astronomy/astrology were of great significance to the civilizations of Europe and Asia. To understand the reason for this and the meaning of the stories about the stars we need to be able to see the world through the eyes of people who lived long ago.

Let's start by piecing together the nature of the universe as understood by our distant ancestors. As we do this it should be borne in mind that these people had a very holistic view of life and the world around them. Today we divide knowledge up into different categories. We have religion and science, and science itself is divided up into different disciplines. People didn't think like this long ago. Everything, science and religion, was interwoven.

Undoubtedly the first thing the stars were used for was navigation. By navigation I don't mean just sailing around in a boat - I mean traveling any great distance. There was a time when there were no roads and no maps and, when someone traveled somewhere they had to get there and back on their own. It is interesting to note that the first trade routes were established 40,000 years ago. That's 30,000 years before the appearance of the first town, which as far as we know is Jericho in the Jordan Valley. Using the stars our ancestors navigated their way across the greatest oceans on earth, across flat featureless plains, and across the greatest deserts – the Sahara, the Kalahari and Gobi deserts. Today, few people would know how to navigate by the stars. But I suggest that once, this was must-know knowledge for every individual.

Of equal, if not of greater importance was time-keeping. We don't tend to think of time keeping as vital information but in reality our society depends upon it and would collapse into chaos without it. Think about it – what would it be like if you didn't have a watch, clock or a calendar? You wouldn't even know when it was your own birthday!

For our distant ancestors, the hunter-gatherers, time keeping wasn't a matter of convenience, it was a strategy for survival. In the continental regions of the northern hemisphere, where most of the early human drama was played out, they experienced extreme changes in climate between winter and summer. Food supplies in any one area varied with the seasons. The wild-life, particularly the big game would migrate backwards and forwards with the coming and going of the seasons. We always use to say that the hunter-gatherer followed the game. But, that's not true. The hunter gatherer preceded the game.

This all makes good sense if you have ever witnessed the seasonal migrations of wild-life. The first time I experienced this was when I was at school as a small child. This was in England, the opposite side of the world to where I now live. Outside of our classroom was a large grove of trees and, one morning, we noticed that there were a lot of birds in these trees. We noticed them because they were making a lot of noise. As time went on more and more birds arrived. It was bit like one of those scenes from Alfred Hitchcock's movie, 'The Birds'. By about ten in the morning you couldn't see the trees for the birds. And then, as if someone had given a signal, the whole lot flew up into the air. I remember all of us running out into the playground. Above was a vast cloud of birds. And then, off they went, and for some time we stood and watched as the great dark cloud of birds slowly disappeared into the distance. It turned out that these birds were migrating to Africa for the winter. Trouble is, if I were a hunter-gatherer and these birds were my tucker, I would be standing there watching my food supplies disappear at a great rate of knots.

The lesson is this, when animals migrate they do so on mass and they move a lot faster than human beings can on foot. So it made good sense that the hunter-gatherer, to ensure that food supplies did not run out, would need to move before the migrations began. In a like manner, if they had traveled far to the north for the short but very fertile summer but left their return too late, the snows would come down and rivers would rise, cutting off their retreat. They would perish in the bitter northern hemisphere winter. How did they know when to move? When considering this keep in mind that they would probably need to move well before there were signs in the trees around them that the season was about to change. The answer is that did so by observing the stars. We know this from living cultures such as the Sioux that lived on the North American plains and depended upon the buffalo. They had their summer camps and their winter camps and the time to up-camp and move was written in the heavens.

When the first farming settlements appeared, the oldest known being found in the Jordan Valley that date back 19,400 years, having some form of calendar was essential. Plant at the wrong time of the year and your crop could fail, and your family would be facing starvation. The harvest was all important, and to this day people in the Andes still plant by the stars.

Now, if you work on the land you may think that you could foretell seasonal events simply by observing the weather conditions, plants and animals. But, you still have a modern calendar in your head. Even if you don't know the day you will know the month you are in and would judge events that you observe accordingly. Unless you are counting days, or observing the sun, moon or stars you would soon loose track of the cycle of the seasons, particularly at the times of El Nino and La Nina.[3]

Long ago, tens of thousands of years before the dawn of civilization, people discovered a connection between the stars and what we now call seasonal events. A bright star would rise up in the dawn twilight and something in the environment would begin to change. A rising star was a portent in the sky – it foretold of a coming event.

Why a rising star? And, how are the stars related to seasonal changes? First I had better explain something the ancients didn't understand – the cause of the seasons.

3 El Nino and La Nina is an oscillation in the surface water temperature in the tropical eastern Pacific Ocean and changes in surface atmospheric pressure in the tropical western Pacific. It generates significant changes in weather patterns and is associated with floods and droughts. These oscillations occur at irregular intervals of between 2 and 7 years and can last from 9 to 24 months.

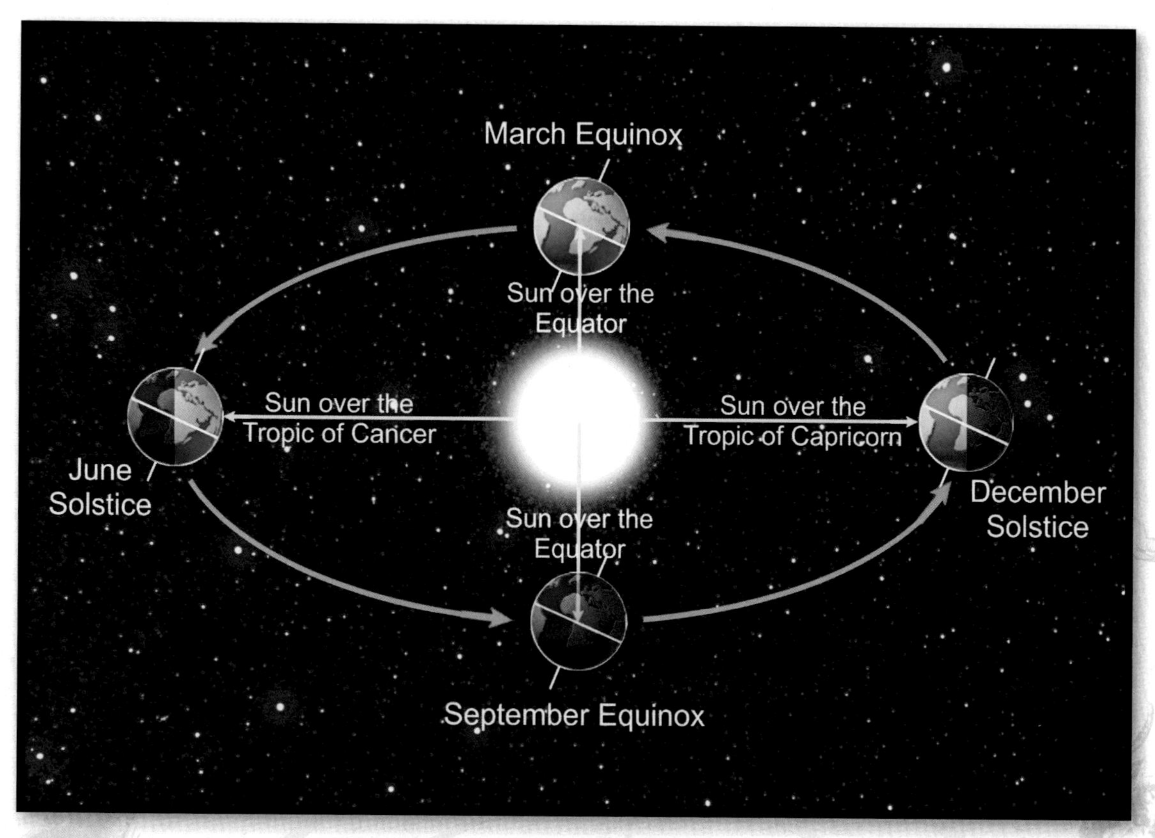

Fig. 3: *The seasons are due to the tilt of the Earth's axis.* R.Hall

We experience seasons because the Earth's axis of rotation is tiled 23.5 degrees to the plane of its orbit around the Sun. It maintains this tilt with the poles of the Earth aligned towards the same point in the sky as it orbits around the sun. The tilt of the earth in respect of the Sun is shown in figure 3. We see that in December the southern hemisphere is tilted towards the sun. Because of this, in the southern hemisphere, the sun is high in the sky at midday and is above the horizon for more than 12 hours. It is summer in the southern hemisphere. The opposite applies to the northern hemisphere which is tilted away from the Sun. It is winter and the sun is low in the sky at midday and above the horizon for less than 12 hours. Six months later, in June, the situation is reversed. Now it is the northern hemisphere that is tilted towards the sun. It is summer in the northern hemisphere and winter in the southern hemisphere. The two points at which one hemisphere is at maximum tilt towards the sun are called the solstices. In our present calendar these occur each year around the 21st of June and December.

There are two midway points where the sun is over the equator and neither hemisphere is directly pointed towards the Sun. These are the two equinoxes that occur around March 21st and September 22nd. The March equinox brings spring to the northern hemisphere, autumn in the south. The equinox in September brings autumn to the northern hemisphere, spring in the south.

As the Earth orbits around the Sun the backdrop of stars that we see at night slowly changes. Because the solstices and equinoxes occur at the same points in the Earth's orbit each year, each season has a different backdrop of stars. For example, when the Earth is located at the December Solstice the stars in the opposite direction to the Sun are those that will be seen at night. These will be summer stars for those in the southern hemisphere, winter stars for those in the north. Six months later, at the June solstice, the Earth will be on the opposite side of the Sun and we will look out on a different panorama of stars. Most of the stars that we could see at night in December will now be in our daytime sky and no longer visible. Thus, each season has its own pageant stars.

As the Earth turns on its axis from west to east celestial objects, the sun, moon and stars, appear to move from east to west. Superimposed upon this daily motion is the slow seasonal change due to the orbital motion of the Earth. This can be observed from the surface of the Earth as a steady change in the rise and set times of the stars. Each night any given star will rise or set 4 minutes earlier than the night before. Over time the entire panorama of the heavens is observed to slowly drift westward.

Our distant ancestors were not aware that the Earth was rotating or that it was orbiting around the Sun. It appeared to them that it was Sun, Moon and stars that were moving. Consequently they had a different explanation of what was happening. If you watch a particular star over the months its westward drift will eventually take it into the western evening twilight where it is eventually lost entirely from view. Today we know that the star is still there but in our daytime sky, on the far side of the Sun. But, in ancient times many people believed that the star had been consumed by the eternal fires of the Sun. A couple of weeks later the same star will reappear but this time in the dawn twilight – rising up just before the Sun. This, the first re-appearance of a star following its passage behind the Sun, is known as its 'heliacal rising'. The star had reappeared, it was believed, because it had been resurrected from the eternal fires of the Sun. Here in the heavens they saw a cycle of birth, life, death and the resurrection.

Certain stars were seen as heralds or portents of seasonal changes and events. These portents were delivered at their heliacal rising, as they were sent forth, reborn from the eternal fires of the Sun (the symbol of God). Thus, the shaman or wise man or woman who studied the stars to foretell the future, was looking not at the evening stars but those which were rising just before dawn.

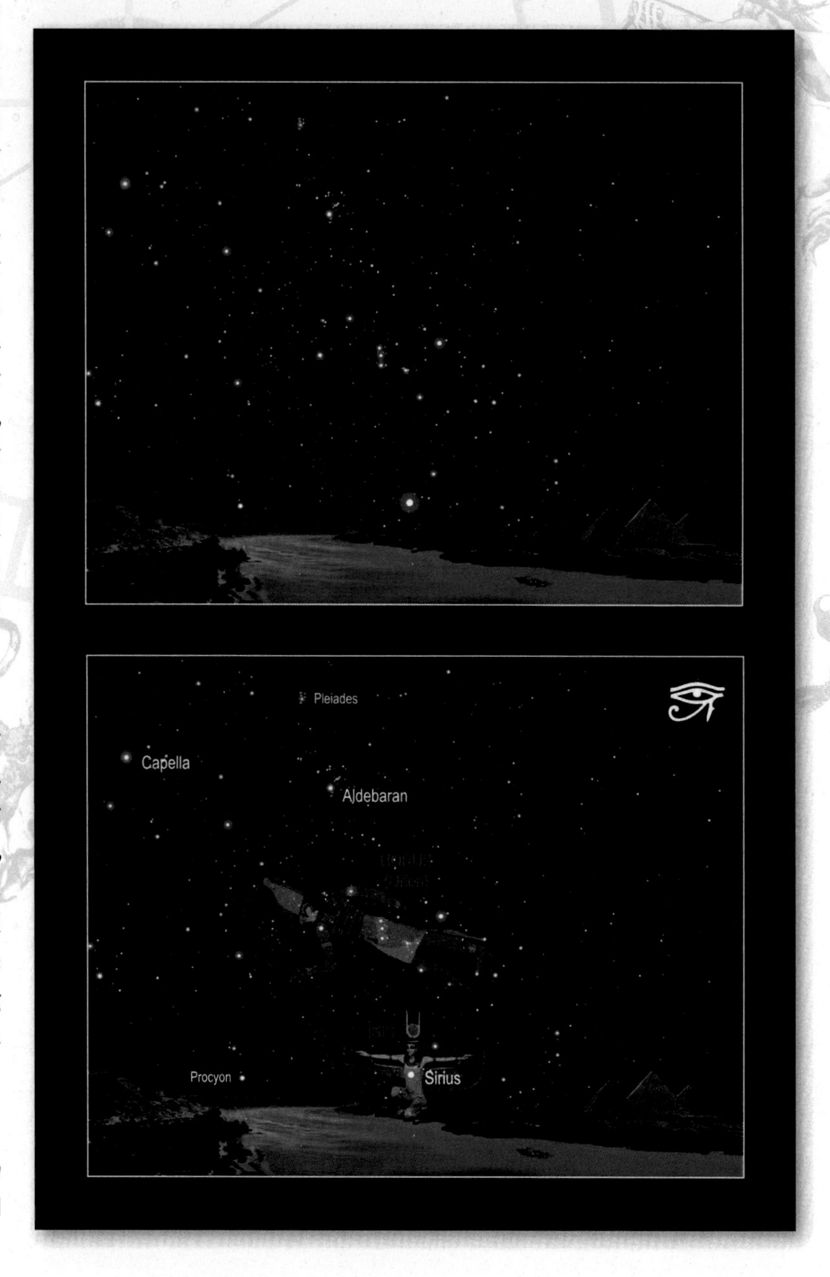

Fig. 4: *The stars rising above the Nile just before dawn 5,000 years ago.* R.Hall

One of the most significant of these celestial portents in ancient times was the heliacal rising of the brightest star, Sirius, pronounced Syrius (Figure 4). Sirius is the brightest true (fixed) star and its brilliance is only outshone by the planets Venus, Jupiter, and sometimes Mars. It is also known as the 'Dog Star'.

As a child, looking at this star from my grandmother's place in the countryside, I can remember being told that it was the 'Dog Star'. But no one seemed to know why it was called a dog. Later, when I became interested in astronomy, I discovered that it was the brightest star in the constellation of Canis Major (the Great Dog) and assumed that was the reason for its name. Not so, later I discovered that it was called the Dog Star long before a constellation was created around it.

Five thousand years ago, when this star rose up in the dawn twilight it signaled the inundation of the Nile. Now, the flooding of the Nile was the most important seasonal event to the Egyptians because it brought fertility to the Nile valley. The Egyptians observed the star rise and then the Nile flood and they thought that they saw cause and effect. They thought it was the star that caused the Nile to flood. It was of course, simply a coincidence that the star rose at that time. None the less it was still good science in its day – to try and find connections between events in the complex world around them.

However, it was from observations such as these that led to the belief that the stars were controlling things down here on earth. The stars were seen as divine beings, or rather the symbols of divine beings, the messengers of the gods. Indeed, Sirius began appearing on Egyptian monuments and temple walls and was worshipped throughout the Nile Valley from 3285 B.C. It was said to be the "Heart of Isis" the great goddess of fertility. [4]

Now back to that question - why is Sirius called a dog? There are in fact two dog stars in the sky. These are Sirius, the Greater Dog, and Procyon the Lesser Dog. Long before the rise of the Egyptian civilization shepherds used dogs to guard their flocks and to warn of approaching danger. Sirius and Procyon were the shepherd dogs in the sky. They stood on guard either side of the celestial Nile, the Milky Way. Their heliacal rising warned the shepherds of the coming flood and that it was time to move the flocks to higher ground.

Let's try and imagine the thoughts and conclusions of our ancestors when contemplating the relationships between heavenly and earthly events. If the stars controlled mighty events such as the flooding of the Nile, the changing of the winds and the coming of the rains, surely they must control everything including the destiny of nations and individual people? Perhaps it was this line of thinking that led to the origin of astrology, the belief that human destiny is written in the stars.

In early farming and pastoral societies the knowledge, the ability to read the stars, was the domain of the 'wise' man or woman – the shaman or witch-doctor. With the rise of civilization star lore and astrology became interwoven in the emerging state religions and often became the exclusive knowledge of the priesthood. These priests were able to predict awesome celestial events such as solar and lunar eclipses. They could foretell the changing of the winds, the flooding of rivers and the coming of the rains. It must have seemed to the ordinary person that they were communicating directly with the gods. Indeed, by reading the stars the priests believed they could read the mind of god.

---

4 Isis was married to her brother Osiris, god of the Nile and the afterworld, and it has been suggested by some that the name Sirius may be derived from his name. However, the Egyptian name for this star was Sothis. It seems more likely that the name Sirius originates from the Greek seirios (σειριοσ), meaning "sparkling" or "scorching". To the Greeks Sirius was the "Scorching One" who, in the 3rd century B.C., rose in the dawn at the hottest time of the year.

3. The Zodiacs

The stars are mansions built by nature's hand,
And, haply, there the spirits of the blest
Dwell clothed in radiance, their immortal vest.

Wordsworth
Poems of the Imagination (c.1810)

The concept of a Zodiac, a band of twelve constellations that mark the path of the Sun, Moon and planets, is found throughout ancient Europe and Asia. However, zodiacs from different traditions differ markedly. Let us now have a look at the major zodiacs as we know them today and where we find them in the sky.

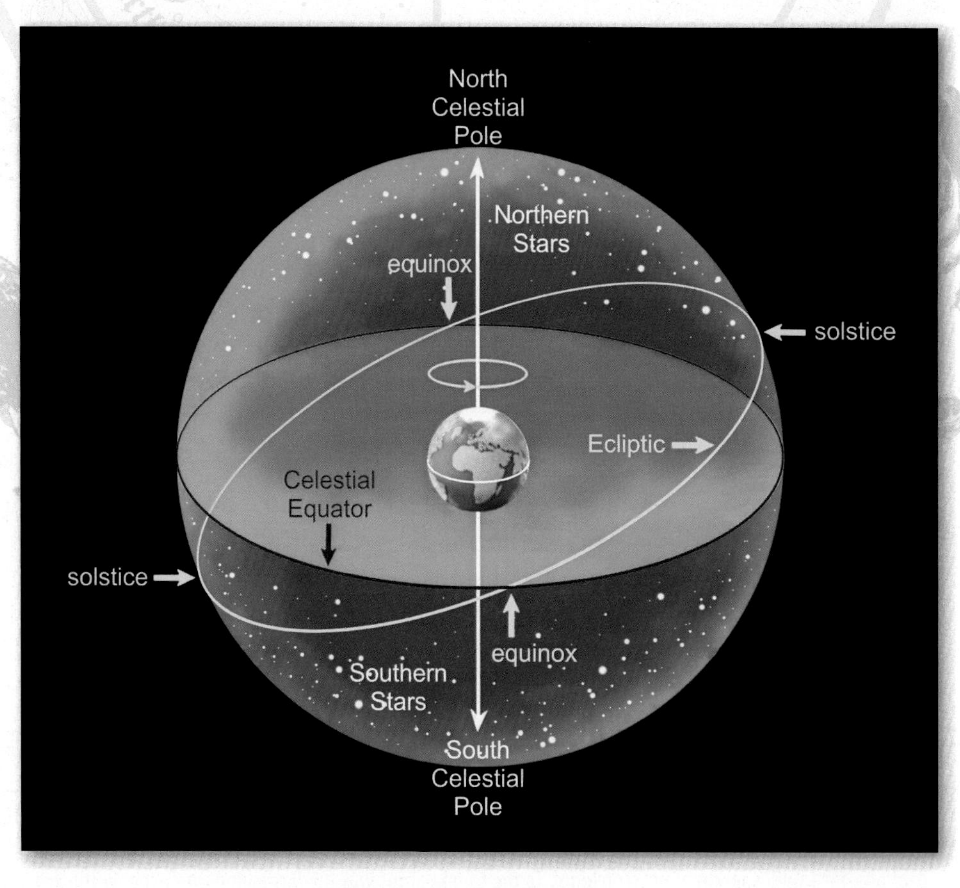

Fig. 5: *The Celestial Sphere.* R.Hall

When we look at the sky it appears to be a great dome over our heads. Indeed, our ancestors believed it to be just that, and that the stars were attached to this invisible sphere. This illusion is due to the immense distances of celestial objects. They are so distant that our sense of perspective fails us. Consequently they all appear to be at the same distance, namely infinity. While this is an illusion it is useful to use this ancient concept to understand celestial mechanics – how celestial objects move across our skies. In figure 5 we see our world as a sphere rotating on its axis within a larger stationary sphere of fixed stars – the celestial sphere.

Although in reality it is the Earth that is turning, we at its surface have no sensation of this motion. To us it seems that it is the celestial sphere that is moving. Consequently we still talk of the sun, moon and stars rising and setting.

Superimposed upon the stellar celestial sphere is the motion of the sun, moon and planets. In the early models of the universe these were believed to be attached to additional transparent spheres. There were spheres within spheres and their motions were linked by wheels within wheels. To this day astrologers still use the word 'wheel' to describe the celestial circle of constellations that make up the zodiac.

The entire celestial sphere appears to rotate around two points, the celestial poles. These celestial poles are a projection of the earth's poles onto the celestial sphere (the line of axis about which the Earth rotates). We can also project the equator of the Earth out onto the celestial sphere. This great circle marks the celestial equator and is exactly 90 degrees from each celestial pole. The celestial equator divides the heavens into two great hemispheres – the northern stars and the southern stars.

Because the Earth's spin axis is tilted 23.5° to the plane of its orbit, the apparent path of the Sun, the ecliptic, is tilted 23.5° to the celestial equator. Consequently, the Sun's annual path around the celestial sphere, takes it back and forth over the celestial equator, spending half of the year in the northern hemisphere, the other half in the south. The two points at which the ecliptic crosses the celestial equator are the equinoxes. The northern spring equinox, the point at which the sun crosses the celestial equator on its journey north, is known as the vernal equinox. This equinox is the point of zero longitude on the celestial sphere from which the positions of stars are measured and time is calculated. The two points at which the Sun is at its most distant from the celestial equator are the solstices.

By international agreement the celestial sphere, the entire heavens, is today divided up into 88 constellations. Each constellation is a region of the sky with fixed borders. In ancient times a constellation was a pattern of fixed stars but the borders were not clearly identified. Consequently modern constellations, the region of sky each includes, are only an approximation of those of antiquity. Furthermore, different cultures often divided the sky up into entirely different constellations. The constellations that lay along the path of the Sun, the ecliptic, are among the most ancient. These, the constellations of the Zodiac, were known as the 'Highway of the Sun'.

The word Zodiac comes from the Greek "Zodiakos Kyklos" which means a 'circle of the animals'. Each sign was a 'zoidion', meaning 'animal figure'. As has often been pointed out, Libra (the Scales) is different in that does not represent a living creature. But, as we shall see later, Libra was not part of the Greek Zodiac. The Babylonian Zodiac, from which the Greek Zodiac was almost certainly derived, did have a set of scales (Libra) but the scales were held by a human figure.

The traditional Zodiac divides the path of the sun into 12 constellations or signs (because each was considered to be a sign in sky), each averaging 30 degrees in length. There are 12 signs because the year is divided into 12 months (the reason why there are 12 months will be explained later). In the ancient calendars, at the beginning of each month, the Sun would move from one sign to the next. Those familiar with star charts will know that modern astronomical constellation boundaries incorporate a 13th constellation along the ecliptic, Ophiuchus, the Serpent Holder. But these are modern constellation boundaries and Ophiuchus, although an ancient and important constellation was not originally a sign of the Zodiac.

In the western world, the zodiac we are familiar with uses the constellation names now adopted by the International Astronomical Union. These are listed below, in their traditional order along the ecliptic, along with the equivalent twelve Indian zodiacal signs, or Rashis.

Western and Oriental Zodiacs

	Western	meaning	Sanskrit	meaning
1.	Aries	Ram	Mesha	Ram
2.	Taurus	Bull	Vrishabha	Bull
3.	Gemini	Twins	Mithuna	Pair
4.	Cancer	Crab	Karka	Crab
5.	Leo	Lion	Simha	Lion
6.	Virgo	Virgin	Kanya	Girl
7.	Libra	Scales	Tula	Balance
8.	Scorpio	Scorpion	Vrishchika	Scorpion
9.	Sagittarius	Archer	Dhanus	Bow, Arc
10.	Capricorn	Sea-goat	Makara	Sea-monster
11.	Aquarius	Water-carrier	Kumbha	Pitcher, water-pot
12.	Pisces	Fish	Meena	Fish

The two Zodiacs appear to be almost identical because, as you may have guessed, they have a common origin. In contrast, the so-called Chinese zodiac originated from a separate tradition and is very different.

Chinese Zodiac

1. Rat	(Aquarius)	5. Dragon	(Libra)	9. Monkey	(Gemini)
2. Ox	(Capricorn)	6. Serpent	(Virgo)	10. Rooster	(Taurus)
3. Tiger	(Sagittarius)	7. Horse	(Leo)	11. Dog	(Aries)
4. Rabbit	(Scorpio)	8. Sheep	(Cancer)	12. Pig	(Pisces)

The Chinese signs and their approximate relation to the constellations of our Zodiac are listed above. The Chinese signs were not originally based upon the 12 solar months of year. Initially they represented the 12 year cycle of the planet Jupiter[5]. Jupiter takes approximately 12 years to orbit the Sun so each year it moves to another sign along the zodiac. The Chinese Zodiac is therefore, symbols assigned to years. Hence, terms such as the 'Year of Tiger'. So, in China your star sign is determined by the year in which you are born. However, it should be noted that in Chinese astrology the Lunar Mansions or Lunar Zodiac (to be discussed later) is far more significant.

Around 500 B.C., the Chinese signs were rearranged to identify the location and path of the Sun. The Chinese called this path, the ecliptic, the Yellow Path. The zodiac symbols were also attached to the lunar months, and two hour periods of the day.

Looking at the correlation between the Chinese symbols and the constellations you will see that they run in the opposite order to other Zodiacs. The signs are numbered from east to west. This is the opposite direction to the motion of the Sun, Moon and planets through the Zodiac.

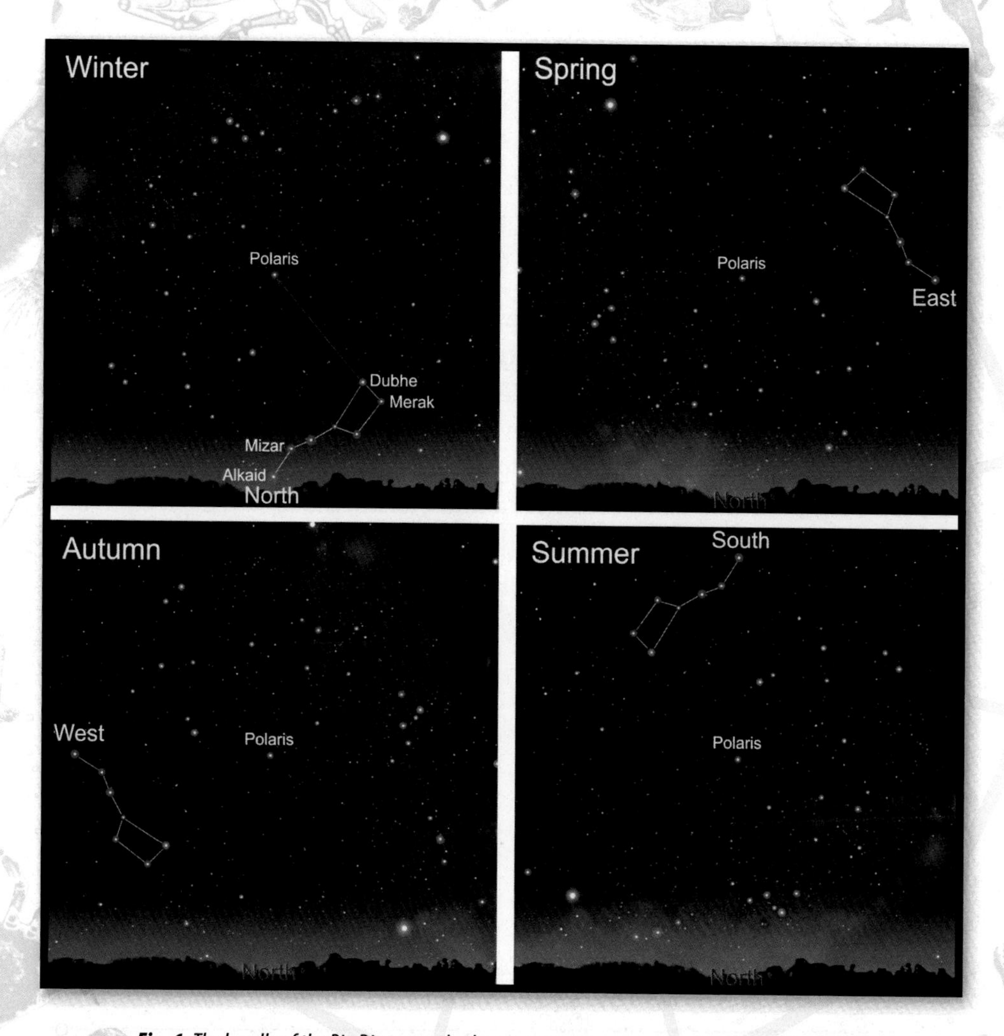

***Fig. 6:** The handle of the Big Dipper marks the seasons as it moves around the North Star.* R.Hall

5 The Mayans of Mesoamerica also used the cycle of Jupiter as the basis of one of their calendars.

Why was the Chinese Zodiac in the reverse direction of other zodiacs? In China the time of seasonal changes were determined by the "Plough" or "Big Dipper". The Big Dipper, seven bright stars in the constellation of Ursa Major (Great Bear), is circumpolar from latitude 41° north. This means that it never sets, it just slowly moves around the North Star, Polaris which marks the north celestial pole[6] (Figure 6). The Big Dipper is a navigational beacon because two of its stars, Dubhe and Merak, point to and therefore identify Polaris. If you were to observe the position of the Big Dipper every evening at the same time you would notice that it is slowly moving in an anti-clockwise direction around Polaris. In one year it would complete one revolution around the North Star.

In ancient China, around 450 B.C., the 'Tseih Sing', which means "Seven Stars" (of the Big Dipper), formed a wagon that carried the celestial bureaucrats around the sky. During the early evening the "handle" of the wagon, Alkaid and Mizar (η and ζ Ursa Majoris) pointed north in winter, east at spring, south in summer, and west in autumn. The Chinese divided the horizon into 12 sections according to where the 'handle' was pointing in the twelve months of the year. These divisions, known as the Twelve Earthly Branches, are as follows.

1. Zi (North)	4. Mao (East)	7. Wu (South)	10. You (West)
2. Chou	5. Chen	8. Wei	11. Xu
3. Yin	6. Si	9. Shen	12. Hai

The Earthly Branches were then related to the position of the Sun to create the 12 houses or signs of the Yellow Road. These were arranged in the direction the Handle moved, east to west.

Our Western year of 2008 marked the beginning of a new cycle in the Chinese Zodiac, the Year of the Rat. But, what happened to the cat? Looking at the Chinese Zodiac it is the only domestic animal not in the Twelve. According to legend, the Jade Emperor invited all of the animals to a great race. The order in which the first twelve crossed the finish line is the order in which they appear in the zodiac and, as you can see, the rat won. Before the race the cat and the rat were good friends. Both were poor swimmers and when they came to a river they asked the ox if they could ride across on his back. In mid-stream, the rat in his greed to be first pushed the cat off, who was then swept away. The rat stayed on the head of the ox and, just as the ox was about to cross the finish line, the rat leapt across first. The cat never managed to catch up and make it through to the first twelve and, ever since that day, the cat and the rat have been enemies.

---

6 Polaris lies within one degree of the north celestial pole. It will be closest to the NCP in the year 2100.

4. Turning of the Wheel

From the earliest of times people recognized the sun as the sustainer of life and the primary source of well-being. As the great wheel of the zodiac turned in the heavens there were four points that signaled major changes in the climate and, above all changes in food supplies. These were the solstices and equinoxes that marked the turning points of the seasons. All of the signs of the zodiac were important but those that were related to the solstices and equinoxes were the most significant. Furthermore, these signs divided the sky into the four quadrants of the Zodiac. The sun is in the following signs at these turning points of the seasons.

Northern Hemisphere	**Sign**	**Southern Hemisphere**
Spring Equinox	Pisces	Autumn Equinox
Summer Solstice	Gemini	Winter Solstice
Autumn Equinox	Virgo	Spring Equinox
Winter Solstice	Sagittarius	Summer Solstice

Whereas you can physically observe which sign the moon or a planet is in, except at the very rare occurrence of a total eclipse of the sun, you can't see which sign the sun is in. It has to be calculated.

But, to calculate which sign the sun is in at the time of an equinox or solstice you first have to know when they occur. Nomadic people, because they were continually on the move, are unlikely to have understood the complex seasonal cycle of the sun. They relied upon stars to forewarn them of seasonal changes. How did people first work out the cycle of the sun and the times of the solstices and equinoxes?

When the first permanent farming settlements appeared people would have noticed that the rise and set position of the sun along the horizon changed with the seasons. Further, with the passage of time, they would have realized that these changes were cyclic and predicable. Let's see how it works.

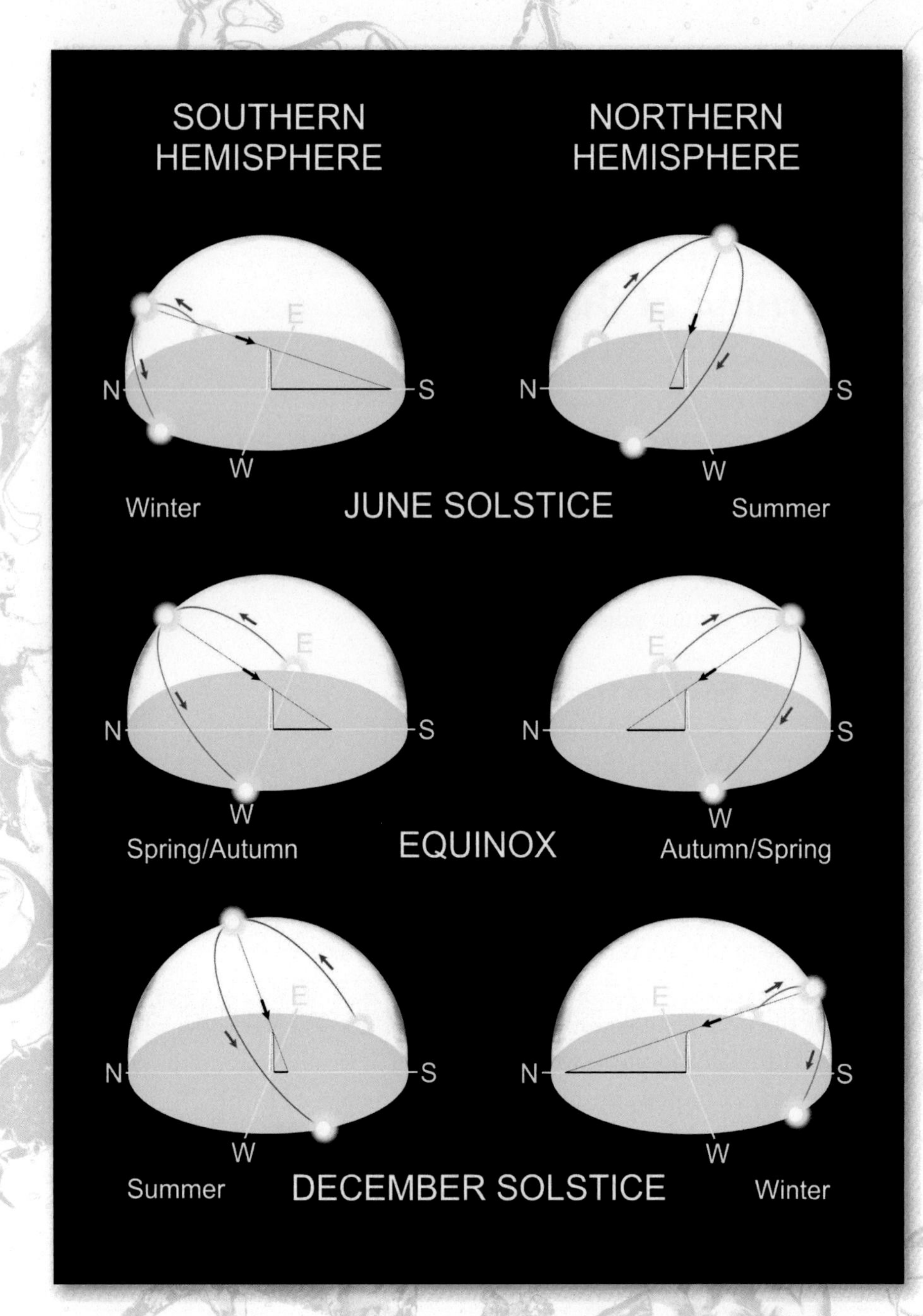

***Fig. 7:** The path of the Sun across the sky at latitudes 45° north and south.* R.Hall

Let's return to our concept of the celestial sphere but this time we will look at what we can observe from the surface of the Earth. Figure 7 shows the sun's path across the sky at the times of the equinoxes and solstices at latitude 45 degrees north (right) and south (left). The diagrams are centered on the observer. The observer's horizon is the edge of the disk and the celestial sphere, the part that can be seen by the observer, is the sky dome above.

Because the Earth axis of rotation has a tilt of 23.5° the apparent path of the sun across the sky varies by 23.5x2 = 47° over the year. From the surface of the Earth we observe the following.

At an equinox, and only at an equinox, the Sun rises due east and sets due west. At latitude 45°, north and south, the sun at noon has an altitude of 45°. It should be noted that the altitude of this path will depend upon the latitude of the observer. If you were standing at the equator the sun would pass directly overhead and cast no shadow at noon. At an equinox the sun is above the horizon for twelve hours. Hence the term equinox, meaning equal day and equal night.

The equinox that occurs around March 21st is the spring equinox in the northern hemisphere, autumn in the south. This is also known as the vernal equinox.

After the March equinox, the rise and set positions of the sun, day by day, move northward. As it does so the length of the sun's path across the sky, its altitude at noon, and hence the hours of daylight all increase in the northern hemisphere. The opposite occurs in the southern hemisphere.

Around June 21st the sun reaches its furthest point north. It is the summer solstice or mid-summer's day in the northern hemisphere. It is the day with the greatest hours of daylight, hence the term the longest day. On this day, at latitude 45° north, the sun is above the horizon for more than 15 hours and at noon has an altitude of 68.5°.

The opposite occurs in the southern hemisphere – it is the winter solstice with the shortest hours of daylight. At latitude 45° south the sun is above the horizon for less than 9 hours and at noon its altitude is only 21.5°.

At midday, at the time of this solstice, the sun will be directly overhead and will cast no shadows at latitude 23.5 degrees north. This latitude is known as the Tropic of Cancer. It is so called because 2500 years ago, when this coordinate system was formalized, the Sun was in Cancer when it reached the northern solstice. This latitude, N23.5, is still known as the Tropic of Cancer even though, due to precession (to be explained in the next chapter), the solstice now occurs in Gemini.

At the time of a solstice the movement of the sun along the horizon halts. Hence the term solstice, meaning "the sun standing still".

After the June solstice the sun reverses its path and its rise and set positions move back towards the east and the hours of daylight decrease in the northern hemisphere but increase in the south. Around September 22nd the sun arrives at the northern hemisphere autumn, the southern hemisphere spring equinox. One again the sun is over the equator and rises due east and sets due west.

After the September equinox the rise and set positions of the sun move to the south until, around December 21st, it reaches the southern solstice. Conditions are the exact reverse of what occurred at the June solstice. It is mid-winter's day in the northern hemisphere, mid-summer's day in the south. At noon on the December solstice the sun passes directly overhead at the "Tropic of Capricorn", latitude south 23.5°. Here again this latitude was named 2,500 years ago when the Sun was actually in Capricornus when it reached the southern solstice. Precession has now moved it into Sagittarius.

After the December solstice the sun again reverses its path, arriving back at an equinox in March. And so the cycle continues.

In Aotearoa, New Zealand, Maori have a special story that illustrates the cycle of the sun and the seasons. This story is of particular interest because in blends the use of stars as seasonal beacons with the cycle of the sun. The people from which this story comes were essentially agriculturists but had recently descended from and continued in part to be hunter-gatherers.

The Maori name for the Sun is Ra; which is interesting because the ancient Egyptians also called the Sun Ra or Re. The story says that Te Ra is a great god with two wives (who are also sisters), the summer maid Hine-Ruamati, and the winter maid Hine-Takurua. Over the year, by dawn's light, you can watch him move from one wife to the other.

The two wives are bright stars that mark the times of the solstices. Hine-Ruamati is the bright reddish star Antares in Scorpius, which rises in the dawn twilight in late December, close to the southern hemisphere summer solstice. Hine-Takurua is the brilliant blue-white star Sirius in Canis Major, which rises in the dawn twilight in mid-June, close to the southern hemisphere winter solstice. Takurua was also known as the "Frost Star" because her twinkling forewarned of a heavy frost. Thus, at a solstice one of the two wives is close to, and rising just before Te Ra (The Sun).

The story goes on to say that "When he is with one of the maids he moves slowly because he is reluctant to leave the comfort of a wife. But, he moves very quickly when traveling between wives, especially when he is visible to both." This is exactly what can be observed with the rising sun. Close to a solstice the sun's daily change in its rise position is very small. Indeed, for a couple of days either side of the solstice it is difficult to detect any change in the sun's rise position without special instruments. However, between solstices the rise position of the sun moves rapidly along the horizon with the greatest changes occurring near an equinox. At an equinox Te Ra is visible to both wives. The two stars reach their highest point in the sky, one in the dawn twilight, and the other in the evening twilight.

From a fixed observation point people, long ago, were able to place posts or stones that marked the rise or set positions of the sun at the solstices and equinoxes. (Figure 8) These structures, like Stonehenge, enabled them to measure the solar year, to predict the times of seasonal change, and to formulate a solar calendar around which they could plan their lives. With the passage of time the solstices and equinoxes evolved into great religious festivals, Holy Days (Holidays) that were celebrated throughout the ancient world.

Fig. 8: *Heel-stones at Stonehenge Aotearoa mark where the Sun rises at the solstices and equinoxes.* R.Hall

5. Precession of the Equinoxes

"The time has come," the Walrus said,
"To talk of many things:

Of shoes – and ships – and sealing wax –
Of cabbages – and kings –
And why the sea is boiling hot –
And whether pigs have wings."

Lewis Carroll, Through the Looking Glass

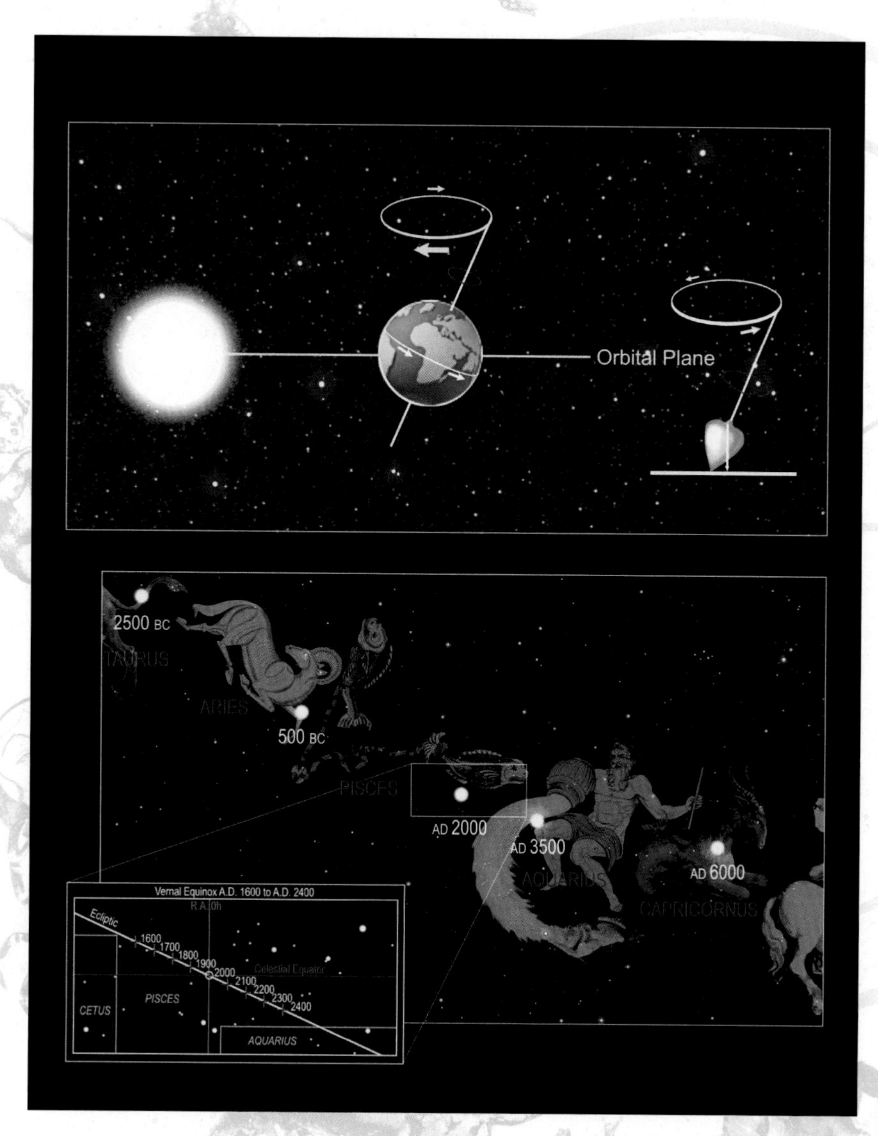

Let's return to that question posed at the beginning of this book - what is your star sign? Well, if your birth date is for example August 1st, and you were to consult most astrologers in the western world or, simply look it up in the newspaper, you would discover that your star sign is Leo. But, if you were to consult a Hindu astrologer he or she would tell you that your star sign is Karka, Cancer. Now, they can't both be right. Is it Leo or Cancer? The answer can be found in something known as 'Precession of the Equinoxes'.

In chapter 2, Signs and Symbols in the Sky, I said that the solstices and equinoxes occur at the same points in the Earth's orbit each year. Consequently, the Sun would be in the same constellation at the same time each year and, we would see the same stars at night at the same time each year. This is not entirely correct.

Fig. 9: *The location of the solstices and equinoxes in the Zodiac slowly change due to a cyclic wobble in the Earth's axis of rotation.* R.Hall.

Like a giant spinning top the Earth is slowly wobbling on its axis (See figure 9). The axial tilt remains the same at 23.5°, but the location in the sky that the poles are pointing is slowly changing. The movement is very slow, one complete cyclic wobble taking 26,000 years. One of the effects of this slow change in the direction of the Earth's tilt is to cause the location of the two equinoxes to slowly move along the orbit of the Earth. Which also means that the location of both the equinoxes and solstices is slowly moving against the background stars, hence the name Precession of the Equinoxes.

The path of the Sun along the constellations of the Zodiac remains the same but, the positions of the solstices and equinoxes changes. Over the 26,000-year cycle each solstice and equinox moves through each of the constellations of the Zodiac. Today the vernal equinox, the northern spring equinox, is in Pisces, 2,500 years ago it was in Aries and 5,000 years ago it was in Taurus. It moves from one constellation to the next approximately every 2,200 years. Figure 9 shows the location of the Sun at the vernal equinox at epochs past, present and in the future. The inset shows the changing position of the equinox and hence the intersection of the celestial equator and the line of 0 hour right ascension relative to the background stars.

The calendar we use in the western world is a solar calendar. Without corrections the dates of the solstices and equinoxes would shift by one day every 71 years. With time the seasons would drift through the calendar so that in 13,000 years from now December would be mid-summer in the northern hemisphere and mid-winter in the south. Today, to keep our calendar in sink with the seasons, astronomers make periodic adjustments to our time-keeping ensuring that the vernal equinox always falls on March 21st.

While the vernal equinox always falls on March 21st its location in the starry background, along the ecliptic, moves by one moon diameter every 36 years. This also means that the date upon which the Sun enters a sign of the Zodiac shifts by one day every 71 years. These changes are then, insignificant over the lifetime of an individual but, compounded over the centuries, they become very marked.

The Greek astronomer Hipparchus discovered precession in the 2nd century B.C. Working from a series of Babylonian star charts that had been drawn up over the centuries he noticed a slow drift in the longitude of bright Zodiacal stars relative to the equinoxes. He was able to explain this as a slow forward motion of the equinoxes through the Zodiac. The Egyptians called this cycle of precession 'The Great Year'.

Hipparchus drew up the first catalogue of stars, accurately defining their brightness and positions, and redefined the boundaries of the constellations. In this new scheme the vernal equinox was, at that time, close to the beginning of Aries. This became known as the "First Point in Aries". Three centuries later the astronomer/astrologer Ptolemy consolidated this system and made the first point in Aries part of orthodox astrology. In Ptolemy's time the vernal equinox was actually very close to the start of Aries and, from that time onward, Aries marked the beginning of the Zodiac.

Around A.D. 400 what was left of the old Roman Empire became a Christian Theocracy. A theocracy is a state run by clerics. All other religions, with the exception of Judaism were outlawed. The temples of other religions were destroyed, their property confiscated and their books burnt. Astrology along with the developing sciences almost died out. This era of cultural decline became known as the 'Dark Ages'.

Astronomy and astrology was reintroduced from Arabia in the 12th century. This time it was embraced by the Church. This was undoubtedly because the Christian world was falling behind the surrounding world in learning and technology. However, astrology was Christianized and carefully controlled by the Vatican. It was used to support biblical prophecy and subject to religious dogma in which *'that which had been created by god cannot be changed.'* Consequently, Ptolemy's astrological star charts were adopted but without correction due to precession. This retained the zodiac as it was at the time of the birth of Christ.

When religious laws were eventually relaxed the old astrology was essentially rediscovered by the general public and underwent a revival. Unfortunately, most of the new breed of astrologers, unlike their ancient counterparts, was not familiar with practical astronomy and celestial mechanics. They adopted Ptolemy's system (the Christianized system) which had the vernal equinox located at the beginning of Aries, without realizing that this was true only in Ptolemy's time. Today, almost 2,000 years after Ptolemy, precession has carried this astrological zodiac hopelessly out of synchronization with the constellations in the heavens. Cyril Fagan, a historian and reputably one of the great astrologers of the 20th century, called this the "greatest blunder that has ever been made in the history of astrology".

In India astrology has always played an important part in peoples lives. Because their traditions were essentially unbroken over time they didn't make the same mistake. So the answer is that it is the Hindu astrologers that have got it right. If you were born on August 1st your star sign is Cancer.

The beginning dates of the Western and Indian or Jyotish astrological signs are given below along with the astronomical dates upon which the sun enters the constellation. The astronomical dates are for epoch A.D. 2000 and are based upon modern constellation boundaries which, as mentioned earlier, are only an approximation of the boundaries of the ancient constellations. In addition, these dates can vary by up to two days depending upon leap years and time zones. However, it can be seen that the oriental dates, because they include precession, are much closer to the modern astronomical boundaries. It should also be noted that, unlike modern constellations, the Indian signs are equal divisions of the zodiac, each 30° in length along the ecliptic.

Western Sign and date		**Indian Sign and date**		**Astronomical**
Aries	March 21	Mesha	April 14	April 18
Taurus	April 21	Vrishabha	May 15	May 14
Gemini	May 22	Mithuna	June 15	June 21
Cancer	June 22	Karka	July 17	July 20
Leo	July 23	Simha	August 17	August 10
Virgo	August 24	Kanya	September 17	September 16
Libra	September 24	Tula	October 18	October 31
Scorpio	October 24	Vrishchika	November 17	November 22*
Sagittarius	November 23	Dhanus	December 16	December 17
Capricorn	December 22	Makara	January 15	January 20
Aquarius	January 21	Kumbha	February 13	February 17
Pisces	February 20	Meena	March 15	March 12

* In modern star charts the Sun passes through the constellation of Ophiuchus from November 30 to December 16.

Most astrologers in the western world still use a zodiac wheel that is almost 2,000 years out of date. You may wonder why this has not been corrected. Mysticism and pseudo-science has grown in popularity in the western world, particularly in the second half of the 20th century, and this has produced vast numbers of self-appointed soothsayers. Most of these 'new-age' astrologers, unlike their ancient predecessors, know little or no astronomy and rely upon information written by others, which, as we have seen, contains a major error. The western astrological zodiac is now so entrenched in the beliefs of the 'new age' astrologers its difficult to see how they could correct it without losing credibility. So, instead of admitting a mistake and correcting things they have invented a new zodiac to conceal the mess.

As an explanation for the anomalies in the western astrological zodiac many astrologers talk about there being two types of zodiacs. There is the "tropical" zodiac used by western astrologers which is based upon fixed calendar dates, and the "sidereal" zodiac used in the orient which is based upon the position of the sun in the constellations. But there is in reality only one zodiac and this is the one they call the sidereal zodiac – the constellations in the heavens along the path of the Sun. The so called tropical zodiac is just calendar dates on a piece of paper. In the ancient traditions of astrology, your star sign was based upon which constellation the sun was in when you were born – the symbol of god and the life giving powers of the sun. One astrologer said to me that for him the zodiac was just a band across the sky and that it didn't matter that the dates no longer coincided with the passage of the sun through a constellation. I found this astonishing. What he was saying was that the real constellations were unimportant. What was important to him was a date in a calendar created by Roman emperors, popes and politicians. Surely if astrology is to have any credibility, a birth sign should be determined by the positions of celestial objects at the time of one's birth not a man-made calendar date.

Another said to me that it was not the Sun or even the stars that were important, it was the energy arriving from that region of space that was important. This really is modern mumbo-jumbo. Apart from the fact that there is absolutely no scientific evidence for the existence of this energy it still begs the question as to why the level of this energy from space would be controlled and dictated by a man-made calendar date.

It is interesting to note that the familiar star signs seen in newspapers and magazines that undoubtedly led to the 20th century popularization of astrology were banned in the British Empire until 1910. When these old laws were repealed one British newspaper decided to publish the 'stars'. The people who put this together were the newspaper staff, who were neither astronomers nor astrologers. They didn't realize that the charts that they were using were those used in the Holy Roman Empire which, by that time, were almost two thousands of years out of date. But it was so popular that it rapidly spread around the western world.

So, if you believe in astrology, and what you have read in the newspapers doesn't always appear right, it's because all your life you've been looking at the wrong star sign.

PART II:

CAVES, PILLARS & GATES

Where did the Zodiac come from and how old is it? In this part of the book we trace the origin of the signs of the zodiac, their relationship with the solstices and equinoxes, and early beliefs on the nature of the cosmos.

6. The Celestial Flocks

Heaven's utmost deep
Gives up her stars, and like a flock of sheep
They pass before our eye, are number'd, and roll on.

Shelley
Prometheus Unbound (c.1820)

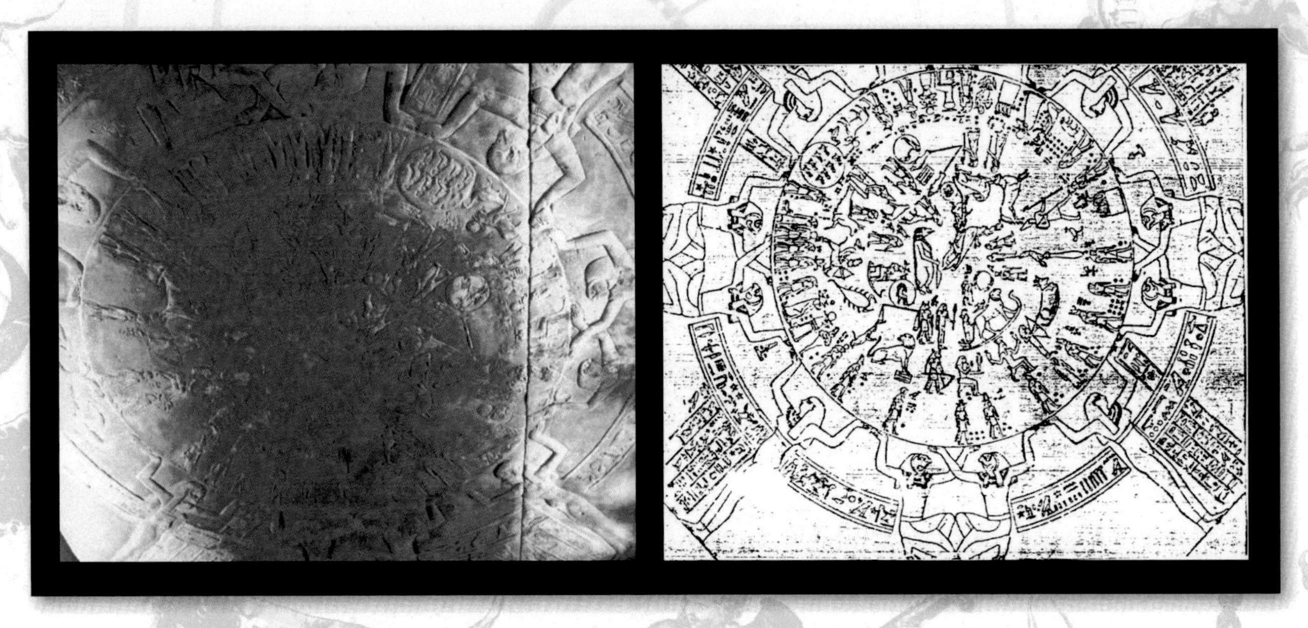

Fig. 10: *The oldest surviving complete Zodiac wheel wasfound in the Temple of Denderah, a little north of Luxor in Egypt. Left: original sandstone sculpture in the Louvre; Right: archeologist's drawing.* Wikipedia.

The oldest surviving complete and detailed representation of a Zodiac wheel is the 'Denderah Zodiac'. The original sculptured sandstone is shown in figure 10 along with a drawing of the figures. This zodiac originally formed the ceiling in a place dedicated to Osiris within the Temple of Denderah, which is a little north of Luxor in Egypt. The vault of the heavens is depicted as a great disc supported four women. Between each woman are two falcon-headed spirits who also support the heavens (4 + 8 = 12). The four women are the 'Pillars of Heaven', the meaning of which will be discussed in the chapter after next. On the outer perimeter of the disc are 36 spirits which represent the Decons, the 36 ten-day weeks of the ancient Egyptian year. Inside the ring of Decons are the 12 signs of the Zodiac and, inside the Zodiac wheel, are figures representing the five known planets. At the centre is a bull's hind leg which represents the stars of what we now know as the Plough or Big Dipper (Ursa Major – The Great Bear).

The positions of the planets relative to the zodiacal constellations in the Denderah Zodiac are for June-August 50 B.C. Although the sculpture is Egyptian the Zodiacal figures are undoubtedly of Babylonian origin. These figures first appear on Babylonian cuneiform horoscopes around 410 B.C.

The zodiac we have today was, save one major alteration, inherited from the Greeks and was formalized about 500 B.C. A Greek zodiac wheel bordering a procession of horses and musicians is found on a 4th century B.C. fresco in a Thracian tomb. This zodiac scheme, which the Greeks were using to establish a solar calendar, was in turn borrowed from the Babylonians.

The Babylonian zodiac, which was established about 700 B.C., did not include Aries (The Ram). In its place was a constellation known as Mul lu Hun ga, the "Hired Man". The Greeks created Aries out of the stars of this constellation which, at that time, contained the vernal equinox. It marked the beginning of the year and the beginning of the Zodiac.

In the folk-lore of early pastoral people the stars were regarded as celestial flocks. Each group of stars or constellation had its own leader. Aries was the ram that led the heavenly flock through the year. In Greek mythology Aries is the sacrificial ram that became the 'golden fleece' in the Grove of Ares (Mars). To the Hebrews Aries represented the 'Altar and the Sacrifice' for it was said that the sun was here when the people were released from their bondage in Egypt. Therefore these stars were known as the constellation of the Ram by other cultures long before the Greeks incorporated it into the Zodiac around 500 B.C.

While the Babylonians had a set of scales (Libra) in their Zodiac it was not included in the Greek Zodiac. The Romans reintroduced Libra during the reign of Julius Caesar around 50 B.C. It was placed at the autumn equinox (northern hemisphere) and symbolized the balance of the seasons and the scales of justice in the Roman Empire.

The origin of the Babylonian zodiac is usually attributed to the Mesopotamians but it may well have been imported from the Indus Valley. The oldest Vedic text is the Rig Veda which dates back to between 1,300 B.C. and 1900 B.C., some argue as early as 3,000 B.C. The Dirghatamas Rig Veda states "With four times ninety names Vishnu sets in motion moving forces like a turning wheel". Vishnu was the sun god and ruler of the heavens. The four divisions are believed to be the four seasons or solstices and equinoxes. Four times ninety is 360, the number of degrees in a circle. It also says "Seven half embryos form the seed of the world. They stand in the dharma by the direction of Vishnu." This is believed to refer to the seven planets.

However, we cannot be certain that any of this refers to a solar zodiac. I use the term solar zodiac to refer to the traditional zodiac of 12 divisions along the path of the sun. Pre-dating the solar zodiac are the so called lunar zodiacs (to be discussed in a later chapter). These are based upon the path and cycle of the moon and were undoubtedly used to create the solar zodiac. The fact is that there is no substantial evidence of the existence of a complete solar zodiac wheel before that developed by the Babylonians around 700 B.C.

Indeed, rather than inheriting it from elsewhere it may have been the Babylonians that invented the solar zodiac. The oldest detailed astronomical texts are Babylonian and these date to around 1600 B.C. They refer to the Bull, the Lion, and the Scorpion, figures built around the 'Royal Stars' – the Pillars of Heaven, but there is no mention of a zodiac as such. It is probable that the first zodiac consisted of just four signs which marked the equinoxes and solstices and divided the year into the four seasons. Then, at a later date, each quarter was divided into three separate constellations to accommodate the twelve months of the year.

While the Zodiac itself doesn't appear to be older than about 3,000 years, many of the individual constellations which became part of it have a much greater antiquity. These constellations had important meanings to the people that created them.

"Constellations have always been troublesome things to name. If you give one of them a fanciful name, it will always refuse to live up to it; it will always persist in not resembling the thing it has been named for".

Mark Twain – Following the Equator

It is a common misconception, even amongst astronomers, that the constellations were named because the patterns they formed resembled something or other. Well, this is true for a small number of constellations such as the Southern Cross (Crux). But, the Southern Cross is a relatively new constellation created by navigators in the 16th century from some of the stars of Centaurus. Most of the ancient constellations were named for their symbolism. Many of them originally had a special meaning relating to seasonal events or tasks of the people. But due to precession their meanings have changed with the passage of time. However, these ancient meanings give us clues to the date of their origin.

Aquarius for example, is so called not because it looks like a man pouring water from an urn, but because when the sun entered this constellation 5,000 years ago, it marked the onset of the rainy season. Aquarius symbolized god pouring water down on the Earth.[7]

The heliacal rising of Leo 7,000 years ago marked the onset of the hottest time of the year. At this time the arid conditions drove the lions in from the wilderness which prayed upon domestic animals. The symbolism was the seasonal appearance of the lion, which was also a warning to the shepherd.

However, due to precession the rains no longer coincide with the sun's passage into Aquarius nor does Leo mark the onset of the hottest time of the year. So, Aquarius probably originates around 3000 B.C., Leo around 5000 B.C. Down through the ages the signs of the Zodiac have meant different things to different cultures. Aquarius no longer marks the onset of the winter rains but, in a thousand years from now, it will be the dawn herald of spring in the northern hemisphere.

Taurus (The Bull) is not only one of the most ancient of constellations it is also perhaps, the most celebrated of zodiacal signs. This is because the vernal equinox lay within its boundaries from 4,000 to 1850 B.C., a period that has been described as the golden age of archaic astronomy. Stories and religious symbolism relating to the Bull date as far back as 6000 B.C., so the constellation may be this old.

Virgo (The Virgin) is another constellation that has a great antiquity and some have suggested that it is the oldest of all zodiacal signs. Virgo has always been an important sign and invariably represented a deity. To the Romans she was Astraea, the goddess of Justice that held the Scales of Justice, Libra. To the Egyptians she is, in the zodiacs of Denderah and Thebes, the goddess Isis clasping in her arms Horus, the infant southern Sun-god. In the Middle Ages this same figure reappears as the Virgin Mary holding the child Jesus.

To the Greeks Virgo was Proserpina (Persephone), the daughter of the great goddess Demeter, or Demeter herself. Sometimes Virgo was called the 'Maiden', which is the old Sicilian name of Persephone. Demeter was the Harvest Queen who at that time rose just before dawn at the autumn equinox. The brightest star in Virgo is the blue-white first magnitude star Spica. The name Spica dates back to around 500 B.C. when the star's heliacal

7 The Islamic figure of Aquarius is just the pitcher. This is because in Islam you are not allowed to portray an image of God.

rising signaled the time to bring in the harvest. It was the spike or ear of wheat held in the hand of Virgo.

In India Virgo was Kanya, mother of the great Krishna. She is also identified as Cybele, the Great Earth-Mother Goddess of Anatolia and Rome. Cybele rode in a chariot drawn by lions. In the heavens Leo immediately precedes her. Thus, although we cannot be certain, the origins of Virgo may date back to as much as 6000 B.C.

There are some people who argue for a great antiquity of the Zodiac and that Virgo, because of its association with the Earth-Mother Goddess, originally marked her most holy day - the spring equinox. If this is so it would imply that the constellation of Virgo is at least 14,000 years old, because that is how far we need to go back in time to find Virgo marking the spring equinox.

Earth-Mother cults almost certainly go back this far in time but it is highly unlikely that any of the constellations are this old. The reason for this is something called 'proper motion', the line of sight motion of the stars.

The stars are distant suns that are so remote that they appear as nothing more than points of light of differing brightness. The stars are in motion but, unlike the planets (which look like stars to the unaided eye), their distances are so great that it takes eons before there is any noticeable change in their position. Consequently, the patterns they form, which we call constellations, remain fixed for thousands of years. The stars we see today and the patterns they form in our night sky are much the same as those seen by the Greeks three thousand years ago.

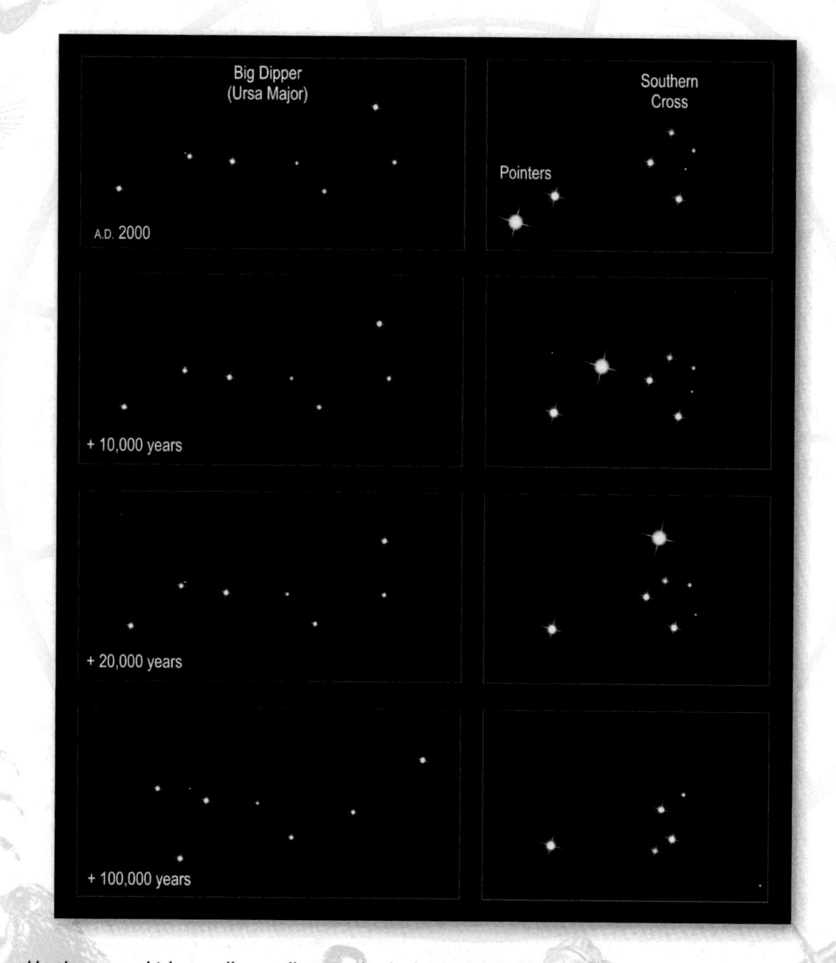

Fig. 11: *The patterns formed by the stars, which we call constellations, slowly change due to the individual motion of the stars through space. The charts show the change in the Big Dipper and the Southern Cross over the next 100,000 years.* R. Hall

However, as soon as we start to talk about ten or more thousands of years we are into that eon region and the motion of the individual stars becomes significant. Take a look at figure 11 which shows how the star patterns that form the *Southern Cross* & and the *Big Dipper* will change with the passage of time.

What we see is that the patterns the stars form are changing and, they become unrecognizable in a time period much over 10,000 years. In addition some stars will fade away as they move off into the galaxy while others will brighten as they approach the Solar system. It is ridiculous to suggest, as some authors have, that some zodiacal constellations were recognized tens or hundreds of thousands of years ago. The patterns of the constellations dissolve with the passage of time. Those that we see today simply didn't exist in those ancient epochs.

What we and our early ancestors could both identify is not the constellations but the brightest of stars. Most of the bright stars in our sky are highly luminous giant stars. They appear bright not because they are close to us but because they are intrinsically bright, thousands, sometimes tens of thousands, of times brighter than the sun. These stars are very distant but shine out over the light-years like cosmic lighthouses. Because of their great distance these stars hardly appear to move at all in 20,000 years. Spica, the brightest star in Virgo, is one of these stars.

From the earliest of times people have used certain bright stars as portents of seasonal events. The present constellations were built around these bright stars at a later date. Whereas the constellation of Virgo had probably not been formulated at that time, the heliacal rising of its bright star Spica would have signaled spring 15,000 years ago.

Imagine how important this star was at that time. The world was still in the grip of the last ice-age and, when this star rose up in the dawn twilight, it signaled that winter had come to an end, life on earth was about to be re-born. That would have been a pretty significant portent. The Babylonians called this star 'Emuka Tin-tir-Ki', which means the 'Might of the Abode of Life'.

7. Clan of the Cave Woman

Virgin august! come in thy regal state
With soft majestic grace and brow serene;
Though the fierce Lion's reign is overpast
The summer's heat is all thine own as yet,
And all untouched thy robe of living green
By the rude fingers of the northern blast.
R. J. Philbrick's Virgo

Let us now take a journey back in time to when I believe it may all have begun. To a time and place which gave birth to many of our fundamental spiritual beliefs, from which emerged the symbolism behind the signs of the zodiac.

Our species, Homo-Sapien, first appeared in North Africa, Ethiopia – Sudan region, 130,000 years ago.[8] From there they migrated into Europe and then into Asia along the tropics. These early homo-sapiens differed little in their technology from that of their predecessors Homo Erectus, from whom they evolved. They used fire, built temporary shelters, and used crude stone tools.

Around 40,000 years ago an amazing transformation took place which was both cultural and technological. A new breed of people appeared in what is now Asia Minor (Anatolia). They invented the harpoon and the bow and arrow, and they used needles and thread to fashion footwear and garments of clothing. With these new crafts they were able to hunt big game and colonize temperate, even arctic environments. In a relatively short span of time they spread throughout the old world and then, 20,000 years ago, made their way into the Americas.

Fig. 12: *Paleolithic cave paintings in the Lascaux Caves, Dordogne, southwest France. Photo by Sisse Brimberg, National Geographic. Insert: Venus figurine*

8 This suggests that everyone alive today descended from people with black skins.

These people also produced works of art and buried their dead with ceremony. Graves have been found that contain artifacts which suggest that these people had some concept of an afterlife. Artwork consisted of the famous cave paintings in France (Lascaux, La Marche, Chauvet and others) and figurines carved out of stone or ivory (Willendorf, de Sireil and many others in Russia).

We use to call these people cavemen and cavewomen because it was originally thought that they lived in caves. But, as hunter-gatherers who lived a nomadic way of life in an environment in which food supplies in any one area varied markedly with the seasons, it is far more likely that caves would have been temporary shelters occupied only during certain seasons. Furthermore, the caves that contain the elaborate paintings show no sign of domestic use. Some of them are found in deep caves, at a depth far beyond what would be practical as a shelter or home. It is more than probable that these caves were sanctuaries that had a spiritual significance. The paintings depict the basic needs of the people – food supplies and procreation (Figure 12). Many anthropologists have suggested that the paintings were ritualistic or formed part of a ritual that would bring success in hunting and childbearing. Now, this theory gains weight from the fact that, with the rise of civilization, the earliest religions on earth that we know of were Earth-Mother cults and that caves were sacred to the goddess and used as temples.

Stones were said to be the bones of Mother Earth and caves were her womb. It was a common belief in times of old that when the sun set it was plunging into the earth, into some hidden passage. The early Greeks believed that the sun was placed in a chariot each day and driven across the sky by the sun god Helius (later Apollo). But for many early cultures it was far more basic. The setting sun was the penis of the sky father entering the earth-mother.

Fig. 13: *Tiki.* R.Hall

In New Zealand many people wear a carving of the Maori 'Tiki' around their neck (Figure 13). You will find Tikis for sale in just about every New Zealand tourist shop. However, few people know that Tiki is the personification of the penis of the sun god as it penetrates (sets) in Papatuanuku, the earth-mother. A common expression in New Zealand, particularly when talking about tourists, is to take people for a 'Tiki Tour'. What they mean by this is to take a tour that makes short visits at lots of interesting (tourist) spots. Few if any of the people who use this expression really know what it means. In pre-European times a Maori chief would sometimes tour the lands of neighboring tribes taking as wives women of high rank. The purpose of this was to produce children that would have blood ties with his neighbours. He was taking his tiki for a tour.[9]

In addition to paintings on the walls of caves our early ancestors also left behind carvings. Perhaps the most famous of these artifacts are the so called "Venus figurines" (insert on figure 12). They are all very similar; each figurine portrays a grossly overweight female. There has been much debate about what these figurines represent. As has been pointed out by many anthropologists they are unlikely to be depicting the body build of a woman of that time. Whereas what is considered beautiful in the female form varies from culture to culture and, in some cultures obesity is a sign of wealth and rank, hunter-gatherers, no matter what their rank, couldn't afford to be obese. A grossly obese person could not survive amongst a people who were constantly on the move and who gained a living by hunting and gathering. Further more, none of the skeletal remains of these people show any signs that they were overweight.

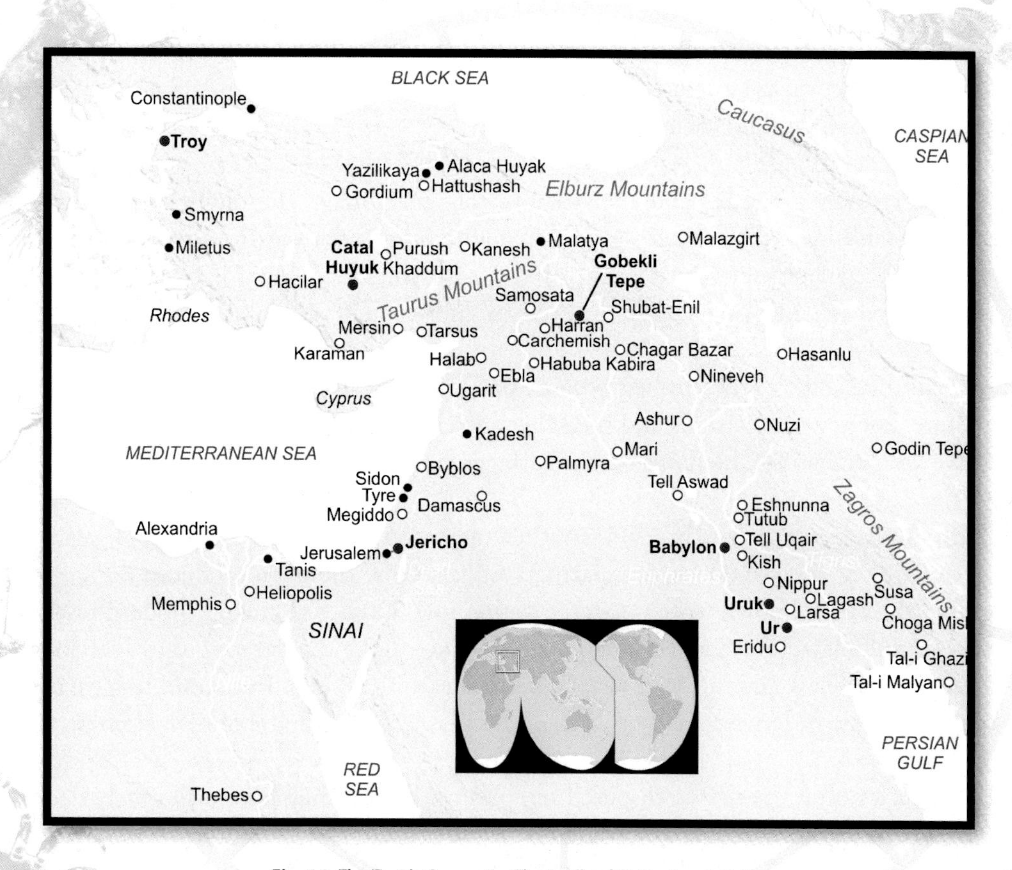

Fig. 14: *The 'Fertile Crescent' – The Cradle of Civilization.* R.Hall

---

9 Another common phallic symbol that most people don't recognize as such is the steeple on a church. Originally they adorned pagan temples throughout the Roman Empire. The early Christians, particularly those in Europe, rather than compete with pagan symbolism sometimes adopted it. The steeple (phallus) was placed so as to be the highest point on the building so that it could be seen from a great distance. Its meaning: By displaying a phallus for the entire community to see it demonstrated and pointed to the direction of ownership.

Looking at the figurines we see that they all depict a pregnant woman and that the breasts and genitals are exaggerated. The figures have no face or feet. Now, this isn't because they didn't have the skills. Carvings have been found in which exquisite facial features have been carved. Some anthropologists have suggested that they have all been carved by women and it is the exaggerated view a pregnant woman sees of herself when she looks down over her naked body. Standing up a woman about to give birth cannot see her feet and, of course, she cannot see her own face. However, others believe that they are artifacts from a prehistoric earth-mother fertility cult. The figurines, which are all small enough to be carried or held in the hand, may have been fertility charms, thus the exaggeration of the reproductive features of a women's body. If they did belong to an earth-mother cult they are the relicts of the oldest known religion on earth. The figurines vary in age from 15,000 to 27,000 years.

The change from the nomadic way of life of the hunter-gatherer to settled agricultural communities was gradual, occurring in different places at different times. The first farming settlement that we know of appeared in the Jordan Valley at the height of the last ice age, 19,400 years ago. Farming communities also appeared in Japan around 17,000 years ago. As the ice-age waned small farming communities began to appear throughout what is known as the Fertile Crescent (Figure 14). This region is also referred to as the 'Cradle of Civilization'.

11,600 years ago hunter-gatherers built a great temple, the oldest known on Earth, at a site known as Gobekle Tepe in southern Turkey. The elaborate structure covers 22 acres and consists of massive concentric stone walls with carved limestone pillars 18 feet high and weighing 16 tons. Some archaeologists believe that spiritual beliefs and the love of spectacular structures gave rise to civilization. In other words the temples came first and then communities were built around them.

Some villages became trading centres and evolved into the first towns. The oldest known town on earth is Jericho, which was established 10,500 years ago. The name Jericho is derived from the Hebrew 'Yerah", which means 'Moon'. There is evidence that the people of Jericho worshiped a Moon deity and that they established the very first agricultural calendar based upon the cycle of the Moon.

With the passage of time some of these towns became powerful city states. With the rise of civilization came a new social structure. There were divisions and specializations of labour, an established aristocracy, and the new rising middle class of merchants. Chieftains became princes and shamans became priests.

Catal Huyuk in Anatolia (present day Turkey) is another ancient town founded about 9,500 years ago. By 6,000 B.C. it had a population of about 6,000 traders, farmers and artisans. The town produced pottery and textiles. It is here that we find the oldest known woven material dating to 7,200 B.C. Unlike a modern town Catal Huyuk had no streets. One building adjoined another and people walked along the flat roof tops. Trapdoors and ladders gave access to the houses below and the dead were buried beneath the floors. This seems to have been a common practice because the people of Jericho also buried their dead beneath the floors of their homes.

Here at Catal Huyuk we find the first undisputed temples on Earth. They have elaborately decorated shrines with walls painted with scenes that depict hunting and ritual burials. In the temple are a number of statues of women that have the same physical form as those found in caves 20,000 or more years ago. One of these statues is of a grossly overweight woman sitting on a throne with two leopards on either side of her. The woman is giving birth with the head of the baby clearly visible. In addition to leopards, which some claim are female lions, the other animal that is found in the temple is the bull. It is worth remembering at this point that the woman, the lion and the bull are three very significant signs of the Zodiac, Virgo, Leo and Taurus.

The temple with the female effigies at Catal Huyuk is believed to be dedicated to a Mother Goddess and may represent the earliest known shrine to what would become the Sumerian Goddess Inanna. The Sumerians are a people that settled in Mesopotamia from about 5,500 B.C. Mesopotamia means 'between two rivers' and is the fertile plain between the Tigris and Euphrates Rivers, in what is now Iraq. The Sumerians developed writing (on clay tablets) and invented the wheel around 3,100 B.C.

Inanna was the Goddess of the great city state of Uruk; which incidentally, some historians believe is the origin of the word Iraq. Her name may be derived from Nin Anna, which means Queen of Heaven and or Nanna, the moon god. She was the goddess of love and fertility. Her priestesses practiced ritual prostitution at her temples celebrating love, sensuality and the mysteries of sex. But she was also a ferocious goddess of war. With the passage of time her cult spread throughout the whole of the Middle East and beyond. She became the Babylonian Ishtar and the Greek Aphrodite. She was the goddess of the seasons and her great festival was the spring equinox.

The origin of the constellation of Virgo is believed to be associated with this ancient goddess and was probably built around its brightest star Spica, which rose just before dawn at the spring equinox 14,000 years ago. Virgo means virgin. But, why was a fertility goddess called a virgin?

A major problem with interpreting ancient writings is that languages evolve. The first time you listen to Shakespeare it sounds strange, but that was how English was spoken in those days. I'm told by a linguistic expert that if I were to travel a thousand years back in time I would have difficulty understanding a word of so called 'common English'!

The meanings of words can also change. Let me give you a few examples. If I were to ask you whether or not you were a pagan most readers would probably say not. But, the word pagan comes from the Latin *paganus*, which means peasant or country folk. So, if you live in the countryside you're a pagan and, probably a villain or a heathen. A *villain* is a rural resident, someone who lives in a village and a *heathen* is someone who lives on a heath – a shepherd or a gypsy.

How did country folk and villagers become pagans, heathens and villains? Well, in Europe during the first millennium, as the power and influence of Christianity spread, the Church managed to convert the people in the towns and cities and almost eradicate the old religions. But they were not so successful in rural areas where the people continued in the 'old ways'. Consequently, the word pagan began to be associated with people who continued to worship the pre-Christian deities. It was the policy of the Church hierarchy to demonize the gods of other religions so that people who followed a different faith were said to be bad or evil people, villains. The word 'wicked' is a debased form of the word *wicca*, which is an old English title for a priest or priestess of the old religion. A person who preached the old ways was a wicked person. One of the most well known characters in mythology is Merlin. Do you think of Merlin as a wicked person? Merlin was a wiccan, which means wizard.

Another, more recent example is the word gay. The meaning attached to the word gay today is very different to that of forty years ago. In 2,000 years from now how would a researcher interpret a 20th century writing in which someone is described as gay?

The word virgin didn't always mean what it does today. In Europe, a few centuries ago, it simply meant a maiden or unmarried woman. Further back into antiquity its meaning often had nothing to do with a woman's sexual activity. A virgin was 'a wise woman of high rank'. Priestesses were often called virgins. As priestesses they were certainly considered to be wise and of high rank. But, most weren't virgins as the term is used today.

It should also be noted that virgin births, immaculate conceptions are a common feature in religious stories of old. Zeus often sired children upon mortal woman. The Judea-Christian stories are little more subtle, the woman in question is visited by the 'Holy Spirit'. According to the Bible John the Baptist was the result of an immaculate conception and it could be argued therefore, that he too was a son of God. It may come as a surprise but virgin births are not uncommon. If you have been brought up in a 'Christian' society you will know the story of the virgin birth of Christ[10], and possibly that of his mother, Mary. If however, you have read the Qur'an (Koran) you will also know that Muhammad was born from a virgin. Hebrew writings tell us that Moses was born from a virgin and that Abraham was born from a virgin. In fact anyone who was really important was born from a virgin. I will leave it to you the reader to decide whether all these women in history were giving birth without being visited by a man, or did it really mean that the person in question was born from a very wise and important woman.

10 According to Paul Morris, Head of Religious Studies at Victoria University, Wellington, New Zealand, the Virgin Birth is based on the words of Isaiah 7:14, and was a mistranslation of a Hebrew passage in the Bible. When it was translated into Greek, the Hebrew word, "almah" (young woman), was translated into Greek, "parthenos", which means "virgin".

8. The Pillars of Heaven

Canst thou bind the chains of the Pleiades, or loose the bands of Orion?

Job 38:31

For perhaps 30,000 years people had used certain bright stars to forewarn them of seasonal changes. They were signs or portents in the sky. Four of these bright stars formed the framework around which the Zodiac was constructed. These stars were known as the 'Pillars of Heaven'.

The solstices and equinoxes, the four seasonal turning points of the year, were the most important days in the calendars of the early civilizations. With the passage of time they evolved into the Great Festivals that were, and still are, celebrated around the world. The Pillars of Heaven were four bright stars which, 5,000 years ago, marked the points in the sky where the sun was at the times of the equinoxes and solstices. They were also known as the 'Royal Stars'.

We can easily identify three of the Pillars of Heaven. The golden-yellow star Aldebaran, the Heart of the Bull (Taurus), marked the spring equinox. The blue-white star Regulus, Heart of the Lion (Leo), marked the summer solstice. And, the bright reddish star Antares, the Heart of the Scorpion, marked the autumn equinox. The winter solstice was in Aquarius but there are no first magnitude stars in this constellation. However, it seems almost certain that this fourth Royal Star is the bright white star Fomalhaut. This star is at the correct longitude, approximately 90 degrees from Aldebaran and Antares, but is about 20° south of the ecliptic. In modern star charts Fomalhaut is the brightest star in the constellation of Pisces Austrinus, the Southern Fish. It is the fish that drinks the water that flows from Aquarius. In Persia, 5,000 years ago, Fomalhaut was known as Hastorang and was said to be one of the four Guardians of Heaven, the sentinels that watched over other stars.

However, some authors have suggested that the winter solstice pillar was Altair, the first magnitude star in the constellation of Aquila, the Eagle. The reason for this is that in Persian temples (now Iran), the sun-god Mithra is depicted fighting a griffin. A griffin (back cover upper right) combined the features of a bull (Taurus at the vernal equinox), a lion (Leo at the summer solstice), a scorpion (Scorpius at the autumn equinox), and an eagle. Altair is 28° north of the ecliptic and is therefore easier to see than Fomalhaut from northern latitudes but, it is some 45° longitude (R.A. 3 hours) removed from the then location of the winter solstice. Fomalhaut, which was described by the Persians as one of the four Guardians of Heaven, is the more likely candidate.

Furthermore, there is uncertainty as to which constellation the Eagle referred. Ancient astronomical/astrological projections of the heavens divided the celestial sphere into four quadrants. These quadrants were identified with the figures of a bull, a lion, an eagle and a man pouring water from a jar. In this representation we have an eagle rather than a Scorpion identifying the autumnal equinox. So which is it, a scorpion or an eagle?

The answer is of course, that both could be correct. Different cultures made different figures out of the stars. To the Chinese Scorpius is a dragon; to the Babylonians Aries was a man. In addition, with some cultures, the figure and symbolism of a constellation changed with the seasons. In Maori astronomy for example, we find the same stars being put together in different ways at different times of the year to represent different seasonal events.

Falcons (or eagles) were sacred to the Egyptians. The falcon is identified with a creator god who, in the form of a falcon, flew at the beginning of time. The sun-god Horus is depicted as a falcon or a man with the head of a falcon and, his mother Isis, who often took the form of a golden falcon, is depicted as a woman with wings (Figure 4). Horus is usually identified with the constellation of Orion, which is close to and rises and sets with Taurus.

We know that the stars at the autumnal equinox represented a scorpion, or rather a Scorpion Man, to the people who lived along the Euphrates 6,000 years ago. It was a symbol of darkness marking the decline of the sun's power. Perhaps at the spring equinox, when these stars were opposite the sun and due south in the night sky, they represented an eagle symbolizing the infant Horus and the re-birth of the Sun.

Fig. 15: *The Zodiac 3,000B.C. Atlas was guardian of the Pillars of Heaven, four bright stars which marked the location of the solstices and equinoxes.Farnese Atlas – MuseoArcheologico, Nazionale di Napoli, Naples, Italy.* R.Hall

The Zodiac signs (as the figures we identify them today), along with the location of the solstices, equinoxes and the Royal Stars, for 3000 B.C. is shown in figure 15. Note however, that at this time, the Zodiac probably consisted of just the four signs marking the seasons, the solstices and equinoxes.

The signs that contained the equinoxes and solstices were of the greatest significance. The most important of these was usually the sign that marked the spring equinox for it was the 'leader of the heavenly host' and marked the 'rebirth' of life on earth. It was a sign that the food supplies of the people would soon be restored. But, due to precession the sign of spring changes with the passage of time. Today the spring or vernal equinox is in Pisces. 2,500 years ago it was in Aries and 5,000 years ago it was in Taurus.

For more than two thousand years (4000 B.C. to 1850 B.C.) the Sun was in Taurus at the time of the spring equinox. The Bull was 'the prince and leader of the celestial host' and held as a sacred constellation by most of the nations in Europe and Asia. Images of The Bull adorn temple walls. We find statues of gods with the head of a Bull or bulls with a human face. Statues of the great Egyptian fertility goddess Isis show her wearing a headdress in which the sun is held between the horns of a bull (Figure 4).

The Bull is an important figure in early creation stories. In the ancient Persian/Hindu story the god Mithra, the Son of Light, fought and killed the 'primeval bull' which released the life forces of nature. (front cover lower left) From the bones of the bull he made bread and from its blood came the wine.

Symbols from this era are still with us today. The letter 'T' is derived from Tau and is believed to represent the head and horns of the Bull. It was regarded as a symbol life and resurrection. The ankh, the Crux Ansata (symbol upper right figure 37), is also a symbol of life and fertility and originally represented Taurus (Tau) with the Sun above. It is an emblem of eternal life and was adopted by the early Christians in the Eastern sector of the Roman Empire as the symbol of their faith.

Fig. 16: *Left: The Seven Sisters, ElihuVedder 1885, set against the Pleiades, montage by R.Hall. Right: The Pleiades Star Cluster.* Photo by John Drummond

On the tail of Taurus the Bull is the Pleiades star cluster that sparkles like a casket of jewels[11]. These are the Seven Sisters, the daughters of Atlas. (Figure 16) The cluster contains many hundreds of stars of which six can be easily seen with the unaided eye. A total of nine can be discerned with a keen eye from a dark sky site. In mythology the Seven Sisters were the daughters of the titans Atlas and Pleione (Phoebe) (2+7 =9).

Atlas was guardian of the 'Four Pillars of Heaven' which held up the sky. This symbolism of pillars holding up the sky is also found in the New Zealand Maori story of Ranginui (the Sky Father) being held up on four posts. Like the pillars these posts are bright stars in the sky that identify the solstices and equinoxes.

11 The cluster is on the shoulder of the Bull in the modern representation of the figure of Taurus which shows only the head and forequarters of the bull (figure 37).

Atlas incidentally, did not hold the world on his shoulders. The idea that Atlas held up the world probably comes from the Greek statue in which he is holding a globe on his shoulders. However, a close look at the globe reveals that it is not the world but a celestial globe of the heavens, which embodies the constellations of the astronomer Hipparchus. (Figure 15)

Of the four pillars one was more significant than the others. This was Aldebaran, the Pillar of Spring. The Babylonians called it Marduk, the Spring Sun. Aldebaran marked the point in the sky where the sun was at the time of the spring equinox. But no one could see this happen because it occurred in the daytime sky. But, what they did observe was this… as the sun moved to Aldebaran the Seven Sisters rose in the dawn twilight. Indeed, Aldebaran means "The Follower" because, after the equinox, it followed the Pleiades into the pre-dawn sky. Consequently it was the heliacal rising of the Seven Sisters which marked the time of the equinox and the beginning of the year for most of the peoples of Europe and Asia.

In Greece the Seven sisters were associated with Athene, the Great Goddess of Civilization. Their heliacal rising marked the beginning of the year and the opening of the sea lanes. This was the time when it was safe to sail so it signaled the beginning of trade for that year, and if they were inclined, it was the time to make war. Great temples such as the Parthenon were aligned to their rising.

In India the Pleiades were known as the Krittikas, who were the six nurses of the motherless infant god of war, Skanda. He had six heads for his better nourishment. As we shall see later, the Krittikas marked the beginning of the Hindu Lunar Zodiac and the beginning of their year.

Down here in Aotearoa New Zealand we know the Seven Sisters by a different name; Maori call the star cluster 'Matariki', or rather Matariki and her six daughters. In Maori lore Matariki is one of the most ancient of goddesses and, for many tribes, it is the dawn heliacal rising of Matariki that heralds the Maori New Year.

How is it that Maori, who inhabit a land on the opposite side of the world, have the same traditions as the great civilizations of antiquity? Well, if we roll time back 5,000 years we discover that the ancestors of the Polynesians were, at that time, sea traders who lived and worked along the coastline from Korea to Indonesia. Along this coastline were city states that traded with China to the east and India, Egypt and Greece to the west. Most of these states used a calendar in which the year began with the heliacal rising of the Seven Sisters. When these Asian city states declined is when we believe the migrations into the Pacific began. The sea traders became hunter/gathers of the sea and they carried their ancient star lore with them. So, to this day, Maori will say that when Matariki (Seven Sisters/Pleiades) rises, it is the dawn of a new year – a tradition that began with civilizations lost in the well of time.

Due to precession the heliacal rising of Matariki, which once marked the northern spring equinox, now occurs in early June, close to the winter solstice in the southern hemisphere.

The signs of the Zodiac, it was believed, influenced the lives of people and the fortunes of nations. In addition, it was also believed that certain signs dominated epochs of time influencing the entire world. Most people have heard of the 'Age of Aquarius' or will remember the song 'The Dawning of the Age of Aquarius'. What does this mean? Well, the astrological age or epoch depends upon where that all important spring equinox is. At present the vernal equinox is in Pisces so we are in the Age of Pisces. 2,500 years ago it was the Age of Aries and 5,000 years ago the Age of Taurus. Each age lasts for about 2,200 years.

The epoch of each 'Age' is given below along with the signs that contained the equinoxes and solstices. The dates are based upon the Indian or Jyotish astrological signs because the boundaries of these constellations appear to be unaltered since the zodiac was formalized with each sign extending 30 degrees along the zodiac. The Ages are placed in two groups. The second, prior to 4000 BC, are calculated backward in time to periods before the solar zodiac was created.

	AGE			
Epoch	**Spring**	**Summer**	**Autumn**	**Winter**
AD 300 - present	Pisces	Gemini	Virgo	Sagittarius
1850 BC – AD 300	Aries	Cancer	Libra	Capricornus
4000 -1850 BC	Taurus	Leo	Scorpius	Aquarius
6150 - 4000 BC	Gemini	Virgo	Sagittarius	Pisces
8300 - 6150 BC	Cancer	Libra	Capricornus	Aries
10450 - 8300 BC	Leo	Scorpius	Aquarius	Taurus
12600 - 10450 BC	Virgo	Sagittarius	Pisces	Gemini

It is believed in astrology that each age will bring about a new order with major social changes. Looking at the dates of the ages given in the table above can you see major changes in world history from one to another?

In A.D.391 Christianity was established as the State Religion of the Roman Empire and, it is claimed by some, that Pisces is the 'Age of Christianity'. Indeed, Pisces, the two fishes, was a sign used by early Christians.

"I shall make ye fishers of men"

Perhaps you know the New Testament story of the two fishes and the five loaves - these were said to be the two fishes in the sky. However, Pisces did not originate with Christianity. As we shall see later the story of the two fishes goes back to the time of the Great Flood.

The Age of Aquarius will begin when the vernal equinox moves into Aquarius and this will occur, using original classical constellation boundaries, in the year A.D. 2451 (A.D. 2589 using modern astronomical boundaries). This will be the dawn of the Age of Aquarius. Thus, despite what some people will tell you, the dawning of the Age of Aquarius is still a long way off.

Why is the Age of Aquarius a long awaited and celebrated time? The answer is that the great wheel, the great year, will have turned the first quarter. The peak of the Age of Aquarius will occur in A.D. 3450. At this time the 'Pillars of Heaven', the 'Royal Stars' will once again be located at the equinoxes and solstices. This is the ancient astrological importance of the Age of Aquarius. It also reminds us that the ancients, in order to predict the Age of Aquarius, were aware of precession – even if its significance is lost to many modern day astrologers. The conjunction of the Royal Stars, the Pillars of Heaven, with the solstices and equinoxes (northern hemisphere seasons) will be as follows:

A.D. 3450

Spring Equinox:	Fomalhaut (Aquarius)
Summer Solstice:	Aldebaran (Taurus)
Autumn Equinox:	Regulus (Leo)
Winter Solstice:	Antares (Scorpius)

9. Gate of the Gods

Gilgamesh, whither rovest thou?
The life thou pursuest thou shalt not find.
When the gods created mankind,
Death for mankind they set aside,
Life in their own hands retaining…
Do we build a house for ever?

Epic of Gilgamesh (third millennium B.C.)

Our concept of the nature of the universe has evolved with the passage of time. In order to understand the symbolism behind the signs of the zodiac and the solstices and equinoxes we need to be familiar with early beliefs of the structure of the universe.

With our present wealth of accumulated knowledge it is very easy for us to say that this or that is obvious. But things that are obvious to us were not necessarily so to people who lived long ago. For example, for most of human history, the vast majority of people would have assumed that the earth was flat. Furthermore, the observational evidence they had pointed to that conclusion. In the absence of the knowledge of gravity or of the immense scale of the earth the reasoning may have been as follows. If the earth was round, and you were at any distance from the top, you would find the entire terrain tilted as if it were going down hill towards the horizon. As no traveler had experienced this it would be an unnecessary complication to consider the earth as anything but flat. Now you may say that, standing on a coastline, you have observed a ship rise over the horizon. But, they didn't have ships of that size long ago. If someone traveled out to sea in a small craft it would have been no different to observing people travel away from you across a great plain. As they moved further and further away they appeared smaller and smaller until they were lost from sight. Oceanic seafarers may have suspected something by observing mountains or islands rising or sinking below the horizon. But most people weren't oceanic seafarers and, with the exception of the nomads, never traveled more than about 30 kilometres from where they were born.

A common belief in Europe and Asia at the time of the early civilizations was that the (flat) earth was encircled by a vast ocean. At the centre of the landmass were the tallest mountains. Above the earth was the sky which has the appearance of a vast dome over our heads. The sky dome rested on the horizon at the edge of the world. In the ancient Hindu model of the universe the world rested upon the back of three giant elephants which in turn stood on the back of an enormous turtle the size of the earth. A fundamental question is, what's the turtle standing on? If you can answer this question you will have discovered one of the most awesome characteristics of the Universe.

There is something very different about the sky, particularly the night sky, which has a property unlike anything found on earth. It is something that would have been obvious to our early ancestors but which today, few people would notice. The world around us is continually changing. Nothing around us appears to have any permanence. People are born, live their lives and die. But there is one place in our environment that does have permanence, and that's above our heads. As our ancestors sat by the camp fire generation after generation saw the same familiar stars, the same familiar patterns they formed which we call constellations. The stars appeared to be immortal. Not only are the stars immortal but they are untouchable by mortal beings. It is little wonder then, that just about every culture has placed their god or gods in a place called heaven. The heavens above, that realm that is both immortal and untouchable by mortal beings.

The ancient Egyptians believed that the earth, the mortal realm was nothing more than a reflection of the immortal heavens. The Milky Way was the celestial Nile and, at the beginning of the Egyptian year, the two Niles merged at the horizon. The stars of the celestial Nile were reflected in the earthly Nile. Some archaeologists believe that the three great pyramids at Giza, built on the western side of the Nile, represent the three bright stars in the "Belt" of the constellation of Orion that can be seen on the western side of the celestial Nile. To the Egyptians the constellation we know as Orion represented the falcon-headed god of the sky, Horus. Sometimes it was identified with Osiris, god of the Underworld. See Figure 4.

We have then the earth, which is the realm of mortals, and the heavens which is the immortal realm of the gods. There is a third realm, just mentioned above, the Underworld. We all know what is in the underworld from tales we have been told as kids – fire and brimstone. Anyone who lives near Rotorua or Mount Etna can testify to this. These old beliefs stem from common things that people observed – fire comes from below. While they couldn't see it, the Underworld was also believed to be an immortal realm. Incidentally, the Underworld was not Hell; it was the abode of the dead. Hell is essentially a Judai-Christian concept.

The Heavens, the Earth and the Underworld, these were the three great realms of existence. No one, not even gods could just pass from one realm to another. There were gates or portals through which all beings must pass when moving from one realm to another.

The Hebrews believed that there were two gates to the Heavens and that these were located in the signs which housed the solstices. At that time, around 500 B.C., the Summer Solstice was in Cancer. This was known as 'The Gate of Men', through which souls descended from heaven into human bodies. In Christian times this gate was called the 'Manger', being the portal in the heavens from which the spirit of god descended into the infant Jesus.

The Winter Solstice was in Capricornus. This was known as 'The Gate of Gods', through which the souls of men, after death, ascended to heaven. You must have heard of the "The Pearly Gates", this is where this concept comes from. Incidentally, I wasn't being sexist when I said "...the souls of <u>men</u> ascended to heaven". Until comparatively recently the fiercely patriarchal religions of Judaism, Christianity and Islam, didn't recognise that women had souls.

The Gate to the Underworld was believed to be located on the Earth, usually in a deep cave on the far side of a river.

The gates to Heaven and the Underworld are locked so that no mortal may wander into heaven nor the dead enter the mortal realm.

But, once each year, at the time of a solstice or equinox, one of the gates would open. It was at these times that god (or the gods) entered the mortal realm. Consequently, it was believed that all important divine events would occur at either a solstice or an equinox. The Gate of Gods in Capricornus opened at the time of the Winter Solstice. No living mortal could peer through this gate into heaven because when it was fully open, at the time of the solstice, the sun stood before it.

Whereas the Earth was thought of as a single plane of existence, Heaven and the Underworld were believed to be built on a series of levels. A common belief was that there were twelve levels of Heaven and seven levels of the Underworld. The supreme god of each realm resided at the highest (twelfth) level of Heaven or the lowest (seventh) level of the Underworld. The Egyptians however, believed that the Underworld was divided into 12 realms, which leads us to an intriguing tale.

If you watch the sky for a while you soon become aware that it is moving. The sun, moon and stars appear to rise in the east, slowly move westward and eventually set. This daily motion is simply due to the rotation of the earth. But our ancestors were not aware that the Earth was moving. To them, each day the Sun, Moon and stars rose from the underworld, traveled across the sky, and then descended back into the earth, into the underworld.

The ancient Egyptians believed that during the night the sun was traveling through the Underworld. At the end of each day the sun-god Re descended into the underworld and, if he defeated death once again, would arise triumphant at dawn.

While today we know that night and day is a natural cycle people didn't always think of it in that way. Everything in nature was believed to be controlled by the gods. And, like people, the gods were not always predictable. Perhaps the sun wouldn't rise from the underworld!

Consequently, the Egyptians were very interested in the sun-god's passage through the underworld and his daily struggle and victory over death. They believed that the underworld consisted of twelve regions and, to monitor the sun-god's journey, the priests kept watch for the rising of twelve equally spaced stars in succession. It was the sun-god that turned the wheel of the heavens and, so long as the heavens continued to turn, they knew he was moving. The rising of each star signaled the successful passage of Re through one of the realms of the Underworld. These stars were called the 'Hours' and the time interval between the rising of each star became known an hour. Later this concept of twelve equal divisions of the night was extended into the daylight to give us our day of 24 hours. So, if you have ever wondered why we have 24 hours in a day it all comes down from monitoring the journey of an ancient sun-god through the underworld.

The cultural significance of the solstices and equinoxes is then, that they were the original holy days upon which significant spiritual/religious events took place. These were the times of the 'act of god'. The gates to heaven, the portals used by god (or the gods) to enter the mortal realm, were located in the two signs which contained the solstices (at that time). These were the Gate of Gods in Capricornus and the Gate of Men in Cancer.

PART III:

THE GREAT FESTIVALS

The first Zodiac probably consisted of just four signs that were built around the four 'Pillars of Heaven', four bright stars that identified the location of the Sun at the solstices and equinoxes. In this section we discuss the myths and symbolism built around the equinoxes and solstices and their associated zodiacal signs. Here we find the origins of some of the great stories of antiquity that became incorporated into world religions.

10. Dance of the Seven Veils

"I am all that hath been, and is, and shall be, And my veil no mortal hath yet removed."

Inscription on the temple of Sais,
Lower Egypt

There is one
race of men, one race of gods;
both have breath of life from a single mother.
But sundered power holds us divided, so that one is nothing, while for the other the brazen sky is established
their sure citadel forever.
Yet we have some likeness in great intelligence, or strength, to the immortals,
though we know not what the day will bring,
what course after nightfall
destiny has written that we must run to the end.

Pindar, Sixth Nemean Ode

In ancient times, throughout Europe and Western Asia, there were four great festivals that were celebrated each the year. These were the two equinoxes which were the times for planting and harvesting, and the two solstices which were the longest and the shortest days (hours of sunlight). In some calendars they were separate days that were not attached to a month. They were the four days marked zero, they were the days of the gods. We still celebrate these festivals today but you may not think that you do. Of the four there was one that was more important than all the others. This was the Spring Equinox.[12] For most of the people of Europe and western Asia, prior to 45 B.C., the spring equinox marked the beginning of the year. Even in Britain, until1752, the year began not on the 1st of January but on March 25th, close to the spring equinox. Today we call this great festival of the spring equinox, Easter.

Most people tend to be very narrowly focused in their own culture. We tend to think that stories that are special to us are unique. But when we look further a field we often discover that other cultures have similar traditions. Most of the cultures that arose from Mesopotamia, the great civilizations of Europe and Asia Minor, all have an Easter story. It is the story of the resurrection of a male god. And, these stories pre-date Christianity by thousands of years. I'd like to tell you one of the oldest of stories that we have of the resurrection, from which the festival of Easter arose.

12 The Spring Equinox was in Taurus from 4000 B.C. but spring itself was identified with the heliacal rising of the Pleiades. From about 2500 B.C. the heliacal rising of Aries, which became identified with a Babylonian fertility god, was the herald of spring. The equinox moved from Taurus into Aries in 1850 B.C. From about 500 B.C. the heliacal rising of Pisces became the sign of spring. Pisces was the sign of Israel and, as mentioned earlier, it became a symbol of the early Christians. The spring equinox moved from Aries into Pisces in A.D. 300.

The story comes from Babylon (c. 2000 B.C.) and was undoubtedly inherited from the Sumerians (c.3000 B.C.). The Babylonian and Sumerian stories are very, very similar but the Babylonian story is more complete.

But, first some background to the story. For most of human history, in the spiritual beliefs of people, the 'Creator' was not a 'He' but a 'She', a goddess. This should not be a surprise because it is an observed fact that it is the female that brings life into the world. Further, it is 'she' who generally protects and nurtures her young. So, if there were a creator wouldn't it be female?

There is one race of men, one race of gods; both have breath of life from a single mother.
Pindar

This creator goddess gave birth to a family of 'gods'. Initially they were simply identified as the forces of nature. With time they evolved to become a reflection of the society that worshiped them. The individual gods were models of the moral values, arts and skills of the people and, like a human family, there was rivalry between gods.

You will recall that it was a common belief that the universe consisted of three realms – the Heavens, the Earth, and the Underworld. Each realm had its own gods which was ruled by a king or queen. The rulers of the three realms formed 'The Trinity'. The concept of the Trinity is an ancient belief and pre-dates all existing religions. Most people in the western world will know of the Christian 'Trinity' - God the Father, God the Son and God the Holy Ghost; or the Hindu Triad – Brahma the Creator, Vishnu the Preserver, and Shiva (Siva) the Destroyer.[13] Brahma was identified with the rising sun, Vishnu with the sun at noon and Shiva with the setting sun.

The Trinity is also reflected in religious orders. Some religions were ruled by a Triumvirate of powerful priests known as the Priest (Pope), the Prophet and the King. So too were religious orders of knights – the Grand Master, the Seneschal, and the Marshal.

Incidentally, one of the major differences between Islam and Christianity, and between different factions within Christianity, is the Catholic recognition of the Trinity. Muslims argue that instead of the "One God", recognition of the Trinity follows the old pagan beliefs and pays homage to three different gods. Catholics however, argue that the Trinity is three different aspects of the One God. In much the same way it could be argued that the Hindu Triune is also three different aspects of a single deity. Personally, I think that it's just a matter of terminology. The role of the Angel Gabriel seems to me to be very similar to that of Hermes (Mercury). Whether you call them gods or angels is just a matter of labeling. Had the Old Testament been written by the ancient Greeks, Satan would have been called a god rather than a fallen angel.

Ishtar and Tammuz

In the Babylonian tradition the three great divinities of the Trinity were brothers and sisters. The Great Goddess of the Earth, ruler of the mortal realm was Ishtar (to the Sumerians she was known as Inanna). Her brother Shamash (the Sun) ruled the Heavens, and her sister Ereshkigal, Queen of the Dead, ruled the Underworld.

13 'The Destroyer' is not a negative label. Shiva is also known as 'The Auspicious One' and is often depicted with his foot on a demon. He was also the Lord of Time and held a drum which he used to bring the universe into being. The beat of the drum is time and ultimately will sound the destruction of this universe and the birth of the next.

Ishtar was also the Babylonian name for the planet Venus, with which the goddess is identified. While the Sumerians knew this goddess as Inanna they called her "The message of Istar" (Venus?).

The planet Venus is seen either just before dawn, when we call it the morning star or, just after sunset when it is called the evening star. Ishtar was the Two-faced goddess. When she was the evening star she was the goddess of sex, love, desire and fertility. Eternally promiscuous, she bathed in a sacred lake each evening to restore her virginity. She was the goddess of music, and her slaves played instruments and sang wherever she went. As the morning star she was a blood thirsty goddess of war. Her songs of lust became war-cries which froze the enemies blood, her arrows of desire became weapons of destruction. This warrior aspect was her primary attribute to the war-like Assyrians. To the Babylonians she was the "Lioness", the mother goddess who protected the city.

Now for the story: This Great Goddess had a male companion, a young man who, although a mortal, was the love of her life. His name was Tammuz (in the Sumerian story he is a shepherd by the name of Dumuzi). One day he is out hunting and is mortally wounded. Following his death the Great Goddess pined for her lost love, she ceased to work and the world began to die. The leaves shriveled and fell from the trees, the sun began to disappear from the sky, and the food supplies of the people began to disappear.

Things got worse and worse on Earth until eventually the goddess decided to do that which no mortal woman can do. She would descend into the Underworld and get her lover back.

Now, because she was the goddess of the upper world she was forbidden access to the Underworld. Her sister Ereshkigal ruled the Underworld. But, the two sisters didn't get on. When Ishtar asked for her lover back Ereshkigal said "No way, all those that come to me never return to the mortal plane".

Ishtar flew into a rage and threatened to break down the gates of the underworld and release the dead into the mortal realm, who would devour the living. This panicked the other gods who then brokered a deal between the two goddesses. Her sister, Queen of the Underworld, agreed to allow Ishtar to descend through the Seven Levels of the Underworld, on the condition that at each of the Seven Gates she part with an item of clothing. This is the ancient story enacted by the "Dance of the Seven Veils".

Now, if your mind is conjuring up the image of the perfect woman carrying out the original striptease, you have lost the plot. In New Zealand, Maori will say that when a man loses his mana or power he is naked. The removal of her garments, the vestments of her office, is symbolic of her relinquishing her powers. Consequently when she arrived before her sister she was naked, she was without power. The evil Ereshkigal set Namtar, the plague demon, loose upon her sister and Ishtar was covered with diseases and then imprisoned in the darkest depths of the Underworld.

Because Ishtar was absent, the earth was without fertility. The Babylonian texts explain, in some detail, that when the goddess is absent from the earth all male creatures, including man, become totally impotent. She is remember, the goddess of sex and fertility. When her brother Shamash (the Sun) realized what had happened he ordered Ishtar's release and threatened war on the Underworld. A new deal was brokered. The demon Namtar, fearing reprisal, released Ishtar and sprinkled her with the waters of life restoring her health and beauty.[14] The spirit of Tammuz was given to Ishtar and, at each of the seven gates, her vestments of office were returned. Ishtar returned to the mortal plane and resurrected her lover from the dead, who then became a god.

---

14 It was said that the waters of life flowed through the Underworld and that this was the reason why life sprung from below.

There was however, a price to pay. In the deal brokered with Ereshkigal, Tammuz would, from that day forward, spend half of the year on Earth and the other half in the Underworld. Each year, on the anniversary of his death, Tammuz returns to the Underworld. This is the Autumn Equinox. And, when he does so, Ishtar refuses to work. That is why the leaves once again fall from the trees and it becomes cold and miserable. When he is returned to her she resumes her work and life on earth is born again. This is the Spring Equinox. The Easter story is then, the story of the seasons, and the resurrection of life following the harsh winter.

The first sign in the Babylonian Zodiac was "Mul lu Hun ga", the "Hired Man" which was equated with Dumuzi, the shepherd lover of Innana. In the Babylonian tradition Dumuzi became Tammuz the God of Spring, the consort of Ishtar. At that time (c. 2000 B.C.) the 'Hired Man' rose in the dawn at the time of the spring equinox. Around 500 B.C. the Greeks transformed this constellation into Aries the Ram, the leader of the celestial flock. Its brightest star Hamal, was called 'Dil-kar' the Proclaimer of the Dawn, the Son of Light, the Son of God.

With the passage of time the cult of the Mother Goddess spread throughout the ancient world and was adopted or assimilated into the belief systems of many cultures. She has gone by many names. As already discussed she was Innana to the Sumerians and Ishtar to the Babylonians. To the Mesopotamians she was Astarte, she was the Egyptian Isis, the Anatolian Cybele, the Greek Aphrodite, the Roman Venus, the Germanic Eostre and the Nordic Freyja. That they are all essentially the same goddess is testified by the similarities of the stories and ancient writings:

"Astarte, The one called Aphrodite by the Greeks."

Aphrodite and Adonis

Aphrodite is today perhaps the most well known of these characters, undoubtedly because of her association with beauty, love and sexual desire. She is sometimes portrayed as flighty but usually she is sweet, coy and sexy, and above all beautiful, the perfect woman in man's eye. This however, is a medieval re-write of her traits.

It is often said that god created man in his own image. I rather think that it was the opposite way around. I believe that the personality and behaviour of god or the gods reflect the culture from which they arose. Thus, the present concept of Aphrodite reflects a medieval (male) perception of the role and status of women within society.

In ancient cultures, in addition to fruitfulness and sensuality, Aphrodite also represented human emotions, passions and creativity. Consequently, she also had a dark side and was subject to jealous rages and wanton destruction. Inanna, the Sumerian goddess of love, was also the goddess of war. The Babylonian Ishtar was known as "The Lioness". In Greek legends Aphrodite is probably responsible for more strife amongst both gods and mortals than any other deity, which is not surprising because she was the Goddess of Desire. She was revered for her power and feared for her temper. She was indeed, "Mighty Aphrodite".

The story of Aphrodite and her lover Adonis is very similar to that of Ishtar and Tummuz. While out hunting Adonis was fatally wounded by a wild boar. At his death, Aphrodite declares:

'Memorials of my sorrow, Adonis, shall endure;
each passing year your death repeated in the hearts of men
shall re-enact my grief and my lament
[in the celebration of the festival Adonia]." –*Metamorphoses 10.724*

From that time onwards, Adonis spent half of the year in the underworld with Persephone, the Greek Queen of the Underworld. This was winter, a time of hardship. But each spring he returned to spend the rest of the year with Aphrodite.

In ancient Italy, women planted Gardens of Adonis in spring. Today, women plant seeds of grains (lentils, fennel, lettuce or flowers) in baskets or pots. When they sprout, the stalks are tied with red ribbons and the gardens are placed on graves on Good Friday. They symbolise life over death.

Demeter and Persephone

Perhaps the most well-known story of the origin of the seasons is that of Demeter and Persephone. Demeter was the earth-mother goddess who spent most of her time doing that which she loved – tending and nurturing the plants and animals. Her daughter Core (Koure), who was said to be the most beautiful being in nature, helped her mother and created and tended the flowers.

Hades, Ruler of the Underworld, fell in love with her and asked her father, his brother Zeus, for her hand in marriage.[15] Zeus agreed but Demeter, horrified at the thought of her daughter dwelling in the darkness of the Underworld, refused.

Hades abducted Core, made her Queen of the Underworld, and renamed her Persephone. Demeter grieved for her lost daughter. She stopped tending the earth, crops withered and the world began to become as barren as the Underworld.

When people began to starve Zeus asked Hades to return Persephone to her mother. Hades agreed but, by this time Persephone had grown to love him and care for the lost souls of his realm. In addition she had eaten six pomegranate seeds and, it was divine law that anyone who ate food in the Underworld had to remain there for ever.

Zeus decreed that she spend six months of the year with her mother as Core and six months of the year as Persephone in the Underworld. Each year, at what is now the autumn equinox, she descends into the Underworld. Demeter once again refuses to work and the world begins to die. At what is now the spring equinox she returns to her mother and the earth blossoms.

In ancient Greece and Rome women worshiped Demeter and Persephone in a mystery cult known as 'The Twain'. Men were excluded.

These then, are some of the first stories of the resurrection at Easter. But, what does the word Easter mean? And what has the Easter Bunny and Easter Eggs to do with Christianity? The answer is that these are ancient pagan symbols and customs that have been assimilated into what is now a Christian festival. The word Easter comes from Eostre (Eastre), who was the Teutonic goddess of spring and fertility. Her symbols were the rabbit which symbolized fertility (any farmer will tell you how potent this is) and coloured eggs for the colours of spring flowers.

Seeds and eggs are often associated with spring festivals because they contain the promise of new life. The tradition of giving Easter eggs comes from the Ukraine, from the Festival of Astarte which was held on March

15 In some account of this story Poseidon is the father of Core.

17th. Astarte is the Mesopotamian name for the Great Goddess. Originally of course, they were not chocolate eggs but real eggs that were dyed red and then decorated with symbols of protection, fertility, wisdom, strength and other qualities. When I was a child, back in England, I remember that on Easter Sunday we always had boiled eggs for breakfast, the shells of which had been dyed red with cochineal.

Spring festivals were dedicated to the great Mother Earth Goddess and celebrated with a feast. People celebrated for the basic reason that their food supplies would soon be restored. Nawriz, Passover, Easter, St Joseph's Day, Maimuna are all variations of the theme of a spring feast.

In ancient Europe, invocations for the fertility of the ground and the abundance in the new crops being planted were often held at the new moon following the spring equinox. At this time young girls danced around the Maypole and couples pledged their vows in hand-fasting ceremonies.

For people in the southern hemisphere the spring equinox occurs around September 22nd. If they were to follow the old traditions September is the time when they should be celebrating spring and giving away 'Easter' eggs. However, these days most of the old festivals have become fixed dates in our calendar and are, for people living in the southern hemisphere, six months out of sink. Easter is one festival that does not have a fixed calendar date. So how does the Pope decide when Easter will fall. In A.D.325 the Council of Nice decreed that "Easter was to fall upon the first Sunday on or after the first full moon on or after the Vernal Equinox."

11. Ops

While writing the book there was a wonderful add on T.V. here in New Zealand in which a child comes home from school and excitedly tells his mother that he has a light-saber. Mum says "that's nice dear" and returns to the kitchen whereupon she hears a great commotion coming from the lounge. Returning to the lounge she discovers that her son really does have a light-saber and is demolishing the furniture. Mum takes the saber from her son and inadvertently cuts off her own arm. Whereupon she says oops! An understatement; actually it's not. We often hear people say oops when they make a mistake. Did you realize that when you say oops that you are calling upon the help of a powerful pagan deity and that the origin and meaning of this word dates back more than 7,000 years?

Ops (Oops), is another name for the Great Earth-Mother Goddess Cybele (Kybele) and her stories play a significant role in the origin of many important traditions and beliefs in Western Society. The worship of Cybele originated in Anatolia in the Near East at the dawn of civilization. Stories of Cybele were brought back from Troy by the Greeks who merged her character with that of Rhea, wife of Cronos. However, the cult proper was introduced into Greece in the 3rd century BC. It was imported to Rome in 204 BC and developed into one of the three major religions of the Roman Empire. The others were Isis and Mithra.

Originally her temples and shrines were always in mountains or caves (Kybele – cave dweller) and her guardians were lions which lived as pets in the temple. She is often depicted as riding a chariot drawn by lions (front cover lower right). In the Zodiac, her chariot is Virgo which is drawn by the lion, Leo.

As the worship of Cybele spread throughout the ancient world, temples were established at each of the great city centres. Her high priestesses, of which there was one at each of the great temples, became known as the Sibyl and were said to have the gift of prophesy. She was the 'oracle' and, when called upon to prophesy, each word or symbol was written on different leaves which were then placed in order on the temple floor. The temple doors were then opened so that people could come in and read the prophecy. However, if upon opening the door there was a gust of wind, Oops, the leaves were scattered making the words unreadable. The prophecy was lost and the only person that could now help you was the goddess herself, which is why her name was called. To this day people still say 'Oops' when they upset something.

Attis, the male companion of Cybele, was born on December 25th. His mother was the Virgin Nana. On the Day of Blood, or Black Friday, Attis dies. There are variants in the stories that remain. He either castrates himself after being driven insane by the goddess or he is castrated by a wild boar. Either way he bleeds to death beneath a pine tree. From that day forward the pine tree was sacred to Cybele.

On Black Friday a pine tree was placed in the temple upon which was hung an effigy of Attis. This is the origin of our Christmas tree. On top of the Christmas Tree is a five-pointed star which, as we shall see later, represents Venus and is the symbol of the goddess.

Also on the Friday the male temple priests (not the Sibyl) gashed themselves with knives as they wept with Cybele. In fact, in pre-Roman times for a male to become a priest of Cybele required a great sacrifice – self castration on Black Friday. The Romans found this distasteful and abolished the practice.

Two days after his death, at dawn on (Easter) Sunday, a priest opens the sepulchre in the temple revealing an empty tomb. The male god had risen from the grave. The day the tomb was opened was known as Hilaria or the Day of Joy. The resurrection of Attis was hailed by the disciples of Cybele as a promise that they too would triumph over death.

Due to the similarities in the stories many researchers believe that this is the origin of some of the tales surrounding the story of the resurrection of Christ. The Christians were not alone in adopting significant stories and symbolism from other cultures. It is a common feature in the evolution of just about every world religion. The story of the "Great Flood' for example is found in Hebrew, Babylonian, Sumerian, Akkadian and Greek folk lore. The stories are all very similar, only the names of the characters have been changed.

In A.D.313 the emperor Constantine legalized Christianity within the Roman Empire. The word 'Catholic' means universal, all-embracing, and tolerant. The early Christians adopted many of the pagan traditions and festivals, although the birth of Christ was, in Egypt and the Eastern Empire, originally celebrated on January 6th (Epiphany).

However, the tolerance and good will to others didn't last long. In A.D.391 the Emperor Theodosius established Nicene Christianity as the State Religion and, with the exception of Judaism, became the only legal religion in the Roman Empire. We cannot be certain whether this was for religious or political reasons or, perhaps both. Whatever the reasons it set the stage for the creation of a Theocracy that would attempt to control the thoughts and knowledge of the people within the guidelines of a single religious dogma. The temples of other religions were sacked and their property seized. Anything associated with other cults that had not been adopted by the Christians was outlawed and in A.D.393 the Olympic Games were abolished. The Sibylline Books were destroyed in A.D.403 by Flavious Stilicho, the commander of the Roman Army under Theodosius. Since that time there has been an attempt to re-write history. Festivals, stories and traditions adopted from other cults were claimed to be purely of Christian origin.

As the Roman Empire evolved into a patriarchal theocracy it would seem at first sight that not only had women been banished from religious office but that the goddess herself had been vanquished. Well, not quite. Try as we may we can never get away from the basic need of that mother figure who looks after us. When we were children it was mum that we ran to when we were hurt or in trouble. It is said that during the First World War, as young men lay dying in the trenches, they called for their mum. Nothing can supplant the love, caring and reassurance we received from our mum. Throughout our entire history no deity has been more enduring than the Mother Goddess. She has been remade with different names and attributes over and over again to suit the needs of different cultural groups. As societies were Christianized people carried on worshiping the Mother as they had done for countless generations, except now she was clothed in Christian vestments. Below is a prayer to the Virgin Mary I found in the Wairarapa Times-Age.

O most beautiful flower of Mount Carmel, fruit of the vine,
Splendorous of heaven, Blessed Mother of the Son of God,
Immaculate Virgin, assist me in this necessity, Oh Star of the Sea.
Help me and show me herein you are my Mother.
O Holy Mary, Mother of God, Queen of Heaven and Earth,

I humbly beseech you from the bottom of my heart, to succour me in my necessity.
There are none that can withstand your power,
O show me here you are my Mother.
O Mary conceived without sin, pray for us who have recourse to thee (repeat 3 times).
Holy Mary I place this cause in your hands (repeat 3 times).
Thank you for your mercy towards me and mine.
Amen.

The Queen of Heaven and Earth! There are none that can withstand your power! O show me here you are my Mother! These are not words to some mere deceased mortal. This is a prayer to, what would be described in any other language, a great goddess.

12. The Son of Light, the Sun of God

The allegory of the death and resurrection of God's heavenly Sun is the key to most religious symbolism and formed the framework around which the Zodiac was created. The names by which the sun was personified were many, but the central feature of these stories is the sun's decline in light and heat at the autumn equinox, its lowest ebb at the winter solstice, and then its renewal at the spring equinox and its glory at the summer solstice. This cycle gave rise to many legends in which the sun-god dies and is then restored to life.

As different tribes mixed and mingled stories of the creation and the seasons were assimilated and then moulded and modified to meet the needs of different people and their environment. We have discussed the symbolism and the festivals associated with the spring equinox. In this chapter we will look at the autumn equinox and the two solstices.

The Lion of Summer

The June solstice, the northern hemisphere summer solstice, was in Leo from 4000 to 1850 B.C. This is the period in which the Zodiac began to be formulated around the Pillars of Heaven that marked the solstices and equinoxes. Consequently Leo was held in great reverence because it was here that the Sun returned to its summit of power and glory. Astrologer-priests distinguished Leo as the sole 'House of the Sun' and taught that the Sun was in this sign when the world was created.

Regulus, the brightest star in Leo, was said to be the 'Leader of the Four Royal Stars'. Its name, which was given by Copernicus, is derived from its earlier name 'Rex' (given by Ptolemy). The Babylonians knew it as 'Sharru' (the King), and in India it was Maghaa (the Mighty). Because kings, pharaohs and emperors were often believed to be living gods or ruled in the name of god, the lion was and continues to be used as a royal emblem.

The summer solstice moved into Cancer in 1850 B.C., and then into Gemini, its current location, in A.D. 300.

In the Celtic calendar dark always preceded light. Therefore the new day started at dusk - the waning of the day followed by dawn, the waxing of the day. This is a tradition that continues in many countries around the world today. The Islamic and Jewish new year begin at sunset. Following the Summer Solstice the hours of daylight decrease. Consequently the ancient Celtic new-year began at the Summer Solstice with the waning of the sun's power. It was known as 'Giamonios' which means the beginning of darkness.

***Fig. 17:** The 'Sunstone', placed at Stonehenge 3100 B.C., marks where the Sun rises at the summer solstice.* R. Hall

Around 3100 B.C, when the summer solstice was in Leo, the first large stone was placed at what would become Stonehenge. This is the Heelstone or Sunstone which marks the point on the horizon where the sun rises at the Summer Solstice (Figure 17). At this time the summer solstice marked the beginning of the Celtic new-year.

It is a common misconception that the druids built Stonehenge. According to Julius Caesar the druids were a Celtic priesthood who flourished in Britain at the time of the Roman conquest and perhaps for a few centuries earlier. At this time Stonehenge was already 2,000 years old and the solstice had moved into Cancer. Furthermore, the Celtic year under the druids began not at the summer solstice but at an autumn festival known Samhain (pronounced Sow-in).

The Harvest Queen

The autumn equinox occurs in late September in the northern hemisphere. The sun was in Scorpius at the time of this equinox from 4000 B.C. to 1850 B.C. It was a symbol of darkness marking the decline of the sun's power.

However, the autumn equinox was a time to celebrate because this was the time to bring in the harvest. The harvest was a very busy time and usually took place at the time of the full moon. The full moon following the equinox provided light after sunset and extended the hours that people could work to bring in the valuable harvest. This was known as the 'Harvest Moon' which, in our calendar, occurs in late September or October.

The autumn equinox moved into Libra in 1850 B.C., and then into Virgo in A.D. 300, where it is to this day. Around 500 B.C., when the equinox was in Libra, it was Virgo that was the sign to bring in the harvest. At the time of the equinox Virgo rose just before dawn. She represented Demeter the Queen of the Harvest, and in her hand she held the star Spica, the ear of wheat. Fifteen hundred years later, when the equinox had moved into Virgo, the dawn rising of the 'Sickle of Leo' became the sign to bring in the harvest (Figure 40).

After all of the work had been done a feast was held. The Druidic-Celtic year ended and was reborn on the festival of Samhain, which was celebrated over a fortnight near the end of October with the new-year beginning at the last quarter moon. Interestingly, it was the rising of the Seven Sisters (Pleiades star cluster) at sunset and their culmination at midnight that gave rise to this great festival.

Samhain means "time of the little sun" or "end of the warm season", and marked the end of summer. In ancient times all the fires of Ireland were extinguished and relit from the druidic fire upon the hill of *Tlachtga*. Later, following the introduction of the Julian calendar, Samhain became a fixed date and was said to occur on October 31st, the day before the first day of winter (northern hemisphere).

At the feast of Samhain food was put outside of the households for family members who had passed on during the previous year and lanterns were placed so that they could find their way. It was therefore a feast for the dead as well as the living.

In the dark ages of the first millenium this great Celtic festival was Christianised by making November 1st "All Saints Day" and turning Samhain, October 31st, into 'Halloween'. Halloween means "All Hallows Eve", the day before "All Saints". The pagan tradition of the feast and honouring the dead was turned into a parody of spooks and ghouls.

Incidentally, the familiar Jack-o'-Lanterns, the ghoulishly carved pumpkins used on the night of Halloween, were not the lanterns used on Samhain. These originate from a time after Samhain was turned into Halloween and are part of Irish folk lore... Jack was a miserable old drunk and a thief who liked to play tricks on everyone. One day he met up with the Devil and tricked him into climbing an apple tree. Once up the tree Jack placed crosses around the trunk so that the Devil couldn't get down. Jack let the Devil go only after he had agreed never to take Jack's soul. Eventually Jack died and went to the pearly gates of Heaven where he was told by Saint Peter that he had lived such a wicked life he would not be admitted into Heaven. Jack then wandered down to Hell but the Devil kept his promise not to take Jack's soul so he was also barred from entering Hell. Poor old Jack had nowhere to go and would have to wander the darkness between Heaven and Hell for eternity. He asked the Devil how he could leave the Gates of Hell as there was no light to find his way. The Devil tossed him an ember from the flames of Hell that would never burn out. Jack placed the ember in a hollowed out turnip, his favourite food, and ever since that day he has roamed the night with his Jack O'Lantern in search of a resting place.

On Halloween the Irish hollowed out turnips and placed a light in them to ward off evil spirits, including Jack. In the nineteenth century Irish immigrants discovered the pumpkin in America. Being larger than a turnip it was easier to carve, so the pumpkin became the Jack O'Lantern.

November the 1st, All Saints Day, was originally the first day of the three-day Festival of Isis, which celebrated the death and resurrection of Osiris. These were perhaps the oldest mystery plays on earth.[16] The worship of Isis, the great goddess of love and destiny, originated in Africa, was nurtured and refined in Egypt, and then spread

16 The last recorded Festival of Isis was held in Rome in A.D. 394.

throughout the ancient world.[17]During the Festival of Isis, professional singers, musicians and dancers, mostly female, performed at the temple. The performance involved actors playing the parts of Isis and Nephthys (her sister) in mystery plays of the murder of Osiris (the male fertility god and husband/brother of Isis) at the hands of his brother Seth (Set).[18]

In this story the body of Osiris is placed in an arc (chest) which is thrown by Seth into the Nile, and swept out to sea. Isis ransacks the earth in search of the body, which she eventually finds. She then transforms herself into a golden kite and, with her wings, breathed life into the phallus of Osiris, and thus she conceived a child with his spirit. Seth returned and ripped the body of Osiris into 14 pieces which he hurled into the Nile. Isis eventually recovered all of his body parts and bandaged them together, forming the first mummy. She then used her powers of magic to resurrect the body. Osiris then travelled over the 'western horizon' to a land that no mortal had seen (Underworld) and settled at the entrance to what would become the 'land of the dead'. He was the first to make this journey and, in doing so, he gave the promise of eternal life to all mortals that followed.

Isis gave birth to Horus, the son of Osiris, the sun of light. From that day forward Horus (the god of light) was the enemy and rival of Seth (god of darkness). In one great battle Seth gouged out the left eye of Horus. It is the reason why the sun-god has a single blazing eye. The lost eye became the moon, which continually winks as it tries to heal itself (the monthly cycle of phases). Isis replaced his lost eye of with the *'Wadjat Eye'*. (symbol on figure 4) This had magical powers and could travel anywhere on its own to keep 'an eye' on the enemies of light.

The never ending struggle between the gods of light and darkness personified the cycle of the seasons. Isis gave birth to the sun-god Horus at the darkest time – the Winter Solstice.

The Holly and the Ivy

The December solstice is mid-winter's day in the northern hemisphere. This solstice was located in Aquarius from 4000 B.C., Capricornus from 1850 B.C., and since A.D. 300 in Sagittarius. The Winter Solstice was, and still is one of the most important festivals of the year. It marked the darkest time of the year. But, it was also a time to celebrate because on this day light defeated darkness and from that day forward the sun would grow in strength. Throughout ancient Europe and Asia Minor it was a festival dedicated to the Sun God. It was the time of the 'Rebirth of the Sun' and, the anniversary of the birth of a divine male child, the Son of Light, the Sun of God.

Throughout the duration of the Roman Empire, up until A.D. 300, the winter solstice was in Capricornus, which was known as 'The Gate of Gods'. In the Roman calendar the Winter Solstice was celebrated on December 25th.

In Rome the winter solstice was known as *'Dies Natalis Solis Invicti'*, the Birthday of the Unconquered One, and the Feast of the Unconquered Sun. The 'Unconquered Sun' is a title given to the Hindu/Persian sun-god Mithra and dates back to a time when these two races formed one people. Mithraism was introduced into Europe following Alexander's conquests. Later, it was adopted by the Roman legionaries and Mithra

17 Isis was the sister-wife of Osiris and the mother of Horus. She was originally the goddess of the Earth and was identified with the star Sirius and the Moon. The Greeks identified her with Demeter, the Roman Ceres, and Io.

18 The Greeks identified Seth with Typhon, a gigantic monster representing darkness and the evil powers of nature. It is suggested by some historians that the Egyptian name for Seth, Sutekh, may have evolved into the word Satan.

became the principle deity of the soldier. The Roman legions carried the religion to the four corners of the Roman Empire.

Mithra was the son of Ahura Mazda, the god of light. According to legend he was born fully grown from a rock. In another legend he was born from a cosmic egg. His mother was the goddess Atargatis (a water deity who was half woman and half fish - the original mermaid). In one sculpture he is depicted emerging from the egg holding the Sword of Truth and the Torch of Light. Surrounding him is the heavens with the twelve signs of the Zodiac. It was Mithra who captured and killed the 'primeval bull' (Taurus) which released the life force of nature (front cover lower left). From the bull's body came herbs and from its blood the vine, and from its seman came the domestic animals.

Mithra was the defender of the just and protector and savior of mankind. He saw the suffering of humanity and decided to be reborn as a mortal. The miracle of his birth was witnessed by shepherds in a field. Mithra was born or re-born on December 25th.

The Feast of the Unconquered Sun, the Winter Solstice, was preceded by the Festival of Saturnalia. Saturn was the Roman god of agriculture and the Golden Age. Temples and homes were decorated with holly which symbolized eternal life (evergreen); and a pine tree which was sacred to Attis (the divine consort of the Earth-Mother Goddess Cybele).

To the Celts the Winter Solstice was the beginning of the light half of the year and was known as Samonios, which means the beginning of light. In Europe the festival of the winter solstice was known as Yule – The Festival of Rebirth. Yule is a Saxon word for light or sun. In Norse it means 'Wheel' or 'The Turning of the Wheel'. The Festival of Yule lasted for 12 days; later these became the 12 days of Christmas. At the beginning of the festival the Yule-Log, decorated with Ivy, was lit in honor of Thor. It symbolized warmth, light and the continuance of life.

In Anglo-Saxon folk lore the Winter Solstice represented the triumph of the Oak King (waxing year) over the Holly King (waning year). The two kings were twins, one light, one dark, caught in eternal rivalry. Holly symbolized eternal life and its berries were said to be drops of blood of Hel (Holli), the Teutonic Goddess of the Underworld. Mistletoe, sacred to the Druids, was placed over the entrance to homes. It was associated with the Oak God and was believed to have magical healing powers.

The Celtic deities differ from those of other cultures in that they spend most of their time on the mortal plane and in close proximity to people. They usually take the form of wild animals, birds or trees. The male counterpart of the Celtic Earth-Mother Goddess was Cernunnos, the Horned God. He was so called because on the mortal plane he often took the form of a magnificent stag. During the Dark ages the Church turned the Horned God into a devil. It is the reason why the Devil has horns.

The birth of a male god, a savior, at the winter solstice (December 25th) is a common theme in the ancient religions. It was the birthday of both Mithra and Attis and, according to the Egyptians, on this date Isis gave birth to the sun-god Horus. In Celtic folk lore Rhiannon gave birth to Pryderio at the darkest time (winter solstice), which was also the birth of the child of light, known as Mabon.

In A.D. 354 the Emperor Augustine adopted December 25th to celebrate the birth of Christ. This, as we have seen, was the day of the pagan rituals of the Sun child – The Sun of Righteousness. The holly became the 'Crown of Thorns' and the berries the 'Blood of Christ'.

In Britain, in 1652, the Puritans abolished Christmas with an act of Parliament:

"No observation shall be had of the five-and-twentieth day of December, commonly called Christmas Day; nor any solemnity used or exercised in churches upon that day in respect thereof."

In presenting the bill its advocates had this to say:

"Christmas Day, the old Heathens Feasting Day in honour to Saturn their idol-God, the Papists Massing Day, taking to heart the Heathenish customs, Popish superstitions, ranting fashions, fearful provocations, horrible abominations committed against the Lord, and his Christ on that day and days following...."

Christmas was restored with the return of the monarchy in 1660.

Part IV:

THE MANSIONS OF THE MOON

While the cycle of the seasons, the climate, weather, and hence the abundance of food, is determined by the Sun, it was the cycle of the Moon around which the first calendars were constructed. The first zodiac, if one can call it that, probably consisted of just four signs that were built around the 'Pillars of Heaven', four bright stars that marked the solstices and equinoxes. Around 700 B.C. the Babylonians formulated the first Zodiac wheel of 12 signs. This Zodiac was almost certainly constructed in an attempt to merge the Solar cycle of the seasons with the Lunar calendar. The Lunar Zodiac, Lunar symbolism, and the calendars are the topics of this section.

13. The Chariot of Artemis

When we talk of the Zodiac as being the path of the Sun we rarely consider how difficult this would have been to identify. The stars cannot be seen when the Sun is above the horizon. In fact stars are not visible unless the sun is at least 10 degrees below the horizon. Where I live in New Zealand the stars begin to appear about 45 minutes after sunset and disappear 45 minutes before sunrise. However, the constellation the Moon is in can be observed, and the path it takes against the background stars is similar to that of the Sun. Consequently the first zodiacs were built along the path of the Moon, not the Sun. In addition, the cycle of the Moon with its ever changing cycle of phases, formed the framework of our calendars. However, before we get on to these topics we should first discuss the cycle of the Moon and the symbolism attached to it.

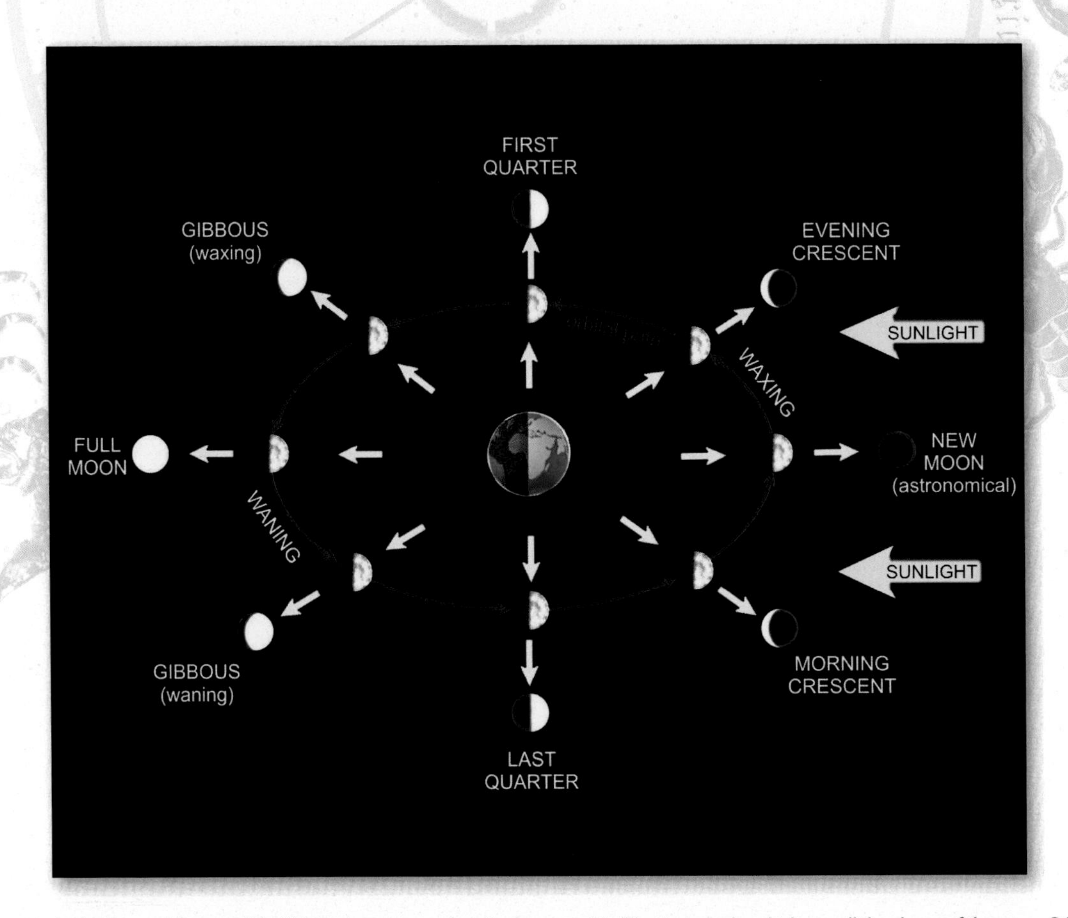

***Fig. 18:** As the Moon orbits around the Earth we see an ever changing amount of its illuminated side, which we call the phases of the moon.* R.Hall

As the Moon orbits around the Earth we see an ever changing amount of its illuminated side, which we call the phases of the moon (Figure 18). At the beginning of each cycle we first see the Moon as a thin crescent low in the west shortly after sundown. We call this the "new moon"[19].

Although the rotation of the Earth carries all celestial objects westward the Moon's orbital motion is slowly taking it eastward against the background stars. Consequently, night by night, it will be observed that the Moon has moved eastward and further away from the Sun. As it does so it sets later and later and waxes – more and more of the Moon is illuminated. After approximately 7 days it reaches first quarter (quarter of its cycle) when it is half illuminated. At this time the Moon, which is now visible in daylight, rises close to midday. After another 7 days the Moon is fully illuminated. The full Moon is opposite the Sun in the sky and rises as the sun sets.

After full, the Moon continues to move eastward but is now waning and moving back towards the Sun. At last quarter, 7 days after full moon, we see another half-moon but this one is in our early morning sky because it doesn't rise until midnight. Thereafter it continues to wane to a thin crescent that is eventually lost in the morning twilight. The Moon is lost from view because its dark side, which reflects little light, is presented towards us and it is close to and rising and setting with the Sun.

The cycle of the moon, from one new moon to the next, takes on average 29.53 solar days but can vary by as much as 12 hours. The lunar cycle is then, approximately one month. This is where the word month comes from. It originates from the old English word monath, meaning moonth. All the first calendars were based upon the cycle of the moon with each new moon marking the beginning of a month. And the weeks of the month, they too come from the cycle of the moon. The weeks divided the monthly cycle of the moon into the four quarters. It takes approximately 7 days for the moon to travel from new to first quarter, another 7 days from first quarter to full, and so on. The word week comes from the old Saxon word wika, which is derived from the Germanic wikon, meaning sequence (of phases).

In ancient times the moon was the foremost symbol of the cycle of birth, life, death and the resurrection. Think of the cycle of the moon like the life of a person. When we first see the Moon it is a thin crescent and sheds little light. Then night by night it grows in brilliance until full (mature). It then slowly fades away and, at the end of the cycle, is eventually lost from view. What is the count in days from the last sighting of the old moon to the first appearance of the new moon? The answer is 3 days. In the Bible, what is the count in days from the crucifixion to the resurrection? This count of three days is important in similar stories from other cultures. When the moon is first seen at the beginning of its cycle we call it a new moon. Why do we call it a new moon? Because in ancient times this is exactly what they thought is was – a brand new moon resurrected from the eternal fires of the sun.

It may seem difficult to believe today that people thought that each new moon really was new and different from the last. But, people use to think that the earth is flat. In Africa, the Xhosa people believed that the world ended with the sea and, at the bottom of the sea, was a vast pit filled with new moons ready for use.

In most ancient cultures the moon was a feminine symbol associated with birth, death and rebirth. The moon was the celestial chariot of the Moon Goddess who was known as the 'Mother of the World' and the patroness of the life forces of Nature. The ability to give birth to a new life was seen as a divine and magical quality invested in women by the goddess. It had not gone unnoticed that the menstrual cycle of women was very similar to the

---

19 It should be noted that in today's astronomical terminology 'new moon' refers to when the moon is in conjunction with the sun (between the earth and the sun). The moon cannot be seen at this time because it is close to the sun in our daytime sky and its entire illuminated side faces away from us. In ancient times the term 'new moon' always applied to its first appearance following conjunction.

cycle of the moon. Indeed, in many societies, menstrual blood was believed to be such powerful magic that a man would die if he came in contact with it. For this reason, at that time of the month, a woman had to leave the home a stay in a special dwelling.

The Moon Goddess had three aspects; she was the Tri-formed Goddess. When she faced the west, from the new crescent moon to half moon at first quarter, she was the virgin; when she looked to the north (south in the southern hemisphere), she became full and she was the mother; and when she looked to the east, from last quarter to the last sighting in the dawn twilight – the death of the moon, she was the hag. Sometimes these three aspects of the one goddess were given different personalities and different names – Artemis (Diana), Hestia (Vesta), and Hecate.

As well as birth and life the moon goddess was also associated with death and the Underworld. It was believed that the powers of the Underworld rose and fell with the cycle of the moon. At full moon these powers reached a peak, which is the reason why the full moon is an essential ingredient in horror movies and stories about ghosts and vampires. Hecate was the Queen of Ghosts, who roamed the night at the time of full moon accompanied by her legions of ghosts and vampires. She was also the goddess of magic and, like other deities of the Underworld she was a goddess of fertility. It may seem strange that gods of the underworld and death were also fertility gods. But it was an observed fact that new life sprouted from below. It is an ancient tradition, but to this day many people still plant beneath the light of the full moon.

While the Sun and its annual cycle has a monumental effect on living creatures many people believe that the Moon also has a profound effect. The Moon has been held responsible for insanity (lunacy), violence, drunkenness, suicide, and child birth. For example, it is or was believed that children born when the full moon is in Aries will be male and those born when the full moon is in Taurus will be female. Extensive studies carried out in France and the U.S.A. found no correlation with the cycle of the Moon for any of the above.

Without doubt the light shed by the Moon plays an important role in the life cycle of many creatures. For example, the Great Barrier Reef always spawns at full moon; many flying insects, such as moths, emerge at full moon. The light of the Moon helps them to find their way and each other. In a like manner, finding your way in the night is almost certainly the reason why the number of burglaries in some places in the 19th century increased around the time of full Moon.

Other than being a lantern in the night there is little scientific evidence that the Moon has a direct physical effect on living creatures. A common misconception is that because the Moon lifts the waters on Earth generating tides, it must have an effect on living creatures… because they are made mostly of water! This is bad science. The tidal effect of the Moon on the Earth is caused by the difference in the pull of gravity of the Moon from one side of the Earth to the other, a distance of 12,756 kilometres. The tidal effect on a human body would be the difference in the gravitational pull over a distance of less than two metres – an amount too small to measure.

It has also been suggested the Moon is responsible for triggering earthquakes, volcanic eruptions and influencing the weather. While there is some evidence to suggest that large basaltic eruptions in Hawaii may be influenced by the tidal effects of the Moon, there is no correlation with earthquakes and the vast majority of volcanic eruptions. As far as the weather is concerned the tidal effect of the Moon produces an atmospheric temperature variation of 0.02 to 0.03 degree C. – far too small to have any significant effect on the weather. I'm sure that if you could predict the weather by the cycle of the Moon meteorologists would have been on to it a long time ago.

While most of the folk-lore beliefs about the influence of the Moon turn out to be false the Moon was and still is of significant spiritual importance to people around the world. There is no denying that this silver lantern of the night still has magical qualities. Wolves howl at it, lovers meet beneath it, and poets write about it.

The Moon played a significant role in the development of the Zodiac and the calendars. Its path across the heavens is very similar to that of the Sun and is therefore along the Zodiac. The reason we have twelve constellations in the Zodiac is that there are on average 12 lunar months (12 cycles of the Moon) in a solar year. Each month, in the old calendars, the Sun moved through a different constellation of the Zodiac. In a year the Sun moves right around the Zodiac passing through all 12 signs. The Moon does the same thing in one month, moving through a different sign approximately every 2.5 days.

14. The Lunar Mansions

The first calendars and astrological signs were constructed around the cycle and path of the Moon, not the Sun. Each ancient civilization had its own system, but essentially the path of the Moon was usually divided up into 28 signs, sometimes 27, which is the number of days it takes the moon to travel once around the earth measured against the background stars. Thus, each sign marked the daily passage of the Moon against the background stars. These were known as the Moon Signs or Lunar Mansions and they formed a band around the sky that is often called the Lunar Zodiac. The study of the Lunar Zodiac played an important part in the science of observational astronomy in Arabia, Persia, Coptic Egypt, Euphrates valley, India and China.

It should be noted that there are two types of Lunar months. The *sidereal* month is the time it takes the moon to complete one circuit around the sky measured against the fixed stars. This is the Moon's orbital period and takes 27.3 days. Then there is the time it takes the moon to complete one cycle of its phases, which could be measured from one new moon or full moon to the next. This is the *synodic* month and takes 29.5 days. The difference between the two is due to the orbital motion of the Earth around the Sun. Because the position of the Earth in space relative to the Sun is changing, the Moon must complete slightly more than one orbit around the Earth to return to the same phase. The 27 or 28 mansions of the Moon are obviously related to the sidereal period of the Moon, not its phases which were used for the calendars.

The Lunar Zodiac was usually correlated to the seasons or solar cycle by placing the first sign or start of the zodiac at the vernal equinox. The oldest Lunar zodiacs have the Pleiades star cluster as the first sign. This establishes a great antiquity for these zodiacs because we have to go back about 5,000 years before we find the Pleiades located at the vernal equinox. In this chapter we will take a look at three of the oldest Luna Zodiacs, the Arabian, Indian and Chinese.

The Greek Alphabet

To aid in the identification of the stars in each Lunar Sign a system of symbols is used in the following charts. In modern star charts the brighter stars in each constellation, in addition to having a proper name, are labeled with a letter of the Greek alphabet followed by the name of the constellation. The brightest is usually called Alpha, the second brightest Beta, the third brightest Gamma, and so on. The Greek alphabet and the symbols used are as follows.

α alpha	η eta	ν nu	τ tau
β beta	θ theta	ξ xi	υ upsilon
γ gamma	ι iota	ο omicron	φ phi
δ delta	κ kappa	π pi	χ chi
ε epsilon	λ lambda	ρ rho	ψ psi
ζ zeta	μ mu	σ sigma	ω omega

The Arabic Manazil (Lunar Zodiac) has 28 mansions. Each is identified by a star, star cluster, or asterism (mini-constellation). Looking at these star names you will discover that not all are within the modern constellation boundaries of the solar Zodiac. This is in part because the orbit of the Moon is tilted by 5° to the plane of the ecliptic. Consequently the Moon sometimes wanders beyond the Zodiac in modern star charts into neighboring constellations. It occasionally ventures into Cetus (The Whale) instead of Pisces, Orion (The Hunter) instead of Gemini, Auriga (The Charioteer) instead of Taurus or Gemini, Sextans (The Sextant) instead of Leo, and it can just touch Corvus (The Crow) on its journey through Virgo. Sometimes however, in barren regions of the sky, bright stars that are close to, but not on the path of the moon are also used.

	Manazil	Meaning	Star or Asterism
1.	*Al Thurayya*	The Many Little Ones	Pleiades (cluster in Taurus)
2.	*Al Dabaran*	The Follower	Aldebaran (α Tauri)
3.	*Al Hak'ah*	White Spot	Stars around Heka (λ Orionis)
4.	*Al Han'ah*	Mark/Brand (on the Camel)	Alhena (γ Geminorum)
5.	*Al Dirah*	The Forearm (of the Nile)	Castor and Pollux (Gemini)
6.	*Al Nathrah*	Gap (in the hair of the Lion)	Praesepe (cluster in Cancer)
7.	*Al Tarf*	Glance (of the Lion's eye)	λ Leonis and ξ Cancri
8.	*Al Jabhah*	The Forehead (of the Lion)	Sickle of Leo
9.	*Al Zubrah*	The Main (of the Lion)	Zozma & Coxa (δ & θ Leonis)
10.	*Al Sarfah*	The Changer (of weather)	Denebola (β Leonis)
11.	*Al Awwa*	The Barker (the Kennel)	β, γ, δ, ε, and η Virginis
12.	*Al Simak*	The Unarmed	Spica (α Virginis)
13.	*Al Ghafr*	The Covering	Syrma (ι, φ, and κ Virginis)
14.	*Al Zubana*	Claws (of the Scorpion)	α and β Librae
15.	*Iklil al Jabhah*	Crown of the Forehead	Graffias (β Scorpii)
16.	*Al Kalb*	The Heart	Antares (α Scorpii)
17.	*Al Shaulah*	The Sting	Shaula (λ Scorpii)
18.	*Al Na'am*	The Ostriches	γ, δ, ε, and η Sagittarii
19.	*Al Baldah*	The City	Stars around π Sagittarii
20.	*Al Sa'd al Dhabih*	Lucky One of the Slaughterers:	α and β Capricorni
21.	*Al Sa'd al Bula*	Good Fortune of the Swallower:	ε, μ, and ν Aquarii
22.	*Al Sa'd al Su'ud*	The Luckiest of the Lucky	β & ξ Aquarii & ι Capricorni
23.	*Al Sa'd al Ahbiyah*	Lucky Star of Hidden Things:	Sadachbia (γ Aquarii)
24.	*Al Fargh Al Mukdim*	Fore-spout (of water-bucket):	α and β Pegasi
25.	*Al Fargh Al Thani*	The Lower-spout	Algenib (γ Pegasi)
26.	*Al Batn Al Hut*	The Belly of the Fish	16 stars in a figure of 8 from Pisces to Andromeda
27.	*Al Sharatain*	The Two Signs	β and γ Arietis
28.	Al Butain	The Belly (of the Ram)	δ, ε and π Arietis

The names of the mansions, which identify the daily location of the moon, are usually related to ancient Arabian constellations. We have camels and ostriches and a huge lion. The Arabian Lion approximates the position of Leo but covers a much larger region of the sky. In place of Virgo we have the Kennel of dogs that bark at the lion.

The Hindu Nakshatras or Lunar Mansions consist of 28 signs. The Sanskrit word Nakshatra means "that which does not decay." The earliest of these lunar zodiacs start with Krittika, the Pleiades, which dates them to about 3000 B.C. Later zodiacs start at Ashvini (27 in Aries) and therefore demonstrate that astronomers had made adjustments for precession of the equinoxes. These date to about 1000 B.C.

This lunar zodiac uses a bright star or conspicuous cluster or asterism to identify each mansion. Often these stars, like those in the Chinese zodiac, are not within the path of the Moon. Rather, they are bright stars or lines of stars that point to and identify the position of the Moon when both are on the meridian (an imaginary line that runs from due north to due south).

	Nakshatra	Star(s)
1.	*Krittika*	Pleiades (star cluster In Taurus)
2.	*Rohini*	Aldebaran (α Tauri) & Hyades star cluster
3.	*Mrigracirsha*	Heka (λ Orionis)
4.	*Ardra*	Betelgeuse (α Orionis)
5.	*Punarvarsu*	Castor & Pollux (α & β Geminorum)
6.	*Pushya*	Praesepe (star cluster in Cancer)
7.	*Aclesha*	ζ, θ, ε, δ, σ, & η Hydrae (Head of the Hydra)
8.	*Magha*	Regulus (α Leonis) & the Sickle of Leo
9.	*Purva Phalguni*	Zosma (δ Leonis)
10.	*Uttara Phalguni*	Denebola (β Leonis)
11.	*Hasta*	Constellation of Corvus
12.	*Citra (Chitra)*	Spica (α Virginis)
13.	*Svati*	Arcturus (α Bootis)
14.	*Vicakha (Vishakha)*	Zubenelgenubi (α Librae)
15.	*Anuradha*	Graffias (β Scorpii), π Scorpii & Dschubba (δ Scorpii)
16.	Jyestha	Antares (α Scorpii), σ and τ Scorpii
17.	*Vicritau (Mula)*	Shaula (λ Scorpii)
18.	*Purva Ashadha*	Ascella (ζ Sagittarii),τ and π Sagittarii
19.	*Uttara Ashadha*	Nunki (σ Sagittarii)
20.	*Abhijit*	Vega (α Lyrae), ε and ζ Lyrae
21.	*Cravana (Shravana)*	Altair, Alshain, & Tarazed (α, β and γ Aquilae)
22.	*Dhanishta (Cravishtha)*	Constellation of Delphinus
23.	Catabhishaj (Shatabhishak)	Al Thalimain (λ Aquilae) plus 100 surrounding stars
24.	*Purva Bhadra*	Markab and Scheat (α & β Pegasi)
25.	*Uttara Bhadra*	Algenib (γ Pegasi)
26.	*Revati*	32 stars north from ζ Picium north, including δ Piscium
27.	*Acvini (Ashvini)*	Sheratan & Mesarthim (β & γ Arietis)
28.	*Bharani*	33, 35, 39 & 41 Arietis*

* These stars originally formed a separate but small constellation called Musca Borealis (the Northern Fly).

The Chinese Lunar Mansions were originally called the 28 Xiu (Sieu). In the west the lunar mansions are reference points along the path of the Moon or Sun (ecliptic). The Chinese called the ecliptic the

'Yellow Path'. In contrast to the Western system, Chinese mansions are reference points along the celestial equator, called the 'Red Path', measured from a point of celestial longitude (right ascension) determined by the 'Handle' of the 'Big Dipper'.

The mansions are divided into four seasonal groups of seven which also use animals and colours. Each group is also associated with a cardinal point which, as we discussed earlier, is related to the position of the handle of the Big Dipper.

The Green Dragon of Spring	East.
The Red Bird of Summer	South.
The White Tiger of Autumn	West.
The Black Tortoise of Winter	North.

The cardinal points may also refer to the seasonal locations of the full moons. During the northern hemisphere summer the full moons are south of the celestial equator, while during winter they are north of the equator. At the equinoxes the full moon rises due east and sets due west. East and West may symbolize moons of the spring and the fall.

The Green Dragon of Spring is associated with the Chinese constellation *Tsing Lung*, the Azure Dragon. This occupies the region of Scorpius and Libra. Naturally, the Chinese didn't have a Lion (there are no lions in China) and in the place of Leo we have a Horse [20]. Above the horse, occupying the constellations of Hydra and Crater, is the Red Bird.

In the 2nd century A.D. Buddhists introduced the Indian lunar zodiac into China. Where the Chinese Xiu were identified with the Indian nakshatras they were renamed by translating the Indian name into Chinese. Other Chinese Xiu which could not be identified with the Indian zodiac became transliterations of the Nakshatra. The table below shows the relationship between Xiu mansions and the Nakshatra. There are many variations in the names of each Xiu, those in bold are the most common. Stars that are different to those of the Nakshatra are in italics.

	Xiu	**Nakshatra**	**Star(s)**
	Quadrant of the Green Dragon		
1.	***Jiao***, *Keok, Guik, Kio*	(12) Citra	Spica (α Virginis) *and Heze (ζ Virginis)*
2.	***Kang***	(13) Svati	κ, ι, υ *Virginis*
3.	***Di***, *Ti, Dsi, I shi*	(14) Vicakha	Zubenelgenubi (α Librae), γ & ι *Librae*
4.	***Fang***, *Fong*	(15) Anuradha	Graffias, Pi & Dschubba (β,π,δ Scorpii)
5.	***Xin***, *Sam*	(16) Jyestha	Antares (α Scorpii), σ and τ Scorpii
6.	***Wei***, *Vi, Mi*	(17) Vicritau	Shaula (λ) + ε, μ, ζ, η, θ, ι, κ, υ, *Scorpii*
7.	***Ji***, *Ki, Kit*	(18) Purva Ashadha	Ascella (ζ Sagittarii*)*, λ, μ *and* π *Sagittarii*
	Quadrant of the Black Tortoise		
8.	***Dou***, *Dew, Tew, Nan*	(19) Uttara Ashadha:	Nunki (σ Sagittarii)
9.	***Niu***, *Ngu, Gu, Keen*	(20) Abhijit	*Algedi* (α) + ν, ο, π, ρ *Capricorni*

20 A lion can be found in some Chinese charts but this was introduced by Jesuit missionaries in the 16th century.

10.	***Niu***, *Nok, Mo,*	(21) Cravana	*Albali (ε Aquarii)*
11.	***Xu***, *Heu*	(22) Dhanishta	*Sadalsuud (β Aquarii)*
12.	***Wei***, *Gui, Goei*	(23) Catabhishaj	*Sadalmelik (α Aquarii) & θ Aquarii*
13.	***Shi***, *Sal, Ying, She,*	(24) Purva Bhadra	Markab and Scheat (α & β Pegasi)
14.	***Bi***	(25) Uttara Bhadra	Algenib (γ Pegasi) *and Alpheratz (α Andromedae)*
	Quadrant of the White Tiger		
15.	***Kui***, *Goei or Kwei*	(26) Revati	*Mirach (β Andromedae)*
16.	***Lou***, *Leu, Low*	(27) Acvini	*Hamal,* Sheratan, Mesarthim (α, β γ Arietis)
17.	***Wei***, *Vij, Oei*	(28) Bharani	33, 35, 39 & 41 Arietis
18.	***Mao***, *Mol*	(1) Krittika	Pleiades star cluster in Taurus
19.	***Pi***, *Peih, Pal*	(2) Rohini	Aldebaran (α Tauri) & Hyades cluster
20.	***Zi***, *Tsee, Tsok, Keo*	(3) Mrigracirsha	Heka (λOrionis)
21.	***Shen***, *Sal, Tsan*	(4) Ardra	*The Seven brightest stars in Orion*
	Quadrant of the Red Bird		
22.	***Jing***, *Tsing, Tiam*	(5) Punarvarsu	Castor & Pollux (α & β Geminorum)
23.	***Gui***, *Kwei, Kut*	(6) Pushya	Praesepe star cluster in Cancer
24.	***Liu***, *Lieu*	(7) Aclesha	ζ, θ, ε, δ, σ, & η Hyd (Head of the Hydra)
25.	***Xing***, *Sing, Tah*	(8) Magha	*Alphard (α Hydrae),* σ *and* τ *Hydrae*
26.	***Zhang***, *Tjung, Chang*	(9) Purva Phalguni	κ, ν1, ν2, λ, μ, & φ *Hydrae*
27.	***Yih***, *Yen*	(10) Uttara Phalguni	*Cons. Crater plus & 22 in Hydra*
28.	***Zhen***, *Kusam, Tchin*	(11) Hasta	Constellation of Corvus

Around 500 B.C. the first sign Jiao, culminated (it was due south and at its highest point in the sky) at midnight at the time of the northern spring equinox. In other words it was directly opposite the sun, due south. During the evening, as Jiao steadily moved towards the meridian due south, the Green Dragon of Spring rose in the south-east.

15. Shadow Reading

As we have seen, lunar zodiacs which have their own asterisms predate the solar zodiac. The oldest of lunar zodiacs may date back to 3000 B.C. or earlier but 'The Zodiac', the solar zodiac as we know it, was not fully formulated until about 700 B.C. To create a solar zodiac and calendar that defined the seasons and the times of festivals our ancestors had to correlate the lunar zodiac and lunar calendar with the seasonal cycle of the sun. To do this they had to work out the precise annual path of the sun against the background stars. You cannot see the stars when the sun is above the horizon, so how did they work this out? The answer is that they used shadows.

From the careful observation of stars in the vicinity of the north celestial pole early astronomers were able to identify the north cardinal point on the horizon. This provided an accurate compass for that location. Using the compass they were then able to draw a meridian line on the ground, a line of longitude that runs geographically north to south. (Figure 1) By placing a gnomon (a vertical pillar) on the meridian line the Chinese were able to work out the times of the solstices.

When a celestial object crosses the sky it is at its highest point when it crosses the meridian. At noon, when the sun is at its highest, the shadow of the gnomon falls onto the meridian line. Now, the altitude of the sun at noon varies throughout the year. It is at its highest on mid-summer's day, the summer solstice, and hence the shadow cast is the shortest for the year. At the winter solstice we have the reverse. The sun is at a noon-time low and the shadow cast is the longest for the year. Thus, by measuring the lengths of shadows at noon the times of the solstices is revealed.

Fig. 19: *The Zodiac and the cycle of the Moon.* R. Hall

The Moon also casts shadows. Of particular importance was the shadow of the full moon as it crossed the meridian. At the time of full moon it will be observed that, as the Sun sets the Moon rises in the opposite direction. This is because when the Moon is full it is directly opposite the Sun. Even though our ancestors couldn't see the stars in the vicinity of the Sun they could work out where it was. On the celestial sphere, or zodiac wheel, the Sun would be in the constellation directly opposite the location of the full moon. Therefore, each month as the Sun moves from one constellation of the Zodiac to the next so does the location of the full moon. Take a look at Figure 19. In January the Earth is at point 'A' in its orbit around the Sun. The Sun is in the constellation of Capricornus and the full moon is in Cancer. A lunar month later the Earth has moved to point 'B'. The moon has traveled around the Earth and is again opposite the Sun. The moon is full again but is now in the constellation of Leo. Meanwhile the Sun has moved into the sign opposite the moon, Aquarius.

Because the Moon's path is close to the ecliptic, when the moon is full and therefore directly opposite the sun, it will be located in the approximate seasonal opposite position to that of the sun. In other words, when the sun is close to the winter solstice the full moon will be located close to the sun's summer solstice position, and visa versa. Consequently, when on the meridian, winter full moons are high in the sky while summer full moons are low.

The Moon could be used to determine which sign the sun was in but, on its own it was not precise enough to determine the exact position of the sun and hence, the ecliptic. This is because the orbit of the moon is tilted by 5° to the path of the Sun. Consequently, the moon's path takes it above and below the plane of the ecliptic.

This tilt of the moon's orbit is the reason why we don't get eclipses every month. An eclipse can only occur when the Sun, Moon and Earth line up perfectly at the same celestial longitude and latitude (right ascension and declination). This can occur only at one of the two nodes where the path of the moon crosses the ecliptic and, at the same moment, the moon is either new or full. An eclipse of the sun occurs at new moon when the moon passes directly in front of the sun. An eclipse of the moon occurs when the moon is full and the shadow of the Earth falls on the moon.

To determine the precise path of the Sun astronomers first had to produce an accurate chart of the celestial sphere. This was achieved by carefully measuring the altitude of stars and the exact time as they crossed the meridian. Using an accurate clock they also measured the time it took for the celestial sphere to complete one revolution.[21] Then, by measuring the altitude of the Sun each day at noon (by the length of the shadow it cast), and the precise time measured from a reference point on the celestial sphere, they were able to chart the path of the sun, the ecliptic, on the celestial sphere. However, to carry out these observations they needed an accurate clock and, these were provided by what was essentially a dripping tap.

Water clocks were among the first inventions that could measure time independent of the observation of celestial cycles. Indeed, they could be used to test the regularity of the celestial cycles. It turns out that many of the celestial cycles were not, as was originally thought, fixed and unchanging. The length of the day, measured from one noon to the next, varied throughout the year. The length of the lunar month also varied throughout the year. There were, they discovered, cycles within cycles. One of the

21 This is the 'sidereal day – the time it takes for the Earth to rotate once relative to the stars.

oldest water clocks known was found in the tomb of the Egyptian Pharaoh Amenhotep I. This dates to around 1500 B.C. The Greeks began using water clocks around 325 B.C. They called them *clepsydras*, which means 'water thief'.

The Greek philosophers were renowned for attempting to explain the universe in natural rather than supernatural terms. Aristarchus, who lived in the second century B.C., used solar and planetary motions to show that the earth and the planets revolved around the sun. Although based on scientific reasoning his theory gained little support until it was re-introduced by Nicolaus Copernicus almost two thousand years later.

Around 250 B.C., lived one of the greatest astronomers and mathematicians of all time, Hipparchus. Among other things he produced the first known star catalogue and the most detailed and accurate maps of the heavens in the ancient world. With accurate charts, clocks and observations he noted that the motion of the Sun was not uniform, but a little faster in winter and a little slower in summer. This is because the orbit of the earth around the sun is not a circle but an ellipse. Consequently, the speed at which the earth is moving along its orbit is continually changing. The earth is closest to the sun and moving fastest around January 4th. It is most distant from the sun, and slowest, around July 7th.

The variation in the length of a day is another product of the elliptical orbit. The time it takes the earth to turn once in respect of the sun depends upon the earth's orbital location and the speed at which it is moving. Our mechanical watches cannot work with an ever changing length of the day. That is why we use 'mean time' in which the length of a day is an average of all days of the year.

The ever changing length of the day was discovered by measuring time with a water-clock and observing the passage of the sun's shadow over a meridian line. There is a meridian line at Stonehenge Aotearoa set in white tiles south of the gnomon, the obelisk. (Figure 1) If the length of a day were always exactly 24 hours, as measured by the water-clock or today, by our watches, the shadow of the obelisk would always fall on the meridian line at noon.[22] But, 24 hours is just the average length of a day. On short days the shadow crosses the meridian before mean noon, and after mean noon on long days. If you plot the position of the point of the shadow at mean noon every day, over a year it will generate a figure of eight called an analemma. This is the yellow line in Figure 1. The only time a day is exactly 24 hours long is where the analemma crosses the meridian line.

Because the Sun is in the same location at the same time each year its shadow can be used to formulate a calendar. The analemma at Stonehenge Aotearoa has been graduated to show the date and the times of the solstices and equinoxes. Using the calendar dates the analemma can then be divided up into the constellations of the Zodiac. Thus, the noon-time shadow can be used to determine the sign the Sun is in that day.

Once the path of the sun and moon was known astronomers could predict awesome events such as eclipses. Any unpredicted celestial event was seen as a portent of disaster. An unexpected eclipse struck terror in the hearts of people. Solar eclipses were particularly frightening because it appeared as if the

22 Actually, because New Zealand Standard Time is based on the mean-time the sun crosses the 180° longitude, and because Stonehenge Aotearoa is located at 175° 34' east, the sun crosses the analemma at Stonehenge Aotearoa at 12:18pm each day, 1:18pm when daylight saving is in effect.

sun, the source of life and well being, was being devoured. Thus, a very important duty of the astrologer/priest was to predict eclipses and forewarn the monarch or emperor of a potential disaster. The thinking was that if you could predict the coming of a bad omen you could do something about it and avoid the consequences. This was usually accomplished by making some great sacrifice to the gods.

The greater the impending disaster the greater the required sacrifice to enlist the aid of the gods. Nothing could be direr than the sun going out. At these times, in ancient Babylonia, they made the supreme sacrifice. It was the practice to sacrifice the King at a solar eclipse. Naturally the kings were not too happy about this custom. So, in later times, when a solar eclipse was predicted the King abdicated. Some poor person of lowly rank, such as a slave, was crowned king. After the sacrificial death of the pauper king the old king was reinstated.

The ability to predict eclipses reinforced the authority of the astrologer/priest. It did however, have its down side. Two Chinese astronomers who failed to predict an eclipse were beheaded.

On the banks of the Mediterranean, on the edge of the Nile Delta, is the ancient Egyptian city of Alexandria. Here the pharaohs built the 'Great Library' which became the centre of knowledge and learning throughout the ancient world. When a ship arrived in Alexandria it was searched by the pharaoh's soldiers. Unlike today, these ancient customs officers weren't looking for contraband, what they were looking for was books. If a book was found it was seized and taken to the Great Library where it was carefully copied. The original was kept by the pharaoh and the copy given back to the owners. In this and other ways the Great Library collected books from the four corners of the known world. At its height the library is said to have contained more than a million books which, in those days, were handwritten on papyrus and vellum scrolls.

The keeper of all this knowledge, the Chief Librarian, was appointed by the pharaoh himself. In the year 245 B.C. Ptolemy III became pharaoh and he asked a scholar by the name of Eratosthenes to come to Alexandria to tutor his son. Eratosthenes was born in Cyrene, which in now in Libya, but had spent many years studying in Athens. About the year 240 B.C. Callimachus, the Chief Librarian died, and Ptolemy appointed Eratosthenes as Chief Librarian the Third.

One day while reading through one of the many books from the ships, Eratosthenes came across an interesting account. It described a place called Syene, now known as Aswan, which is about 800 kilometres south of Alexandria. Syene was an oasis used by camel trains on route between Egypt and Ethiopia. The writings said that 'on the longest day of the year, as the day approached noon, the shadows cast by the temple columns became shorter and shorter, until at noon, no shadow was cast'. There was a well at Syene and the manuscript went on to say that, 'on this day the image of the sun could be observed to slither down the well until, precisely at noon, the image of the sun could be seen at the bottom of the well.

Eratosthenes realized that for this to happen, the sun at noon had to be directly overhead. Although he had never observed it, he wondered whether shadows at Alexandria did completely disappear at noon on the longest day. So, at the next mid-summers day at Alexandria he used a gnomon to check it out. And, it turned out that the gnomon did cast a shadow at noon! Puzzled by his findings, Eratoshenes wondered how temple columns at Alexandria cast a definite shadow at noon, while at the same time in Syene no shadow at all. After a while he realized that there could be only one possible explanation, the surface of the Earth was curved.

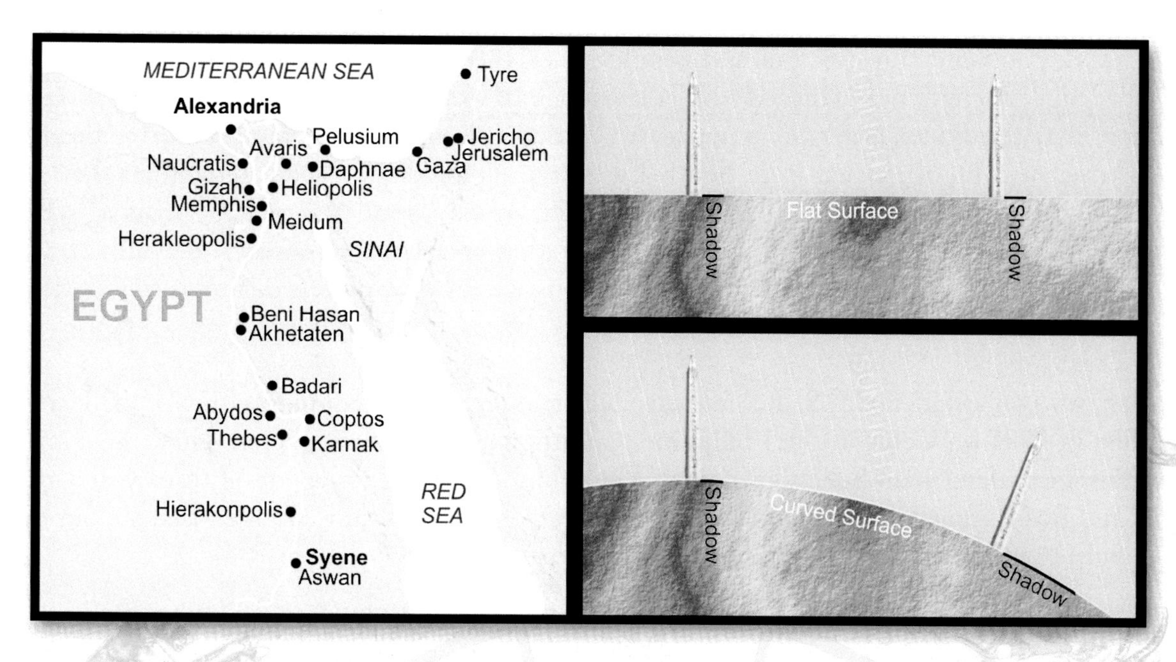

***Fig. 20:** In the 3rd century B.C. Eratosthanes demonstrated that the world was round by the differing lengths of shadows cast at the same time in Alexandria and Syene.* R. Hall

Eratosthenes assumed, and rightly so, that the sun was so far away that light rays, when they strike the earth, are parallel. If the earth were flat, two pillars of the same height, no matter where they are placed, should cast shadows of exactly the same length at the same moment in time. If however, the earth where a sphere, pillars at different latitudes would stand at different angles to each other and, cast shadows of different lengths. (See figure 20) Eratosthanes was not the first person to suggest that the earth was round. Aristotle argued this in the 3rd century B.C, pointing out, that during a lunar eclipse, the shadow of the earth was curved. However, it was Eratosthenes that backed up his theory with experiment. He moved the concept of the spherical earth from philosophy to science fact and, calculated the circumference of the earth.

The shadows cast by pillars at different latitudes provided more than just evidence that the world was round. Each pillar stands upright relative to the centre of the earth. Pillars at different points will be at an angle to each other and this angle will depend on their distance apart and the curvature of the sphere. Eratosthanes knew that he could calculate the angle of each pillar by the lengths of the shadows cast. If he knew the distance between the two pillars he could calculate the circumference of the earth.

At that time the distance between Alexandria and Syene was not well known. So, to carry out his calculations, he hired a man to pace out the entire distance. This is not as outrageous as it sounds. Following the flooding of the Nile, which was an annual event, there were always disputes on where property boundaries where. To settle disputes the courts employed people who were experienced in pacing out distances and fixing boundary points. Eratosthanes undoubtedly used one of these professional pacers. None the less, from these measurements he calculated the circumference of the earth to be 250,000 stadia. A stadia is 157.2 metres, so he calculated the circumference of the earth to be 39,300 kilometres. A remarkable result because the true value is 40,078 kilometres.

From that day forward ships set sail and people risked their lives to circumnavigate the world. To our knowledge this was not achieved until the voyage of Ferdinand Magellan in the 16th century. We know that the Phoenicians circumnavigated Africa and established trade routes to the Indies. The Polynesians, who originated from South-East Asia, explored the Pacific and, either they, or the Egyptians, or both crossed the Pacific and reached the Americas more than 2000 years ago. Recent analysis of Egyptian mummies dating back to this era has revealed the presence of South American cotton and nicotine. Whether these products were brought back by explorers or simply found their way along the trade routes is not known.

Magellan, who set sail in A.D.1519, did not personally complete the circumnavigation of the Earth. He was killed in 1521 on Cebu, in the Philippines, and the voyage was completed under the command of the Basque seafarer Juan Sebastian Elcano. The round the world voyage took three years and one month and was completed on September 18th 1522. Of the 5 ships and 200 men that set out on the voyage only 18 of the original crew survived and returned home.

"How many discoveries were made, fortunes won, and lives lost on the theories of a man who could read shadows?"

Carl Sagan

16. The Calendars

"Time creates the sky and the earth. Time creates that past and the future. By Time the sun burns, through Time all beings exist, in Time the eyes see. Time is the lord of all."

Atharva-veda (19.54)

Astronomically and astrologically the signs of the Zodiac are linked to calendar dates. How is our calendar related to the stars and what was the thinking behind its construction?

For large and prosperous communities to function efficiently an accurate calendar was required for records, communications, co-ordination and future planning. All calendars are based upon the motion of celestial objects.

The solar year, which is the time it takes for the Earth to orbit once around the Sun, determines the seasons. But, it was the Moon, which marked off the weeks and months, that was used as the first clock and around which the first calendars were constructed. Trouble was that the cycle of the moon doesn't fit into the solar year as a whole number. Neither is the solar year a whole number of days. The solar year is 365 days, 5 hours, 48 minutes, and 45.5 seconds. A lunar year of 12 moonths is 354 days, 11 hours and 15 minutes.

The western world uses a solar calendar and it's easy to fall into the trap of thinking everybody in the world uses the same calendar. At the beginning of this century I can remember T.V. presenters saying everyone around the world is celebrating the new millennium. But they weren't because most people around world use different calendars.

Islamic Year

The Islamic year consists of 12 lunar months of 29 or 30 days each, a total of 354 days. In this calendar thirty years constitutes a cycle in which the 2nd, 5th, 7th, 10th, 16th, 18th, 21st, 24th, 26th, and 29th years are leap years of 355 days. The Islamic calendar is not correlated to the solar calendar and consequently, because their lunar year is 11 days shorter than the solar year, festivals and the beginning of their year slowly slip through the seasons. The first day of the Islamic new-year is celebrated on the first day of Muharram. This is the lunar anniversary of the day Mohammed began his journey from Mecca to Medina in the Islamic Year 1 (A.D. 622). In our solar year 2008 there were two Islamic new-years – January 9th and December 28th. The Islamic year begins at sunset with the actual sighting of the new moon.

Chinese Year

The Chinese year begins on the second new Moon after the winter solstice. This occurs in late January or February. The Chinese year 4,644 began on February 18th in 2007, in 2008 it began on February 7th. Why the second new moon rather than the first? Perhaps this is to do with precession of the equinoxes. The Chinese calendar is very ancient and over time precession would have moved the winter solstice relative to the fixed stars of the lunar mansions. If the start of the new-year was tied to a particular mansion, with time precession would move the beginning of the year from the first to the second new moon.

Maori Year

Many people adopted a calendar with a duel system: The moon was used as the time-piece that provided weeks and months but, to keep the seasons in step with the calendar, the solar cycle was used to mark the beginning of the year. The starting point for the year would be a solstice or equinox, or the annual heliacal rising of a bright star.

An example is the calendar used by the Maori of Aotearoa, New Zealand. Each month began with the new moon but it was the heliacal rising of the Pleiades star cluster, known as Matariki, which was the herald of the new-year. However, the rising of the stars did not themselves mark the beginning of the year. Following the sighting of Matariki, which occurs in early June, a watch was then kept on the western evening twilight for the next new moon. As with the heliacal rising of Matariki the time of the new-moon was not an event calculated in time. It began when the new moon was seen.

With the sighting of the new-moon a festival, the period of misrule began. All contracts, including that of marriage, were suspended. A slave had the same rights as a chief. Anyone could sleep with who they liked irrespective of rank and not be punished. This period of misrule is similar to a Roman festival of Saturnalia in which the roles of everyone in a household were reversed. The slaves or servants became the masters and were waited upon by the lords.

The period of misrule lasted for about 10 days, until the Moon became full. This, the rising of the full moon, was the start of the new-year. Laws and contract were reinstated and the labours of the new-year began. Thus, the year began in the middle of a month.

With the coming of the Europeans the missionaries in particular were uncomfortable with the period of misrule. They encouraged the people to change their ways and, with the passage of time, the period of misrule was cut out of the calendar and the year was then said to begin at the new moon.

Because we (in the western world) use a solar calendar in which the months are simply divisions of the year, it seems strange starting the year in the middle of a month. But it is not as unusual as you may think. For example, in England and Ireland, between the 12th Century and 1752, the year began on March 25th (spring equinox). March 25th, 1301, was the day after March 24th, 1300.

Lunisolar Year

Because the lunar year of 12 months is approximately 11 days short of the solar year many people used a lunisolar year in which an extra month was added when necessary to keep the calendar in line with the seasons. The Jewish calendar used to this day is a lunasolar calendar of 12 lunar months of 29 days alternating with 30 days (354 days). An extra month is intercalated every 3 years, based on a cycle of 19 years. The first month in the Jewish calendar is Nissan and is correlated with the spring equinox and the Passover. However, the new-year begins with the seventh month Tishri, close to the northern autumn equinox. Each month and the year begin with the new moon at sunset. In ancient times this was determined by direct observation.

The Babylonians, as far as we know, were the first to use a lunisolar calendar. They had a year of 12 lunar months of 30 days each (360 days) with an extra month being added every six years.

***Fig. 21:** The ancient Egyptian calendar had only three seasons and was based upon the Sothis(Sirius) Cycle.* R. Hall

Sothic Cycle

The Egyptian civilization was dependant upon the annual inundation of the Nile which brought fertility to the Nile Valley. Consequently they initially had a calendar which was peculiar to them that divided the year into three seasons (Figure 21). At this time the vernal equinox was in Taurus and the summer solstice in Leo. The first season was *Akhet*, the "Waters" which alludes to the flooding of the

Nile. It began in what is now June with the flood waters arriving from late June and running through to September. This was followed by *Peret*, the season of "Plants" or "Growing". Most of the planting was carried out in what is now October and November but the season itself ran through to February. The third and last season was *Shemu*, "Flowering" or "Harvest" which began in what is now mid-February and ran through to the end of May.

Each of the three seasons was ruled by 12 spirits called decons. These decons were 10 day 'weeks' based upon the heliacal rising of specific stars. In addition to the 36 decons there were 5 intercalary days which provided a calendar of 365 days. The decons formed the Sothic Cycle which began with the rising of the brightest star, Sirius. The Egyptians knew this star as Sothis.

The word Sothis means the 'barker', or 'monitor'. Sothis was the shepherd's dog in the sky and its heliacal rising warned the shepherd of the approaching inundation of the Nile. It was the time to move the flocks to higher grounds and, since that time Sirius has been known as the 'Dog Star'.[23] The star was said to be the 'Heart of Isis', the Egyptian goddess of fertility, and the floodwaters were said to be her tears shed in remembrance of the death of her husband Osiris. The star was also identified with Anubis (Fig 21 centre). Anubis was a god of the underworld who weighed the hearts of the dead (judgment) and is represented by the figure of a man with the head of a dog.[24]

Later, with the development of the Zodiac, this Sothic calendar was replaced with a solar calendar of 12 months of 30 days each, with 5 extra days at the end (365 days). The decons became guardians of the months (3 to each) and their respective signs of the Zodiac. In 238 B.C. Ptolomy III ordered an extra day to be added every 4 years (the leap year).

The Julian Calendar

Some early farming communities used a calendar of 10 months, or 10 plus 2. There were the ten lunar months in which the labours of the people were carried out plus two in which there was little activity. Because of their unimportance, sometimes these two months weren't even given a name.

The early Roman calendar, introduced in the 7th century B.C., consisted of 10 months with a total 304 days that began in March (The Ides of March). Two extra months, January and February, were added later. Because the two months were added at the beginning of the calendar the original names of the months, which are still used today, are out of order – the words September, October, November and December mean the 7th, 8th, 9th and 10th months respectively. In this calendar, because each month was 29 or 30 days long, an extra month had to be added approximately every two years.

When Julius Caesar became ruler of the Roman Empire he became so frustrated with the complications of the calendar he asked the Senate how he was expected to run an empire with such an inefficient calendar. To remedy this, in 45 B.C., he introduced the Julian Calendar. This was a solar calendar of 365 days with a leap year of 366 days every fourth year.

23 The constellation of Canis Major, the Great Dog, was built around this star at a later date.

24 Anubis is usually described as having the head of a jackal because of his association with death. However, it is more likely that it represents a dog which, to the Egyptians, was a sacred animal.

This, the Julian calendar, is essentially what we use today in the western world. There were seven months with 30 days and five with 31, a total 365 days. Julius Caesar had a month named after him (July), but this month originally had only 30 days. Now, it was not fitting that the great name of Julius be given to a lesser month so, a day was taken from February and placed in July topping it up to 31 days. The Emperor Augustus also had a month named after him (August) and, he wanted to have the same number of days as Julius. So he also pinched a day out of February. And, that's why February has 28 days – tinkering by Roman Emperors.

The Gregorian Calendar

The Julian year was 11 minutes and 14 seconds longer than the Solar year. By A.D.1582 this discrepancy placed the northern spring equinox out by 10 days. This, in turn, meant that festivals like Christmas and Easter were beginning to be shifted into the wrong time of the year. To remedy this Pope Gregory XIII decreed that ten days be dropped from October that year. This caused riots in Europe because many people believed that he had taken ten days out of their lives.... One's destiny and the date of one's birth and death were believed to be written in the stars.

To prevent further displacements in the calendar Pope Gregory further decreed that century years divisible by 400 should also be leap years. This fine tuning of the Julian calendar became known as the Gregorian calendar, and it is the one we use today in the western world.

So when is or isn't it a leap year? If the year is divisible by 4 then it is a leap year.... so long as it is not divisible by 100, in which case it is not a leap year. That is, unless it is divisible by 400, then it is a leap year.

Finally, it is important to remember that all calendars are just a bunch of names and numbers put together by people. The number of days in a month has been determined by astronomers, clerics, politicians and Roman emperors. A calendar is only an approximation of celestial cycles.

THE PLANETARY POWERS

The central theme in this section is the origin of the horoscope and the historical significance of natal astrology – the belief that the destiny of people and nations is written in the stars. The planets play a major role in natal astrology which also forms a cornerstone to religions around the world. We begin by looking at ancient cosmologies - early concepts on the nature and origin of the universe - from which astrology and ultimately the sciences emerged. Finally we explore some of the ways in which star lore and astrology, particularly the planetary powers, played a major role in the foundation of major religions.

17. Sex, Serpents and Lucky Numbers

I have always found it interesting how so many people in this day and age, have and believe in, lucky numbers. If you carry out a survey the most popular numbers are 3 and 7, with the number 7 way out in front. In 2007 large numbers of people made a point of getting married on the 7th day of the 7th month, of the 7th year. The number 12 is another significant lucky number. Historically each of these lucky numbers has important symbolism. Their meanings originate from ancient astronomy and star lore, particularly the Zodiac.

Incidentally, there also unlucky numbers and the most unlucky is the number 13, particularly when it is attached to a certain day – Friday the 13th. This superstition dates back to the demise of the Knights Templar. The Knights Templar was a religious military order formed in A.D.1118 to protect Christian pilgrims who were traveling to the Holy Land. These knights, who wore a white tunic with a red cross over their chain mail, played a major role in the Crusades. With time, the Knights Templar became so rich and powerful that they were seen as a threat to the French monarchy and the authority of the Vatican. Pope Clement V and the King of France, Philip IV, secretly conspired to eliminate this powerful military force. The order went out from the Pope and, on Friday the 13th, October 1307, every Templar Knight in Christendom was arrested and charged with heresy. The assets of the Knighthood were seized, and the knights were tortured, confessions taken, and then executed. Jacques de Molay, the last Grand Master of the Knights Templar, was burned at the stake on March 19th, 1314. As he burned he cursed King Philip and Pope Clement, inviting them to join him in death. The Pope died one month later and the King seven months after that.

Let's get back to those lucky numbers. The origin of their significance in our folk-lore has a much greater antiquity than unlucky 13, and has much to do with the wandering stars.

The Planets

The word planet means wanderer and in ancient times it referred to the seven celestial objects that continually moved relative to the 'fixed' (true) stars. The seven wandering stars are the Sun, the Moon, and the five planets known to the ancients (which really are what we call planets today but look like bright stars in the night sky). These planets are Mercury, Venus, Mars, Jupiter and Saturn. All seven travel along the Zodiac. We have discussed the importance of the sun and moon in relation to the Zodiac. In astrology and some calendars the other wandering stars, the planets, are also major players.

Today we know that the Solar System, the Sun and its system of planets, is shaped like a large flat disk with the Sun at the centre and the planets orbiting within the disk. The Earth and the other planets orbit around the Sun's equator and in the same direction as the Sun rotates on its axis. This rotation is anti-clockwise when viewed from a point directly north of the Sun. This is the same direction in which the Earth rotates on its axis, and the same direction the Moon orbits around the Earth.

Seen from the surface of the earth the orbital motion of the Moon and planets takes them slowly eastward against the background stars. This is the same direction the Earth is rotating on its axis, but is in the opposite direction to the westward motion of the heavens produced by that rotation.

In other words, if you watch the stars for a while you will notice that they are slowly moving westward and, if the moon is in the sky it too will slowly move westward. This motion is due to the rotation of the earth. But, if you observe the position of the moon relative to the background stars, after a while you will notice that it is slowly traveling eastward relative to the stars. The rotation of the earth also carries the sun from east to west but, if we could see the stars during the day, we would notice that the Sun is also slowly moving eastward relative to the background stars. Much the same applies to the planets except that, due to the orbital motion of the earth, the apparent paths of the planets can appear quite complex.

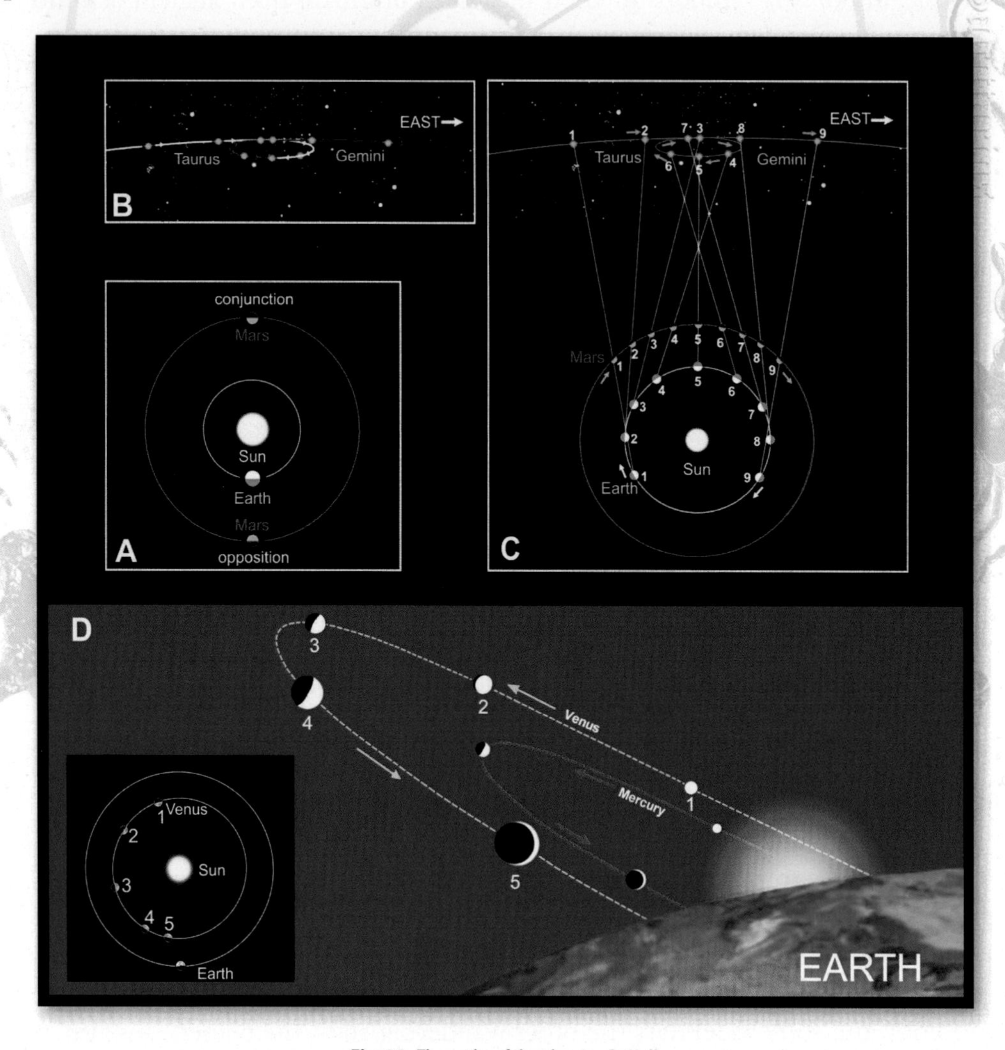

***Fig. 22:** The paths of the planets.* R. Hall

For example, an outer planet is at its brightest when it is closest to the Earth. At this moment it is directly opposite the Sun, what we call opposition (Figure 22A). As it approaches opposition it undergoes a curious change in its path against the background stars. Its eastward motion slows, halts, and then reverses into a westward motion. After opposition the westward motion slows to a halt and the planet reverts to its normal easterly motion. This loop reversal is called *retrograde motion* (Figure 22B).

Retrograde motion puzzled ancient astronomers who believed the Earth to be stationary at the centre of the Universe. The retrograde motion of a planet is in fact the projection of the Earth's orbital motion (Figure 22C). Because the Earth has a smaller orbit and higher orbital velocity it overtakes an outer planet at opposition. Prior to opposition the planet appears to be moving in the same direction as the Earth, eastward. As the Earth passes the planet it appears to move backwards relative to the background stars. After opposition we see the planet behind us and once again it appears to be moving in the same direction as the Earth.

The inner planets, Mercury and Venus, also undergo retrograde motion. But, because their orbits are closer to the Sun than the Earth their apparent paths in our sky are somewhat different. An inner planet will first be seen either just before dawn in the east or just after sunset in the west. Its path then takes it, day by day, further away from the sun – it rises earlier or sets later. Eventually, it reaches its most distant point from the sun (as seen from the Earth) and its apparent motion comes to a halt. It then reverses direction and moves back towards the sun. Eventually it is lost from view in the bright twilight. Figure 22D shows what is really happening and the phases that can be seen with a small telescope. After passing either in front of or behind the sun the planet reappears, but on the opposite side of the sun. Thus, an inner planet is seen either east or west of the sun, which means that it alternates between morning and evening appearances. It alternates between being a 'morning star' and an 'evening star'.

The planets look like stars but are really neighbouring worlds that shine only by reflecting the light of the sun. They have different coloured tints due to differences in their reflective surfaces. And, unlike most of the stars, the planets slowly vary in brightness due to their varying distance from the earth. In addition, the closer a planet is to the sun the faster it moves through space and against the background stars.

The ancients observed these differing characteristics of the planets but were unaware of their causes. They believed that the planets were or represented celestial beings and that the phenomena exhibited by each were related to the sphere of influence of that deity.

Seven Planetary Powers

The ancient stories most people in the western world are familiar with belong to Greek mythology. Many of these tales originated from Pelasgian folk lore. The Pelasgians were an ancient people that inhabited the coasts and islands of the east Mediterranean and Aegaean Sea.

In the early creation stories from this region we learn that 'in the beginning' the universe was without form, a chaos of nothingness, a dreamless sleep. Eventually Eurynome, who was known as the 'Goddess of All Things', awoke. She rose and danced naked across the waters of Chaos, separating the sea from the sky as she went.

She rubbed the North Wind, which begets all creatures with child, between her hands. Then the swirling ripples produced by her dance, brought into being Ophion, the cosmic serpent. The serpent, a symbol of knowledge and wisdom, coiled itself around her and creation began. Eurynome transformed herself into a dove and produced the Universal Egg, from which emerged everything that exists.

As the egg hatched the elements, Earth, Fire, Air and Water, within Chaos began to separate. The aether rose above the air to form the firmament of the heavens. Then Gaia, Mother Earth, emerged out of Chaos surrounded by the embracing arms of the ocean.

From the darkness in the firmament emerged immortal seeds, seven male and seven female seeds, which glittered like gems in the sky. In time they grew into mighty beings that would rule the universe. These, the Titans and Titanesses, were known as the 'Planetary Powers'. They were created and paired by Eurynome to oversee the seven astrological entities. They are as follows:

Titan		Titaness		Planet	Power
Hyperion	+	Theia	=	Sun	Illumination
Atlas	+	Phoebe	=	Moon	Enchantment
Coetus	+	Metis	=	Mercury	Wisdom
Oceanus	+	Tethys	=	Venus	Love
Crius	+	Dione	=	Mars	Growth
Eurymedon	+	Themis	=	Jupiter	Law
Cronus	+	Rhea	=	Saturn	Peace

It should be noted that in all of the early creation stories from this part of the world it is an all powerful female figure that is credited with creation. Later, when patriarchal tribes began to take control, she is replaced with a male creator figure and the Titanesses almost forgotten.

Fig. 23: *The First Lady. Left: Lilith, painting by John Collier 1934. Right: The Temptation of Adam and Eve from the 1427 Fresco of the Brancacci Chapel, Santa Maria del Carmine, Florence.* Wikipedia; Espanol.

The story of Adam and Eve is an example of the influence of patriarchy. In the original Genesis story Eve was not the first woman. The first woman was created in the same manner as Adam and her name was Lileath (Lilith, Lilitu – Storm-Lady). She saw herself as Adam's equal and refused to accept him as her master. She also tried to get Adam to eat the fruit of knowledge to discover sex. For this, in the ancient Hebrew story, Yahweh cast her out of the Garden of Eden. After Eve had been created from Adam's rib Lileath returned and tried to persuade Eve to see herself as an equal to man and to eat from the Tree of Knowledge. By the time the Old Testament was written, around 500 B.C., Lileath had been virtually cut out of the story altogether. She became the serpent that tempted Eve. Indeed, in old paintings the serpent is portrayed with the head of a woman (Figure 23). She was turned into a demon, a succubus who preyed upon men by visiting them while they slept and taking their sperm (to create more demons).[25]

Rather than viewing sex as a sin most of the so-called pagan religions celebrated sex as something natural and wonderful. In an alternative, pagan genesis we find a different explanation of the origin of sex and love. The two planetary powers, Oceanus (male) and Tethys (female) combined to form one – the planet Venus, which was the power of love. When they created humankind they were also one, both male and female. These hermaphrodites had four arms, four legs and two faces. Because they failed to honor the gods they were struck in two, one male, the other female. Ever since that time each half has felt a powerful urge, a deep longing to be reunited with its other self. A man seeks a woman and a woman seeks a man to join together and become one again.

Be warned. It was said that if we displeased the gods again we would again be struck in two. Which I guess would make us hopping mad.

The story of Lileath is unfortunate for snakes as well as women. In the vast majority of cultures across the world serpents (snakes) were associated with healing, good luck, fertility and, above all, wisdom. To this day entwined snakes are the symbol of the medical profession. The Judaic and Christian beliefs are virtually alone in regarding the snake as evil and the enemy of mankind.

The 13th sign along the Zodiac in modern star charts, wedged between Scorpius and Sagittarius, is Ophiuchus – The Serpent Bearer. This is an ancient constellation representing a man entwined by a snake (the constellation of Serpens). Ophiuchus is usually identified as Aeculapius, who is reputed to have discovered the healing power of plants from a snake. He was the forerunner of Hippocrates, and his medical skills were such that he had the ability to do that which was reserved for gods - raise the dead. He was struck down by the gods and placed in the heavens, along with his snake.

Other snakes in the sky include Draco (the Serpent-Dragon) that coils around the north celestial pole and Hydra (the Sea Serpent) the largest constellation in the sky. Hydra lays south of and adjacent to the Zodiac with its head close to Cancer. Its body then coils 100 degrees across the sky, passing Leo and Virgo, with the tip of its tail touching Libra.

Eggs and chaos are, for obvious reasons, common in creation stories. Here briefly, is the creation story from an entirely different tradition, the Chinese.

25 In another account she is the consort of God.

At the beginning of time the only thing in existence was a gigantic egg. Inside this egg were Chaos and the giant, Pan Ku. When Pan Ku broke out of the egg, chaos escaped. The lighter, purer parts (Yang) rose to become the sky, while the heavier impure parts (Yin) sank down to become the earth. The giant kept the two forces apart with his body, pushing them further and further apart. After 18,000 years of pushing the two apart the giant died, but by this time the sky was secure above the earth. From the giant's body were made the sun, moon and stars.

Returning to our lucky numbers, we see that "7" is a significant number in ancient astronomy and astrology that became interwoven with religious beliefs. It is the number of the planets, which were believed to be symbols of the gods. It is the number of days in the week which is a quarter of the Moon's cycle of phases. There are the "Seven Sisters", that signaled the beginning of the year in the ancient world. Seven is also, according to ancient belief, the number of days of the creation and the number of levels of the Underworld.

The number 12 also appears frequently. We have of course the twelve Signs of the Zodiac and the twelve months (moonths) of the year. In the old belief system there were twelve levels of Heaven, twelve ruling Gods of the Pantheon, and twelve Kings before the Flood. Then there are the twelve Apostles and the twelve days of Christmas. The number 12 also became a unit of measure (dozen) and coins of the realm (pennies in a shilling).

The other 'power number' is 3. Three is the number of days of the resurrection (of the moon and the male god). The Moon Goddess has three aspects. We have the Three Realms of Existence (Heaven, Earth and the Underworld / birth, life and death) and the divine Trinity which rules them. Finally we have the Trigons of the Zodiac, which will be explained in the next chapter.

18. The Star Man

So God created man in his own image, in the image of God he created him; male and female he created them.
Genesis 1-27

The Ethiopians make their gods black and snub-nosed; the Thracians say theirs have blue eyes and red hair.... Yes, and if oxen and horses or lions had hands, and could paint with their hands, and produce works of art as men do, horses would paint the forms of the gods like horses, and oxen like oxen, and make their bodies in the image of their several kinds.

Xenophanes

For most of human history physical phenomena has been explained in supernatural terms, the movement of the stars and planets, the seasons, the weather, earthquakes, in fact just about everything was believed to be manifested by the gods. About 600 B.C Greek philosophers began to challenge these beliefs and developed theories that explained the world around them in terms of natural phenomena governed by mechanics that could be explained through mathematics. This point in history marks the origin of modern science.

Initially this brought many of these thinkers into conflict with spiritual leaders. In 450 B.C Anaxagoras of Clazomenae suggested that the sun was not a god but a mass of hot glowing metal. He also suggested that the moon was a second earth and that the stars were burning stones. Sacrilege! Anaxagoras was accused of impiety. Fortunately he lived at a place and at a time when free thought and speech was tolerated.

The new ideology was compelling but, instead of displacing mysticism, it often became assimilated with the supernatural. Instead of a thunderbolt we begin to see images of god using a pair of compasses at the time of creation.

"He set a compass upon the face of the depth." Proverb 9 (27)

The purely mechanical concept of the universe was soon incorporated with and gave way to the theory of universal harmony. Pythagoras, a Greek philosopher, developed a mathematical theory of the cosmos in which the orbits of the planets and stars resonated with each other like musical notes. Indeed, it was subsequently believed that the planets generated musical notes (which mortals cannot hear) and together they played the music of the universe. Like a well orchestrated piece of music everything in the universe should be in harmony.

The new theories of celestial mechanics and universal harmony were related to or arose from the Greek theory of the elements.

What is the Universe made out of? That's a question people have put their minds to down through the ages. Today we would say that the fundamental building blocks are called atoms. Atoms themselves are made up from smaller particles but the atom is the smallest unit that retains chemical properties. There are 92 naturally occurring atoms ranging from hydrogen, the smallest, to uranium the largest. These atoms, the basic building blocks of matter, are known as the 'elements'. In the ancient world it was believed that everything, the stars, the earth and people were made from just four basic elements: Fire, Earth, Air and Water, each of which had a special characteristic. [26]

Element	Characteristic
Fire	Hot
Earth	Cold
Air	Dry
Water	Wet

This gave rise to the concept of the four Humors, from which all living things were made. The humors were directly related to the four elements. In some respects the humors were like small molecules in that each contained the characteristics of two elements. The humors are as follows:

Humor	Characteristic
Coler (yellow bile)	Hot and Dry (fire and air)
Blood	Hot and Wet (fire and water)
Phlegm	Cold and Wet (earth and water)
Melancholy (black bile)	Cold and Dry (earth and air)

Fig. 24: Elements, Humors and Trigons. R. Hall

26 The Chinese had a fifth element, Wood.

The seasons and the climate were also associated with the four humors. This relationship of the elements/humors, the seasons and climate, was represented as an octagonal star as shown in figure 24.

The stars were also associated with the humors. Each humor was associated with three signs of the Zodiac. These were known as the 'Trigons'. The three signs within each trigon are located in the sky 120 degrees away from each other. (See figure 24).

Element	Trigon	Characteristics
Fire	Aries, Leo, Sagittarius	hot, dry, ardent
Earth	Taurus, Virgo, Capricornus	cold, dry, heavy
Air	Gemini, Libra, Aquarius	hot, wet, light
Water	Cancer, Scorpio, Pisces	cold, wet, soft

The earth was made from the elements but the stars and people contained the humors; which meant that there was an intimate relationship between people and the stars. This was the reasoning that underpinned astrology; which was divided into two disciplines, natural and judicial astrology

Natural Astrology

There are two branches of natural astrology. The first is meteorology, the astrological prediction of the weather. I wonder if Jim Hickey[27] has ever thought of himself as an astrologer? The difference is that modern meteorologists depend upon satellite weather maps to make their predictions while the ancient meteorologist used the stars (which were related to seasonal events).

In predicting seasonal events the prophets were on pretty firm ground. However, it should at this point be noted that nothing could be taken for granted. Whereas today we know that the seasonal changes are cyclic and due to celestial mechanics this was not the understanding of the ancients. They believed that seasonal events were brought about by the gods and, like ordinary people; the gods were not always predictable. After all, sometimes the rains didn't come and then there was a famine. Ceremony and sacrifice was introduced to please the gods and encourage them to do what humans wanted.

In one story from antiquity we learn that once upon a time it was the gods who toiled to maintain the earth and looked after the crops and animals. The gods got fed up with all this work so they created people to do it for them. This allowed the gods to relax and do other godly things like drinking ambrosia. However, they still had their heavenly chores, to turn the great wheel of the heavens, keep the sun, moon and stars on their courses, and control the great elements – earth, fire, air and water.

The great celestial wheel didn't just turn on its own - someone had to do it and, in ancient Egypt this was the job of Osiris. The Pharaoh was believed to be Horus, the "living god". The funeral ritual following the death of his mortal body was exceedingly elaborate because the very fate of the world depended upon it. The body of the pharaoh was carried in a boat across the Nile to its final resting place. A large boat was often placed within or buried outside the pyramid because the spirit of the pharaoh now had to cross the Celestial Nile (Milky Way). If he succeeded in doing this he became Osiris who turned the great wheel of the heavens. In the Solar Boat (The Sun), the soul of the pharaoh

27 Jim Hickey is a weather presenter on New Zealand T.V.

made his pilgrimage across the celestial Nile and also through the underworld. He then drove the great cosmic cycles of night and day, of death and rebirth of the stars, and of the 'Great Year'.

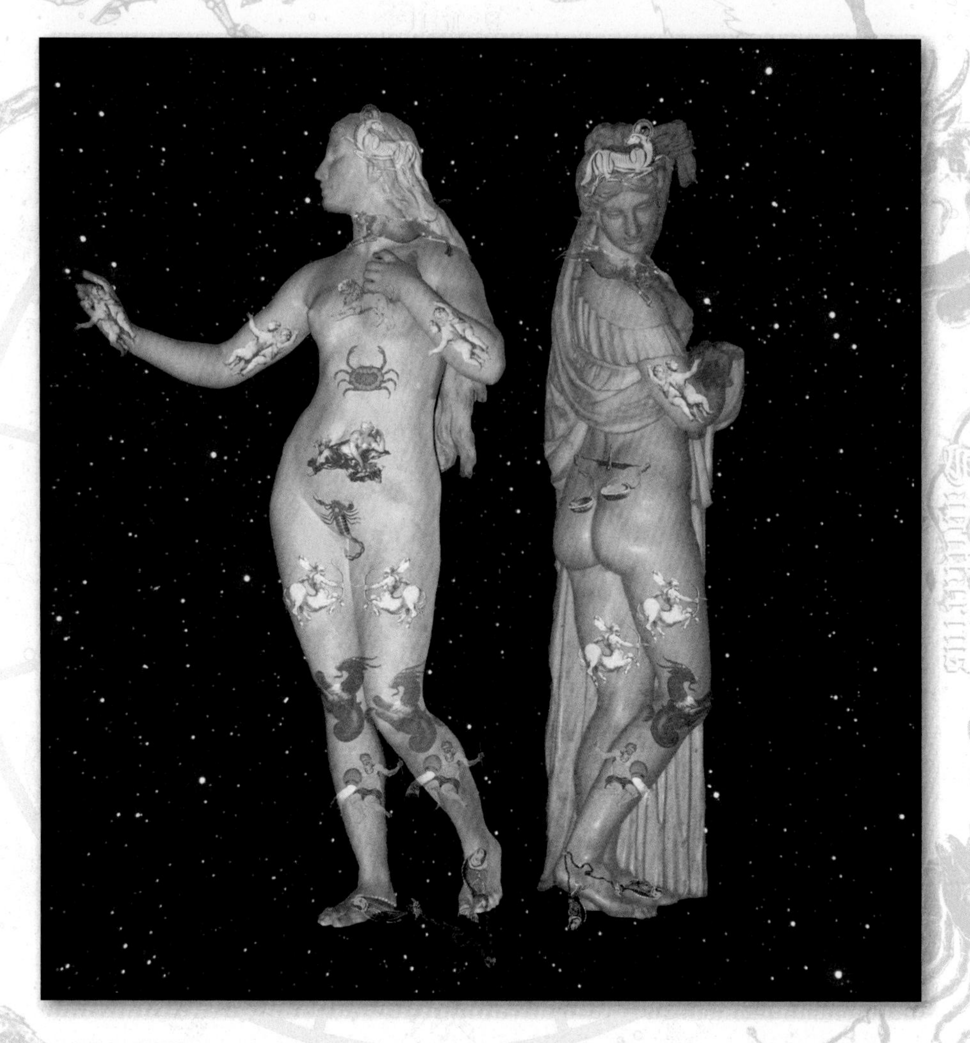

***Fig. 25:** Different parts of the body were believed to be controlled by different signs of the Zodiac.* R. Hall

The second branch of natural astrology deals with medicine. Because human beings were made up from the four humors it was believed that different parts of the body were controlled or associated with different signs of the Zodiac, see figure 25. Right up until the 17th century the practice of medicine was based upon the four humors. In a healthy body everything was in 'harmony'. When a person became sick the humours had become out of balance. It was the task of the practitioner to bring back the harmony. This was done by administering potions (medicine) made from plant or animal extracts that were believed to be related to the zodiacal sign of the afflicted body part. Blood letting was another method of bringing the humours into balance. Sickness, fevers and plagues, were often associated with certain signs of the Zodiac and the positions of the planets. Blood was released from that point on the body associated with the malevolent sign.

God (or the Gods) created man (and woman) in his (or her) own image. That this is almost a universal belief should not come as a surprise. If something doesn't have a human face we can't relate to it and find it difficult to recognize it as either intelligent or having emotions. That's why cartoonists put human faces on animals and inanimate objects.

Not only was it believed that man was created in God's image it was also believed that God created the universe in his own image. The universe it was believed, if we could see it in its entirety, would be in the image of god, which is also the image of man. The universe was the *cosmos* and man was the *microcosmos*. Within each of us, it was believed, is a blue print of the universe. The famous mystic Hildegard of Bingen, who was known as the Sybil of the Rhine, had this to say: "Man carries with him heaven and earth... man's head is spherical, as is the world; it has seven orifices which corresponds to the seven (known) planets, his eyes are the sun and the moon; his chest contains air, his whole body water; the four ages of man correspond to the four seasons..." Everything that happened in the human body, the micrcosmos, was believed to be directly related to the greater cosmos.

Incidentally, the word 'cosmos' (*kosmos*) was adopted into science vocabulary by Plato. This is intriguing because kosmos means women's apparel; it is where the word cosmetics originated. Perhaps the stars are the cosmetics of the universe (of the man in the sky).

Judicial Astrology

This is the astrology that we are familiar with. It asserts that the signs of the Zodiac are the building blocks of an individual's personality and destiny. Its origins are much older than natural astrology and can be traced back to the ancient Egyptians and the Babylonians, perhaps earlier. The personality traits of an individual were determined by his or her natal sign. This was the sign occupied by the sun, the life giving powers of the sun, at either the time of one's birth or the time of one's conception. Obviously the later was generally more difficult to work out with any certainty. In addition, each sign was related to one of the elements and was also considered to be either male or female.

1.	Aries	Fire	Masculine	assertive, impulsive, selfish
2.	Taurus	Earth	Feminine	resourceful, thorough, indulgent
3.	Gemini	Air	Masculine	logical, inquisitive, superficial
4.	Cancer	Water	Feminine	tenacious, sensitive, clinging
5.	Leo	Fire	Masculine	generous, proud, theatrical
6.	Virgo	Earth	Feminine	practical, efficient, critical
7.	Libra	Air	Masculine	co-operative, fair, lazy
8.	Scorpio	Water	Feminine	passionate, sensitive, anxious
9.	Sagittarius	Fire	Masculine	free, straightforward, careless
10.	Capricorn	Earth	Feminine	prudent, cautious, suspicious
11.	Aquarius	Air	Masculine	democratic, unconventional, detached
12.	Pisces	Water	Feminine	imaginative, sensitive, distracted

Some astrologers have said to me that they can pick someone's star sign just by observing their behavior. Can they really? Take a look at the list of personalities and their star signs below. What behavioral patterns does each group have in common and how well do they fit with their zodiacal personality traits? How well do you fit in with your group? Now, you are not allowed to be selective. If for example your star sign is Virgo you might be tempted to believe that you have traits similar to Rutherford or H.G. Wells. But hang on! You've also got Albert De Salva in your group – the Boston Strangler!

Aries
Adolf Hitler
Aretha Franklin
Charlie Chaplin
Charlotte Bronte
Leonardo da Vinci
Pope Benedict XVI

Taurus
Jeffrey Dahmer
Florence Nightingale
Lenin
Mary Wollstonecraft
Pope John Paul
Saddam Hussein

Gemini
Bob Dylan
Che Guevara
John F. Kennedy
John Wesley
Marilyn Monroe
Queen Victoria

Cancer
Edmond Hillary
George W. Bush
Imelda Marcos
Louis Armstrong
Nelson Mandela
Princess Diana

Leo
George Bernard Shaw
John Key
John Howard
Helen Mirren
Madonna
Napoleon

Virgo
Albert De Salva
Ernest Rutherford
H.G. Wells
Michael Jackson
Mother Teresa
Ray Charles

Libra
Dwight Eisenhower
Jesse Jackson
John Lennon
Julie Andrews
Margaret Thatcher
Oscar Wild

Scorpio
Bill Gates
Condoleezza Rice
Edwin Hubble
Joseph Goebbels
Marie Curi
Peter Jackson

Sagittarius
Brad Pitt
Jimi Hendrix
Joseph Stalin
Mark Twain
Ted Bundy
Winston Churchill

Capricorn
Al Capone
Issac Newton
Martin Luther King
Mao Zedong
Muhammad Ali
Tiger Woods

Aquarius
Babe Ruth
Charles Lindbergh
Franklin D. Roosevelt
Oprah Winfrey
Rasputin
Thomas Jefferson

Pisces
Albert Einstein
Elizabeth Taylor
Helen Clark
Osama bin Laden
Robert Mugabe
Yuri Gagarin

The above is of course a very small sample and my astrologer friends will no doubt be saying that I am being flippant about astrology. Maybe, but conjectures based upon these simplified character traits can be found in every newspaper horoscope that you pick up. I am also aware that many leading astrologers also hold these newspaper horoscopes in contempt.

The horoscopes of the ancient astrologer were somewhat different. He or she carefully observed the positions of the stars and planets, their shimmering, colour and luster, and other phenomena in the night sky. From these observations horoscopes were constructed that provided information of what lay in the future. It was an immensely powerful ideology. The Pharaoh or Caesar would not embark upon a campaign without consulting the court astrologer. Astrology played a major role in the politics of Imperial Rome. The Emperor Tiberius for example, surrounded himself with astrologers who advised him on his every move.

Predicting seasonal events was one thing but using the stars to predict the destiny of humans was a little trickier. However, it must be born in mind that many of these celestial portents concerning human affairs would have been self-fulfilling prophesies. If you are firmly of the belief that your destiny is written by, or controlled by the stars, you are more than likely to follow the advice and live by the teachings of the prophet. As a simple example, imagine that it had been prophesized that you would meet a "tall dark handsome/beautiful stranger". Believing this you would be looking out for this person. Every time someone appeared that fitted this description you would be wondering, is this him or her? If you thought it likely that this were the person in the prophecy you would probably behave in a manner that would attract the attention of that person.... And the prophecy would come true.

Fig. 26: *The Bayeux Tapestry (11th Century), Musee de la Tapisserie de Bayeux. Legend above: Isti Mirant Stella, "These men wonder at a star".* Wikipedia

Some self-fulfilling prophesies have changed the course of history. Bright comets have often been seen as harbingers of doom that foretold the "death of princes and the fall of kingdoms". This bad press on comets probably dates back to the fall of Jerusalem. In A.D.66, shortly before the city was sacked by the Romans, a bright comet appeared and hung over the city like the "Sword of Damocles". In 1066 this same comet re-appeared in the sky[28]. The portent was, according to the sages, that there would be the fall of a kingdom. So the Normans turned this to their advantage and promptly invaded England. The famous Bayeux tapestry (figure 26) shows the Normans pointing at the comet which is traveling towards the doomed English King Harold.

Not everyone adhered to the doctrines of judicial astrology. The belief that everyone's destiny had been pre-determined by the stars worried some thinkers. It implied that human beings had no free will. The

28 The comet seen in AD66 and 1066 was Comet Halley which returns every 76 years.

Emperor Claudius rejected this philosophy. What was the point he asked, of motivating your troops to win a battle if their defeat had already been prophesied. Astrology he believed was a barrier to rational thought and undermined his authority, so he banned astrologers from the city of Rome.

However, the belief that one's destiny was influenced by the stars continued to be widely accepted and flourished in Europe until the Roman Empire became a Christian Theocracy in A.D.391. At this time anything associated with the old religions was outlawed. Saint Augustine denounced "the lying divinations and unholy aberrations of the numerologists". Astrology, along with the emerging sciences, was virtually wiped out of existence in Christendom.

Astrology (and astronomy) was reintroduced to Europe in the 12th century by the Arabs. But this time it was adopted by Christian authorities who used the deterministic aspects of astrology to glorify god and support their doctrines and prophesies.

Astrology (and astronomy) was Christianized. Aristotle's 'Prime Mover', who set the world in motion and turned the great celestial spheres, was replaced by angels. God was placed in the highest level of heaven, in the head of the cosmic man.

***Fig. 27:** Map of the Christian Constellations as depicted by Julius Schiller, from 'The Celestial Atlas, or The Harmony of the Universe,' 1661.*

Everything had to be sanitized. The Venerable Bede, along with other theologians, strongly criticized astronomers for continuing to use star charts that include pagan images. To remedy this, in 1627, the Jesuit astronomer Julius Schiller published 'Coelum Stellatum Christianum', a lavishly illustrated star atlas in which the constellations of the zodiac were renamed after the 12 apostles (figure 27). Taurus for example, became Saint Andrew carrying his cross. Other constellations in the northern hemisphere were re-figured and given names from the New Testament – Andromeda became the Holy Sepulcher, Christ's Tomb. Constellations in the southern hemisphere were given names from the Old Testament so that Argo Navis, the Ship of the Argonauts, became Noah's Ark. Not even the Milky Way was spared. It became the path to Santiago de Compostela.

Fortunately these new maps of the heaven were never adopted by the rest of the astronomical community. The reformation was underway and Christianity was divided. Schiller's Atlas was considered to be too Catholic.

On the periphery of this theocratic stage there were still some individuals who continued to try and understand the universe in natural rather than supernatural terms. When the Vatican claimed that it was angels that turned the celestial spheres Maimonides, a 12th century Jewish philosopher, declared that the angels were nothing more than "the laws of the world". But Maimonides thoughts were more than 500 years before their time.

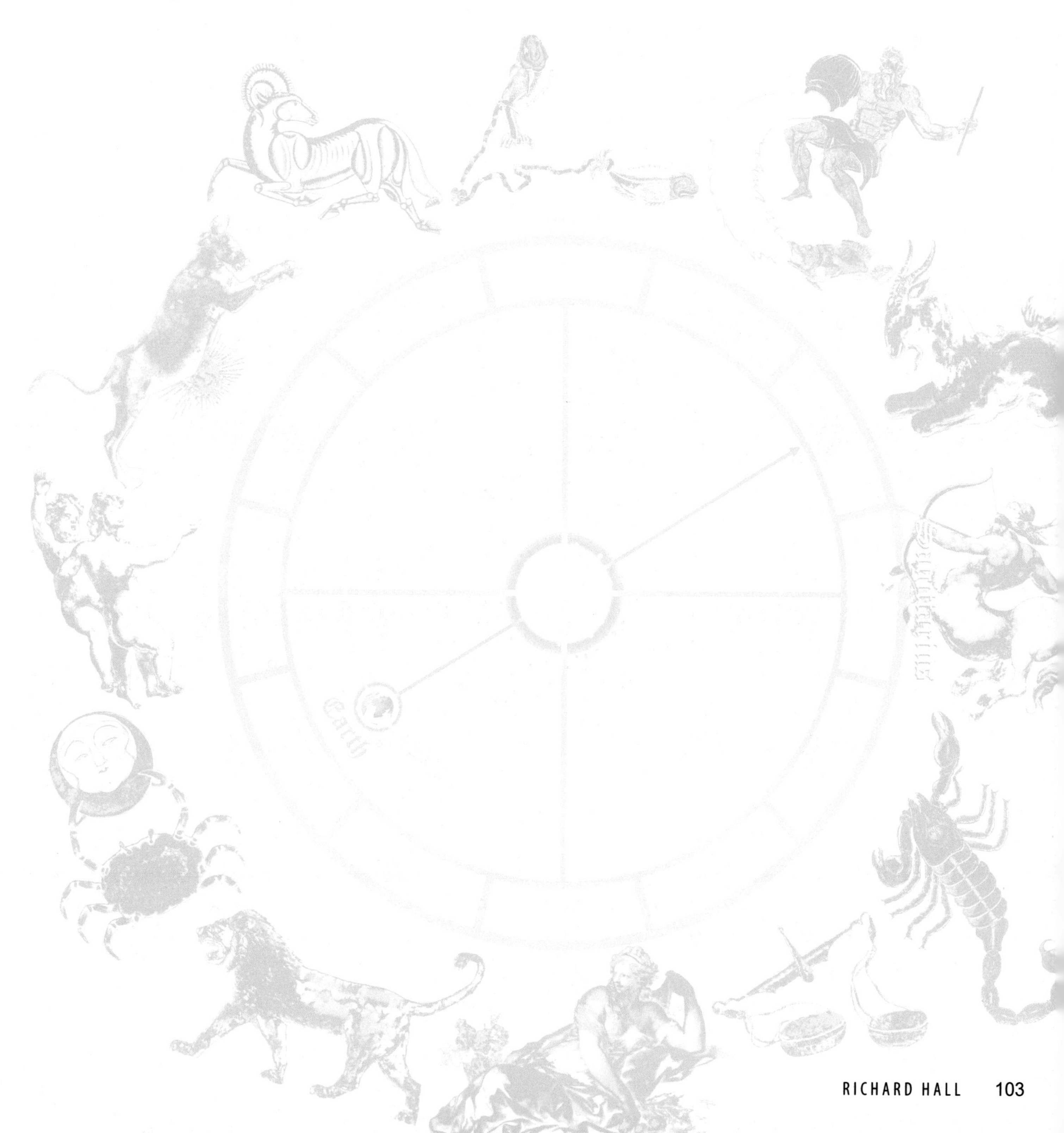

19. The Planetary Houses

***Fig. 28:** Thema Mundi – the oldest known horoscope, 400B.C.* R. Hall

The earliest known horoscope is the **Thema Mundi** which was used by Greek astrologers around 400 BC. (Figure 28) It shows what was believed to be the positions of the seven planets at the time of creation. The birthplace of the Sun was believed to be in the constellation of Leo, close to the star Denebola. The Moon was created in Cancer, Mercury in Virgo, Jupiter in Sagittarius and Saturn in Capricornus. Venus and Mars were born in Scorpius but later, when the Scales were created out of the claws of the Scorpion, the birthplace of Venus was assigned to Libra.

Each planet was believed to rule over one or more of the signs of the Zodiac. These are known as the planetary houses:

House of the Sun:	Leo	(Masculine)
House of the Moon:	Cancer	(Feminine)
Houses of Mercury:	Gemini and Virgo	(Masculine & Feminine)
Houses of Venus:	Taurus and Libra	(Feminine & Masculine)
Houses of Mars:	Aries and Scorpius	(Masculine & Feminine)
Houses of Jupiter:	Pisces and Sagittarius	(Feminine & Masculine)
Houses of Saturn:	Aquarius and Capricornus	(Masculine & Feminine)

The assignment of the planetary houses is based in part upon observational evidence. The observed path a planet takes in our sky is essentially determined by whether it is closer or further from the Sun than the Earth. This divides the planets into two groups, the *inferior* or inner planets Mercury and Venus, which spend most of their time close to the sun, and the *superior* or outer planets Mars, Jupiter and Saturn, who travel right across the vault of the heavens.

In the Thema Mundi the two brightest and most important luminaries are placed side by side with the Moon at the summer solstice in Cancer and the Sun in the following sign Leo, the hottest time of the year. The remaining planets are allocated two houses, one to the solar side and the other to the lunar side of the zodiac. The first to be placed are the inferior planets. The houses of Mercury are located in Gemini and Virgo, the two signs either side of the sun and moon. This is because when Mercury is visible it is either east or west of the sun but always in the sign next to that occupied by the sun. The houses of Venus are located in the next two signs, Libra and Taurus, because Venus is never seen more than two signs away from the sun. Next are the superior planets which are observed to travel all around the zodiac and occupy any sign irrespective of its distance from the sun. These are located in order of the speed in which they move around the Zodiac – Mars, Jupiter and then Saturn (this turns out to be related to the orbital distance of the planet from the sun). The houses of the outermost planet Saturn, which became associated with darkness, are located in Capricornus and Aquarius. When the sun was located in these signs it was the coldest and darkest time of the year.

The Zodiacal Ladder

The planetary houses are also symbolic of the seven ascending steps the sun takes on its journey from the darkest of times to the summit of its power. This refers back to the time when the Summer Solstice was in Leo and the Zodiac was in the process of being formulated. Starting from Aquarius (winter solstice) each step of the ladder takes the Sun higher in the sky – Pisces, Aries, Taurus (spring equinox), Gemini, Cancer, and finally it reaches the summit, Leo. Hence, the rungs of the ladder are in the exact order of the Houses of the Seven Planets.[29]

Each planetary house, and hence each rung or step of the ladder, was also associated with a metal and a colour. These are, from top to bottom:

29 Some historians believe that this is the origin and symbolism of 'Jacob's Ladder' and the Masonic 'Ladder of the Seven Rounds'.

	Sphere	Metal	Colour
7.	Sun	Gold	gold
6.	Moon	Silver	silver
5.	Mercury	Quicksilver	dark blue
4.	Venus	Copper	pale yellow
3.	Mars	Iron	red
2.	Jupiter	Tin	orange
1.	Saturn	Lead	black

Close to what was once the great city of Babylon is the 'Temple of the Seven Spheres'. The temple has a great antiquity and was renovated by Nebuchadnezzar, c1110 B.C. The floor of the temple, like that of the great temple of Belus in Babylon, was built into seven receding stages. On the uppermost stage was the 'arc' or tabernacle. The seven stages were coloured to represent the hues assigned to the 'seven planetary spheres'.

Origin of the Planetary Entities

The planets are named after Roman gods but their mythological and astrological identities were mostly inherited from the equivalent Grecian or Pelasgian deities:-

Roman	Greek
Mercury	Hermes
Venus	Aphrodite
Mars	Ares
Jupiter	Zeus
Saturn	Cronus

Over time the mythologies and symbolism of these deities has become mixed. Mars for example, the planetary power of growth, was originally an agricultural god. The Romans turned him into a god of war. He became the mythical founder of the Roman state and the father of Romulus and Remus. Later he became identified with the Greek god of war, Ares. The stories of Ares became interwoven with those of Mars. However, unlike Ares, Mars was noted for his self control, justice, and fair-dealings.

Saturn, Cronus and Chronos are the most tangled. Saturn was the Roman God of agriculture and the harvest and was identified with Cronus, the male entity of the planetary power of peace. Saturn, being a harvest deity, is often portrayed carrying a scythe. This is the same instrument used by Cronus to castrate Ouranus, the first King of the Gods. The scythe is also associated with the Grim Reaper.

Cronus is often conflated with Chronos, the personification of Time. Chronos emerged from Chaos, you've guessed it, at the beginning of time. He is depicted by the Romans and the Greeks as a wise old man with a long grey beard who turned the Great Wheel of the Zodiac. The planet Saturn was also associated with time and, as we shall see shortly, may represent the original Grim Reaper.

The Planetary Ages

In ancient mythology the history of the universe was divided into four great eras, the ages of gold, silver, bronze and iron.

The Age of Gold (Saturn and Venus)

Gaia, Mother Earth, gave birth to Ouranus (Sky) in her sleep. Ouranus (Uranus) showered his "fertile rain" upon Mother Earth to create all living things. When Ouranus imprisoned her children, Mother Earth asked Cronus (Saturn) to help (in some stories Cronus is the son of Ouranus). He ambushed Ouranus and castrated him, throwing his genitalia into the sea. Aphrodite (Venus), the goddess of desire, rose naked from the blooded foam. When she stepped ashore flowers sprang up where ever her feet touched the ground. Thus began the Age of Gold when Cronus ruled the universe (or was it Aphrodite?). It was a time when the earth provided food without the need for labor, and murder was unknown to humans who lived in peace and harmony (Eden).

In remembrance of the Age of Gold the Romans held a great festival known as Saturnalia. During Saturnalia the roles of masters and slaves were reversed, and morals and etiquette abandoned. Originally it was held on a single day, December 17th. It was so popular that it was extended to the 24th, the day before the winter solstice in the Roman calendar. Everyone joined in the revelry with Caesar playing the role of Saturn. At the time of the festival the sun was in the House of Saturn, Capricornus.

Saturn is the outermost of the planets known to the ancients. In our night sky it looks like a fairly bright yellowish star. Saturn is at its closest to the Earth every 378 days (1.03 years) but, because of its great distance, there is only a small cyclic variation in its brightness. Saturn moves slowly against the background stars taking 29 ½ years to complete one circuit of the zodiac. In days of old this was the average life expectancy of people. It was said that when Saturn returned to the sign that it was in at the time of your birth you were living on borrowed time (Hence the Grim Reaper). Today, most people in the developed world live for at least two, some more than three, Saturnian years.

The House of Venus is in Taurus and, in ancient astrology, Taurus ruled over Arabia, Asia and Scythia. Taurus was also known as *Veneris Sidus, Domus Veneris nocturna* and *Gaudium Veneris*. When Aphrodite (Venus) stepped forth from the water, she came ashore at Scythia. The Scythians were nomadic horsemen from the Russian steppes and their culture and customs eventually became part of Celtic culture.

Venus is the second planet from the Sun and is by far the most brilliant star-like object in our sky. At greatest brilliance she casts a shadow at night and, if you know where to look, Venus can be seen in broad daylight. For this reason, in ancient times, she was known as the "Day Star". Venus is an inferior planet that alternates between morning and evening appearances. She is the celebrated "Morning or Evening Star". Unlike the fleeting appearances of Mercury, Venus remains resident in either our eastern morning or western evening sky for periods of seven to eight months at a time. Her cycle of morning and evening appearances along with changes in her luster repeats every 584 days (1.6 years). The path of Venus takes her up to 48 degrees from the sun or, two zodiacal signs from that occupied by the sun. Thus, Venus can rise before or set after the sun by as much as three hours. In some cultures her morning and evening appearances represented two opposing aspects. As the Evening Star she was the goddess of love and desire, when she was the Morning Star she was the goddess of war.

In pre-Christian times the Morning Star was known as Lucifer. This name is derived from the Latin words lucem ferre, which means "light-bearer", the herald of the Sun and daylight. In post New Testament times Christian theologians debased the deities and symbols of other religions and identified Lucifer as Satan.

Satan (Shaitan) means 'adversary' and in Hebrew folk lore he was originally a 'Lord of Light'. When called upon by Yahweh (God) to knell before his creation, Adam, Satan refused. Adam, he said, was made from clay but he was born of fire. He would knell only before Yahweh. Because he challenged Yahweh he was cast from Heaven into Hell. He then became the Prince of Darkness, the opposite of Yahweh the Prince of Light. Philosophically these two opposites represented the two conflicting aspects of human personality – light and dark, good and evil.

The identification of Lucifer as Satan was based upon a single reference in the Old Testament, Isaiah 14:3-20, in which a Babylonian king had the title of 'Day Star' and was prophesied to fall from heaven and power. Elsewhere in the Bible the use of the word Lucifer is used to identify the Morning Star.

The Age of Silver (Jupiter)

Mother Earth prophesied that Cronus would be overthrown by one of his children, so he devoured each shortly after their birth. Rhea, his sister-wife tricked him with the sixth child. She replaced the child (Zeus, Jupiter) with a carved stone wrapped in blankets, which Cronus swallowed. Cronus thought that he was safe but Rhea had hidden Zeus away. When Zeus came of age he slipped into his father's palace and put poison in his drinking a cup, which caused Cronus to vomit and regurgitate the five immortal children he had swallowed. These were Demeter (Ceres), Hades (Pluto), Hera (Juno), Hestia (Vesta), and Poseidon (Neptune). He also threw up the stone which, according to legend, became the centerpiece of the Delphic Oracle. The children of Cronus (the Olympian gods) made war and defeated Cronus and his Titan allies. When Cronus was cast down Zeus became King of the Gods and the Age of Silver began. It was the age when the seasons came into being.

Jupiter (Zeus) is the largest planet in the Solar System and, despite its great distance from the Earth, is always a conspicuous object when present in our night sky. Jupiter is at its closest point to the earth, and therefore at its brightest, every 399 days (1.09 years). During this cycle, from its most distant to its closest point, it doubles in brightness. However, even at its faintest Jupiter is outshone only by Venus and on rare occasions Mars. Jupiter, King of the Planets, moves sedately against the background stars orbiting around the sun in a period of 11.9 years. This means that it takes approximately 12 years for Jupiter to complete one lap of the Zodiac. Consequently, each year, it moves from one sign to the next. This 12 year cycle in which each year it resides in a different sign of the zodiac formed the framework of calendars created by the Chinese and the Mayans.

The Age of Bronze (Mars)

Hera, Queen of the Gods, had two sons. One, Hephaestus (Vulcan) was twisted and ugly in form but had a beautiful and creative mind. The other, Ares (Mars), was beautiful in form but had a twisted ugly mind. When he was born Ares looked down on the earth and wanted it. But it had already been ruled that no single god could own the mortal plane. When he was denied Ares was infuriated and cursed mankind with death and destruction. Ares became the God of War who loved battle and conflict for its own sake. So began the Age of Bronze which was also known as the Age of War. It was the age when the Trojan War occurred.

Mars is the first and nearest of the superior planets. Most of the time Mars, seen in our night sky, is a bright but not overly conspicuous reddish star. However, every 780 days (2.14 years), the Earth and Mars make a close approach to each other. At these times Mars grows in brightness, up to 45 fold, to become a brilliant star-like object that can outshine all other planets except Venus. It has a strong reddish tint which the ancients said glowed like blood.

The Age of Iron (Mercury)

Hermes (Mercury) was the son of Zeus and Maia, one of the Seven Sisters who were the daughters of Atlas. From the time of his birth he was elusive and fast moving. He was born at dawn, three hours later he was running about; and at noon he slipped away to explore the world. Hermes was the 'Messenger of the Gods' and was the only god that roamed freely between the realms of existence. One of his tasks was to escort the souls of the dead to the underworld. Hermes is the God of Commerce and the Patron of Thieves. He is associated with the Age of Iron, an age of toil and injustice, crime and punishment. It is the age we are now in.

Mercury is the closest planet to the sun and the two are never far apart. It orbits around the sun in just 88 days making on average six fleeting appearances in our sky throughout the year. These alternate between evening and morning appearances in which the planet is never more than 28 degrees from the Sun. This means that Mercury is always in the same sign or a sign directly adjacent to that occupied by the Sun. Consequently, when Mercury is visible, it is usually in a twilight sky, either in the west after sundown or when it rises in the east shortly before the sun. During an appearance its rapid motion against background stars, which takes it first away and then back to the sun, is very noticeable from night to night. Mercury looks like a bright star with a warm, slightly pinkish glow, but its luster is diminished in the twilight sky. Its cycle of morning or evening appearances repeats every 116 days.

The s**even days of the week** were named after the seven wandering stars. Those we use today are derived from the names of Germanic gods or Germanic transliterations of Roman gods:-

	Germanic	**Roman**	
Sunday	*Sunnandeg*	*Solis dies*	Day of the Sun
Monday	*Monandeg*	*Lunis dies*	Day of the Moon
Tuesday	*Tiwesdeg*	*Martis dies*	Day of Mars
Wednesday	*Wodnesdeg, Odin*	*Mercurii dies*	Day of Mercury
Thursday	*Thunres, Day of Thunder*	*Jovis dies*	Day of Jupiter
Friday	*Friatag, Day of Frigg*	*Veneris dies*	Day of Venus
Saturday	*Satern*	*Saturni dies*	Day of Saturn

The Hindu day, *vara*, was also named after the seven wandering stars. In Sanskrit they are in the identical order to the Roman days.

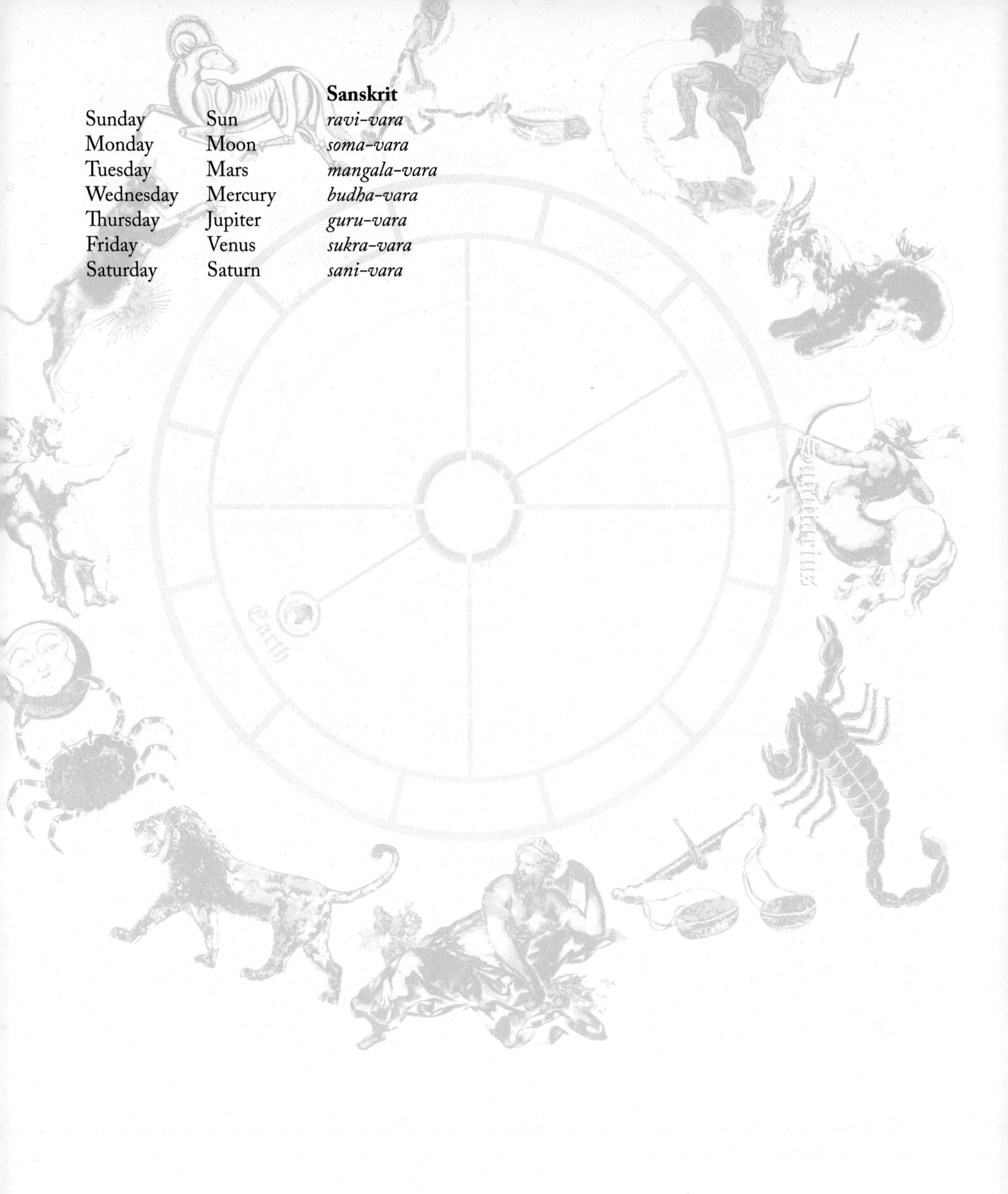

		Sanskrit
Sunday	Sun	*ravi-vara*
Monday	Moon	*soma-vara*
Tuesday	Mars	*mangala-vara*
Wednesday	Mercury	*budha-vara*
Thursday	Jupiter	*guru-vara*
Friday	Venus	*sukra-vara*
Saturday	Saturn	*sani-vara*

20. The Star of Ishtar

Symbols and symbolism become ingrained upon our minds. But like the words in a language their meaning can change with the passage of time. The swastika for example, was originally a sign of good fortune. Because it was adopted by the Nazis it is difficult today to see it as anything other than symbol of hate and racism. Perhaps the most misunderstood symbol is the pentagram which, in modern horror stories is often used as the sign of Satan. In reality the pentagram symbol originally had nothing to do with Satan. It is in fact the symbol of that Earth-Mother goddess we have discussed and was, and still is the foremost religious symbol of the Wicca.

Wicca is an old English name and means wizard or witch (male/female). In medieval times the Christian Church labeled the Wicca as devil worshipers which, in the 16th and 17th centuries, led to the persecution and execution of tens of thousands of people, mostly women. The truth is that the Wicca had nothing to do with Satan, they don't even believe in his existence. They were the followers, priests and priestesses, of the old pre-Christian religion in Europe who worshiped the Mother-Earth Goddess. Incidentally, the last person condemned to death in England on the charge of witchcraft was Jane Wenham at Hartford in 1712. Among the more sensational charges was that she traveled aloft on a broomstick. She was reprieved by Lord Justice Powell who, in his deliberations, uttered the famous statement: "There is no law against flying."

The spiritual significance of the pentagram can be traced back at least 3,000 years to Babylon. The great Babylonian Goddess Ishtar is identified with the planet Venus.

She had two aspects. When she was the evening star she was the goddess of sex, love, desire and fertility. As the morning star she was a blood thirsty goddess of war (see chapter 10). The pentagram is her symbol and it is derived from the motion of the planet Venus in the Zodiac.

The Earth completes one orbit of the Sun in just over 365 days. Venus, being closer to the Sun, completes an orbit every 225 days. Rounding up to whole numbers we find an 8 to 13 ratio in the orbital periods of the two planets. In other words, in an 8 year period the Earth completes 8 orbits of the Sun, Venus 13. This means that, in an 8 year period, Venus laps the Earth (13-8) 5 times. Consequently any configuration of the Sun and the two planets, such as inferior conjunction (when the three are lined up with Venus between the Earth and Sun), will reoccur every 584 days or 5 times in an 8 year period. This 584 day period of reoccurring configurations is known as its synodic period.

Let us take a closer look at a configuration that is easily observed with the unaided eye, the evening appearance of Venus (See Figure 22D). When Venus is first seen she is low in the west and sets shortly after the Sun. Night by night she moves further away from the sun setting later and later. Eventually she reaches her most distant angular distance from the sun, this is called 'greatest elongation east' (at an evening appearance Venus is east of the Sun). Thereafter, Venus appears to move back towards the Sun and, is eventually lost from view. Venus will be an evening star and reach greatest elongation east every 584 days, 5 times in the 8 year period.

If we plot the position of Venus in the zodiac at the time of greatest elongation east, after an interval of 584 days she will be at greatest elongation east again, but this time five constellations further on. However, after 8 years and 5 elongations east, the planet returns close to the point in the zodiac where it started – only a couple of degrees less. If you connect the plots of the position of Venus at successive elongations east, over an eight year period it will draw a pentagram on the zodiac (Figure 29).

***Fig. 29:** The Pentagram and the path of Venus.* R. Hall

Here then is the origin of the symbolism of the pentacle. It is the 'Path of Ishtar' in the zodiac, the cycle of the evening or morning star. Two thousand years later, when Christianity became the state religion in the Roman Empire, Christian theologians demonized other people's gods, Ishtar was turned into the demon Ashtoreth, the Morning star was identified with Satan, and the pentagram was turned upside down and became a symbol of Satanism.

Not all ancient pagan symbols were debased, some were adopted. Whereas Christians adopted the Cross as their symbol, many Islamic cultures adopted the Crescent Moon with a Star (inset upper left Figure 29). This same symbol can be found on Babylonian engravings of Ishtar. The crescent moon is her chalice and the star is the evening star, Venus.

21. The Star of Bethlehem

A question that I have often been asked is 'What was the Star of Bethlehem?' Was it a real event or just mythology? Well, I believe that is was a real event and that it founded one of the world's great religions, Christianity. But before I attempt to explain why and what it was, consider this Babylonian story that predates the birth of Christ by many centuries:

Abraham
Book of Jashar and Maase Abraham (Dead Sea Scrolls)

> *When Abraham was born, a new brilliant star suddenly appeared in the eastern sky and began to move across the heavens.*
> *King Nimrod, astounded by this celestial event, summoned three magicians (magi) who were reputed for their wisdom to explain the significance of this event.*
> *They informed him that it foretold the birth of a male child that was destined to be exalted.*
> *Nimrod fell into a blind rage of terror fearing that this child would one day depose him from his throne.*
> *Nimrod's councillors advised him to kill the son of Terah.*
> *As King Nimrod did not know the specific identity of the child, he gave orders that all male children under the age of two be put to death.*
> *This order was carried out by his merciless soldiers.*

Yuri Koszarycz, School of Theology,
The Australian Catholic University

Like me, the first time you read this story you probably thought, 'I've read that before'! And, the question that arises in one's mind is whether the Biblical nativity story is pure myth that has been adopted from another culture. We know that the gospels were written well after the death of Christ and that the authors were not present at the time of his birth. However, while it is probable that they embellished the nativity story it is also highly probable, due to strongly held beliefs at that time, that there was a special celestial event that triggered the foundation of Christianity. So, if the Star of Bethlehem was a real event, what was it? To attempt to answer this question we need to understand the belief system at that time.

It is written....

As we have discussed in previous chapters it was a common belief that the destiny of people and entire nations was written in the stars. This brings to mind a scene from the film 'Lawrence of Arabia'. Lawrence urgently needed to reach a distant town and, to save time, he decided to take a short-cut by leading his men across a treacherous region of the desert that few had survived passage - the 'Sun's

Anvil'. Under the intense heat one of his men, unnoticed by others, falls from his camel and is left behind.

Upon reaching the other side of the desert he discovers that one of his followers has been lost. Lawrence says "We must go back, he could still be alive". His second in command, an Arab, says "No, he is dead, it was written." Lawrence angrily retorts "Nothing is written". With that he gets back on his camel and rides off on his own into the desert, his companions shake their heads in amazement. Sure enough he finds the lost man staggering deliriously across the scorching plain. When the two eventually get back to the camp Lawrence is welcomed as a returning hero.

A little later, after joining with another tribe, a man is killed in a blood feud. In order to end this feud between the two tribes Lawrence offers to personally carry out the execution of the guilty man. When the man is brought before him Lawrence is shocked to discover that it is the one that he saved from the desert. When Lawrence shoots the man the chieftain of the other tribe says "It was written then".

According to ancient Jewish and Arabic tradition all important events have been prophesied; all important events will occur at either an equinox or solstice; and a portent will herald important events. The coming of the Messiah, the promised deliverer of the Jews, would be a pretty important event and accordingly there would be a portent, a sign in the sky. If there is no sign there would be no Messiah. The sign was his legitimacy.

Fig. 30: *The Zodiac at the time of Christ's birth.* R. Hall

The Gates, Solstices and Equinoxes

Lets start by familiarising ourselves with the Zodiac at the time of Christ's birth, this is shown in figure 30. Of special importance are the signs that contain the equinoxes and solstices, especially as it was believed that all important events would occur at either a solstice or equinox.

The Winter Solstice (Northern Hemisphere) was in Capricornus. This, as mentioned earlier, was known as 'The Gate of Gods', the portal through which the souls of men, after death, ascended to heaven.

December 25th A.D.1 is the *OFFICIAL* birth date of Christ. This would make Christ a Capricorn (prudent, cautious and suspicious)! But as discussed in a previous chapter, December 25th was originally the Roman festival of the winter solstice - the birthday of the sun god 'Mithra'. The Christians adopted this day to celebrate the birth of Christ in A.D. 354

The Spring Equinox was in Aries. The resurrection took place at the spring equinox (Easter), when the sun was in Aries. In Mesopotamia, when the Sun moved into Aries, it was called 'The Sacrifice of Righteousness'. To the Assyrians it was the 'The Altar', 'The Sacrifice'. When Hamal, the brightest star in Aries, rose in the dawn twilight it was called 'The Son of Light', 'The Son of God'.

The Summer Solstice was in Cancer. As mentioned earlier, this was known as 'The Gate of Men', through which souls descended from heaven into human bodies. In Christian times this gate was called the 'Manger', being the portal in the heavens from which the spirit of god descended into the infant Jesus.

The Autumn Equinox was in Libra which represented the 'Balance of the Seasons' and the 'Scales of Justice'. It was also known as the time of the birth of the 'Kings of the House of David', the meaning of which I shall explain later.

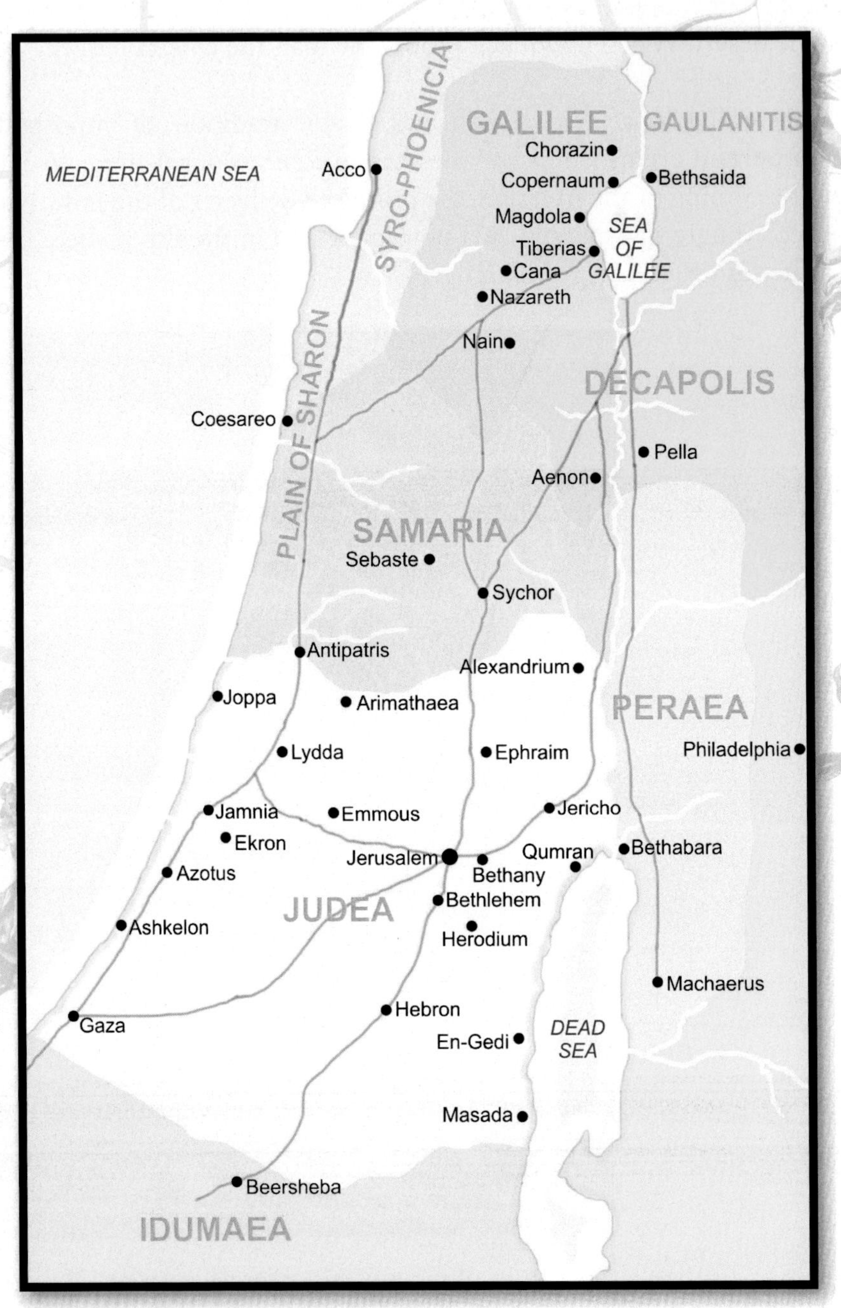

Fig. 31: *Judea at the time of Christ.* R. Hall

The Lamb of God

During the Roman occupation of Judea the Essenes, an ancient-Jewish ascetic sect who claimed to be the descendants of the House of David, established a secret religious centre at Qumran. See the map in Figure 31. Qumran, which was the hiding place of the 'Dead Sea Scrolls', was occupied by the Essenes from 140 B.C. until A.D. 68 when it was destroyed by the Romans. Pliny had this to say of them:

They are a lonely people, the most extraordinary in the world, who live without women, without love, without money.....

While the Essenes lived without women they still married. They supported but lived apart from their wives, whom they only visited to procreate.

Joseph, the father of Jesus, is recorded in the Dead Sea Scrolls as a member of this sect. The linage of Jesus through Joseph back to David is clearly detailed at the beginning of Matthew's Gospel. This explains how the son of a carpenter was able to read and write, a skill learnt by few outside of the aristocracy or priesthood. It also explains the authority he had in the eyes of the common folk – Jesus was almost certainly a Priest of the House of David.

Jesus turned, and saw them following, and said to them, "What do you seek?" And they said to him, "Rabbi, where are you staying?"

John 1-38

This austere religious sect was ruled by a triumvirate of clerics known as the Priest (Pope), the Prophet, and the King (King of the Jews?). Now, to be the king (of the Triumvirate) you had to be born at the Autumn Equinox; the Jewish new-year, which begins at the new moon on or close to the Autumn Equinox. Incidentally, if you were a cleric and you wanted one of your sons to have the opportunity of becoming the King you simply had to plan the visits you made to your wife. If Jesus was the 'King' of the sect he was born on or close to the Autumn Equinox, under Libra.

However, the priests were probably expecting the Messiah to be born on or near the Spring Equinox because it was said that the sun was in Aries when Moses led the people from their bondage in Egypt. So, when was Christ born? Luke gives us a clue in his nativity story:

And in that region there were shepherds out in the field, keeping watch over their flock by night.

Luke 2-8

Whereas today we leave domestic sheep and goats out all night, because of predators this was not the custom in ancient times. For protection the animals were herded and placed in a pen at night; except for one time of the year, the lambing season. If the young were born in the pens they would be trampled. So, during the lambing season the animals stayed in the fields all night and, to protect them from predators, the shepherds watched their flocks by night. This suggests that Christ was born in the spring during the lambing season.

Many years later, when Jesus was a man, he went to see John the Baptist (The Prophet).

The next day he (John the Baptist) saw Jesus coming toward Him, and said, "Behold, the Lamb of God, who takes away the sin of the world!"

John 1:29

The Lamb of God is title given to Aries, the Son of Light, the Son of God. Again, this suggests that he was born at the Spring Equinox. However, it is also possible that both of the above passages from the Bible were created at a later date to legitimize Jesus as the Messiah. There is a lot of important symbolism here. Remember that Hamal, the brightest star in Aries, is the Shepherd that leads the Heavenly Flock. You will also recall that shepherds witnessed the birth of the sun-god Mithra. In addition, of astrological importance, nine months before the spring equinox is the summer solstice. At this time the Sun was in Cancer, the 'Manger', the 'Gate of Men' through which it is said the spirit of god descended into the infant Jesus.

Born Again

In order to find an astronomical explanation for *The Star of Bethlehem* we must fix the absolute date of Christ's birth. We can then look at ancient records to see if something extraordinary happened in the heavens at that time. Working out the year of Christ's birth is almost as difficult as working out what time of the year it occurred.

We learn from Matthew that Christ was born during Herod's reign. Flavius Josephus, a Jewish historian, who was writing around A.D. 80 states that:

Herod the Great died shortly after an eclipse of the Moon. After a period of mourning the Feast of the Passover was celebrated.

It is possible to calculate back using computers and show that a partial (40%) eclipse occurred on March 13th in the year 4 B.C., 29 days before the Passover; and that a total eclipse occurred on January 10th in the year 1 B.C., 88 days before the Passover. Herod died in either 1 B.C. or 4 B.C. Therefore Jesus must have been born before 1 B.C. or, as we shall see, more likely before 4 B.C.

In the nativity story, just prior to the birth of Jesus, Joseph and Mary travel to Bethlehem for a census. We learn from Luke that at this time Augustus was Emperor of Rome and that Cyrenius (Quirinius) was Governor of Syria.

"In these days a decree went out from Caesar Augustus ..."

From Roman records we know that this census occurred in A.D. 6, as much as 10 years after the death of Herod. For many years people studying the Bible have pondered over this inconsistency. However, Barbara Thiering came up with the following explanation:

> *A boy at the age of 12 went through a ceremony equivalent to that of the orthodox Bar Mitzvah, when he was formally separated from his mother; this early initiation was symbolized as a kind of second birth, and he was given a ceremonial vestment to wear.*
> *When Luke says that Jesus was born in the year of the census of Quirinius, which was A.D.6, he was not making an error. Jesus was 12 years old in March, A.D. 6. When Mary "brought him forth" she was*

following the symbolism in which the boy was separated from his mother. When he was "wrapped in cloths" he was being clothed in the ceremonial vestment.

Dr. Barbara Thiering, Jesus, the Man, 1987

In addition to the supernatural all manner of natural celestial phenomena have been suggested as Star of Bethlehem candidates. These include a bright star or planet, a meteor storm, a comet, a conjunction of planets, or a nova. Some old nativity paintings show what looks like a comet overhead which has prompted some people to suggest that it was Halley's Comet. But Halley doesn't fit the time-line; it made an appearance in 12 B.C. followed by another in A.D. 66.

In the time frame that we are looking at, other than planetary conjunctions, the only unusual or spectacular celestial phenomena recorded by the Chinese, who kept meticulous records, was a nova in 5 B.C. But before I explain what a nova is and its significance to the nativity story we need to look closely at planetary conjunctions.

Conjunctions

A conjunction occurs when two or more celestial objects come close together in the sky. While these events are no longer significant in modern astronomy they were immensely important in ancient astrology. Using Kepler's Laws of Planetary Motion we can calculate the positions of the planets for any time in the past or the future. In the time frame of interest the following conjunctions occurred:

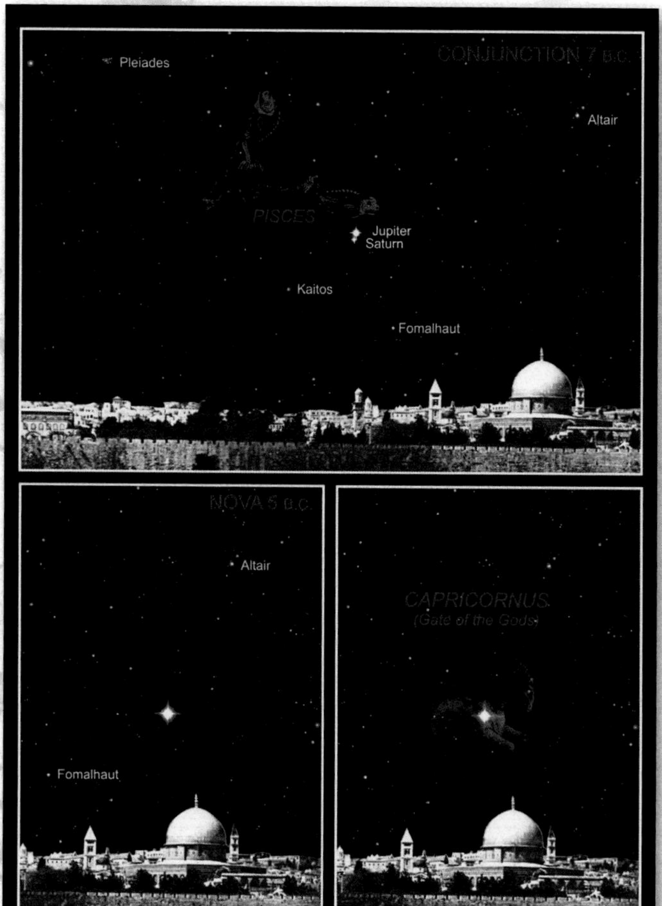

In 7 B.C. there was triple conjunction of the planets Jupiter and Saturn in Pisces.

In September 3 B.C. and February and May 2 B.C. there was a conjunction of Jupiter and Regulus (one of the Royal Stars, the brightest star in Leo).

In August 3 B.C. and June 2 B.C. there was a conjunction of Jupiter and Venus in Leo.

Of the above conjunctions the triple conjunction in 7 B.C. is by far the most significant. This consisted of three close encounters between Jupiter and Saturn in the constellation of Pisces. Seen in the sky the two bright planets performed a dance in which they came very close together and then moved apart three times in succession (See Figure 32). The conjunctions occurred on May 29th, October 3rd and December 4th.

***Fig. 32:** The Stars of Bethlehem* R. Hall

The significance of this triple conjunction is as follows. Jewish tradition regarded Pisces as the sign of Israel. Jupiter was known as Sedeq or "Righteousness" and Saturn was considered to be Israel's Guardian Star. It is said that a triple conjunction of Jupiter and Saturn took place in Pisces three years before the birth of Moses. Consequently, the Rabbis expected another triple conjunction to herald the birth of the Messiah. This was the Sign in the sky that brought the three wise men (astrologers) to Jerusalem (a theory first advocated by Kepler).

The three wise men probably came from Babylon. In 597 B.C. Nebuchadnezzar II occupied Jerusalem, capital of Judah, and carried off many of its leading citizens to Babylon. After a revolt in 587-586 B.C. many Jews were deported to Babylon and their descendants kept there until after the Persian conquest in 331 B.C. If the wise men set out on their journey after observing the first conjunction they would have arrived in Jerusalem around the time of the second conjunction in October. This suggests that Jesus was born in the autumn of 7 B.C.; so perhaps it was his conception that took place at the winter solstice (December 25th). This scenario fits in with the rest of the nativity story because Herod is still alive and well at that time and, the 'star' is still visible in the sky at the time when Christ is born and the wise men arrive in Jerusalem.

Below is the story of the Wise Men as told by Matthew.

2:1 Now when Jesus was born in Bethlehem of Judea in the days of Herod the king, behold, wise men from the East came to Jerusalem, saying, 2:2 "Where is he who has been born king of the Jews? For we have seen his star in the East, and have come to worship him.

The first conjunction would have been observed in the east shortly before the onset of dawn. The later conjunctions would have been observed in the east at the end of evening twilight.

2:3 When Herod the king heard this, he was troubled, and all Jerusalem with him; 2:4 and assembling all the chief priests and scribes of the people, he inquired of them where the Christ was to be born. 2:5 They told him "In Bethlehem of Judea; for so it is written by the prophet: 2:6 'And you, O Bethlehem, in the land of Judah, are by no means least among the rulers of Judah; for from you shall come a ruler who will govern my people Israel.' 2:7 Then Herod summoned the wise men secretly and ascertained from them what time the star appeared;

That Herod had to ask when the star appeared tells us that it was a celestial object and not, as some have suggested, a supernatural object hovering over Bethlehem. It also tells us that the star was not something really spectacular to the untrained eye. It would have been prominent and unusual to the public but it wasn't a headlamp in the sky.

Furthermore, it could not have been at the time of the spring equinox because the spring equinox of 7 B.C. occurred before the portent, and at the following in 6 B.C, the conjunction was over. The 'star' which Herod asked to see, the star the wise men followed to Bethlehem would no longer have been there.

2:8 and he sent them to Bethlehem saying, "go and search diligently for the child, and when you have found him bring me word that I too may come and worship him." 2:9 When they had heard the king they went their way; and lo, the star which they had seen in the East went before them, Till it came to rest over the place where the child was.

The last verse is interesting because Barbara Thiering in her book 'Jesus, the Man' argues that Jesus was born at Qumran and not Bethlehem. Qumran, she says was known to the Essenes as "Bethlehem of Judea", being the Throne of the House of David in exile. She says that it was also known as "the manger". If the star they had seen in the east went before them - they would have traveled east from Jerusalem. If you look at the map in Figure 31 you will see that Qumran is to the east of Jerusalem, Bethlehem to the south. Further, by horse or camel, Qumran is about five or six hours ride from Jerusalem. If the star was seen in the east after sundown and they travelled towards it, after about five hours, due to the rotation of the Earth, the star would be overhead – over the place where the child was.

Nova

In 5 B.C. a nova appeared in the sky and shone for seventy days. The word nova means new star because in ancient times that is what they were believed to be. In reality they are old binary stars that that are subject to violent eruptions. Novae are usually very remote objects and prior to an eruption are well below the visibility of the unaided eye. When an eruption occurs, the star may brighten by a factor of 10,000, sometimes as much as a 100,000 fold. Suddenly a star appears in the sky where none was seen before and at maximum it may outshine the brightest of stars. The star then slowly fades and is eventually lost from view, until the next eruption. But, this may not occur for another 10,000 years or more. Really bright novae are uncommon, maybe one or two a century. Novae brighten and fade in a characteristic way and Chinese records state that the nova of 5 B.C. shone for 70 days. The lengthy duration of visibility tells us that this nova was a brilliant object at peak brightness.

The Nova of 5 B.C. was significant because it occurred in Capricornus, the Gate of the Gods, the constellation where the Sun is at the Winter Solstice on December 25th (see figure 32). Could this have been the legendary Star of Bethlehem? It certainly fits the bill as to what most people imagine the Star of Bethlehem to have looked like.

Two Portents in the Heavens

Revelations 12:1-5 talk of there being two portents in the Heavens.

And a great portent appeared in heaven, a woman clothed with the sun, with the moon under her feet, and on her head a crown of twelve stars; she was with child and she cried out in her Pangs of birth, in anguish for delivery.

And another portent appeared in heaven; behold, a great red dragon, with seven heads and ten horns, and seven diadems upon its heads…And the dragon stood before the woman who was about to bear a child, that he might devour her child when she brought it forth; she brought forth a male child, one who is to rule all the nations with a rod of iron, but her child was caught up to God and to his throne.

It seems more than likely that the Star of Bethlehem consisted of two separate celestial events that have, with the passage of time, been conflated. There was the first portent, the triple conjunction of Jupiter and Saturn in Pisces in 7 B.C., seen by the wise men. This was the sign of the coming of the Messiah. The second portent, remembered by common folk, was the bright nova in Capricornus, the Gate of Gods, in 5 B.C.

I also suspect that Jesus was born around September/October 7 B.C, conceived at the winter solstice under Capricorn, born at the autumn equinox under Libra. Which, I believe is the reason why his persecutors nailed "King of the Jews" to the Cross. At a much later date his followers, to legitimize their claim that he was the Messiah, included passages in the gospels that suggested the time of his birth was the spring equinox. Conceived at the summer solstice, the Gate of Men, and born at the spring equinox under Aries – the Son of Light, the Son of God.

PART VI:

THE SIGNS AND STARS

This part of the book provides historical, astronomical, astrological and mythological information about the individual signs of the Zodiac and their brighter stars. For each sign the Indian or Rashis name is given along with that of the familiar name used in the western world. The following terminology is used:-

Solar Conjunction gives the dates of the passage of the Sun through the sign. Astronomical conjunction is based upon modern constellation boundaries. Rashis is based upon the traditional boundaries of the Indian Zodiac, which western astrologers call the 'Sidereal Zodiac'.

Tropical Calendar gives the calendar dates used by most western astrologers to identify the natal signs, which they call the 'Tropical Zodiac'. When the Zodiac was established over 2,000 years ago the dates for each sign, like the Indian Zodiac, coincided with the passage of the sun. Unlike the Indian Zodiac it was never adjusted for precession so that they no longer do so.

Opposition is the date upon which the sign culminates, is directly opposite the sun and on the meridian at midnight. This is not adjusted for daylight saving.

House and **Element** give the astrological planetary body that is believed to rule over the sign and the element the sign is associated with.

Star Charts In the accompanying star charts the zodiacal constellation name and that of its brightest stars is in gold, the Indian or Jyotish name in red. The modern constellation boundary is in green with the ancient Indian constellation boundary marked by gold arrows on the ecliptic. The names of neighbouring constellations are in dark blue and their bright stars in light blue.

22. PISCES (Meena)

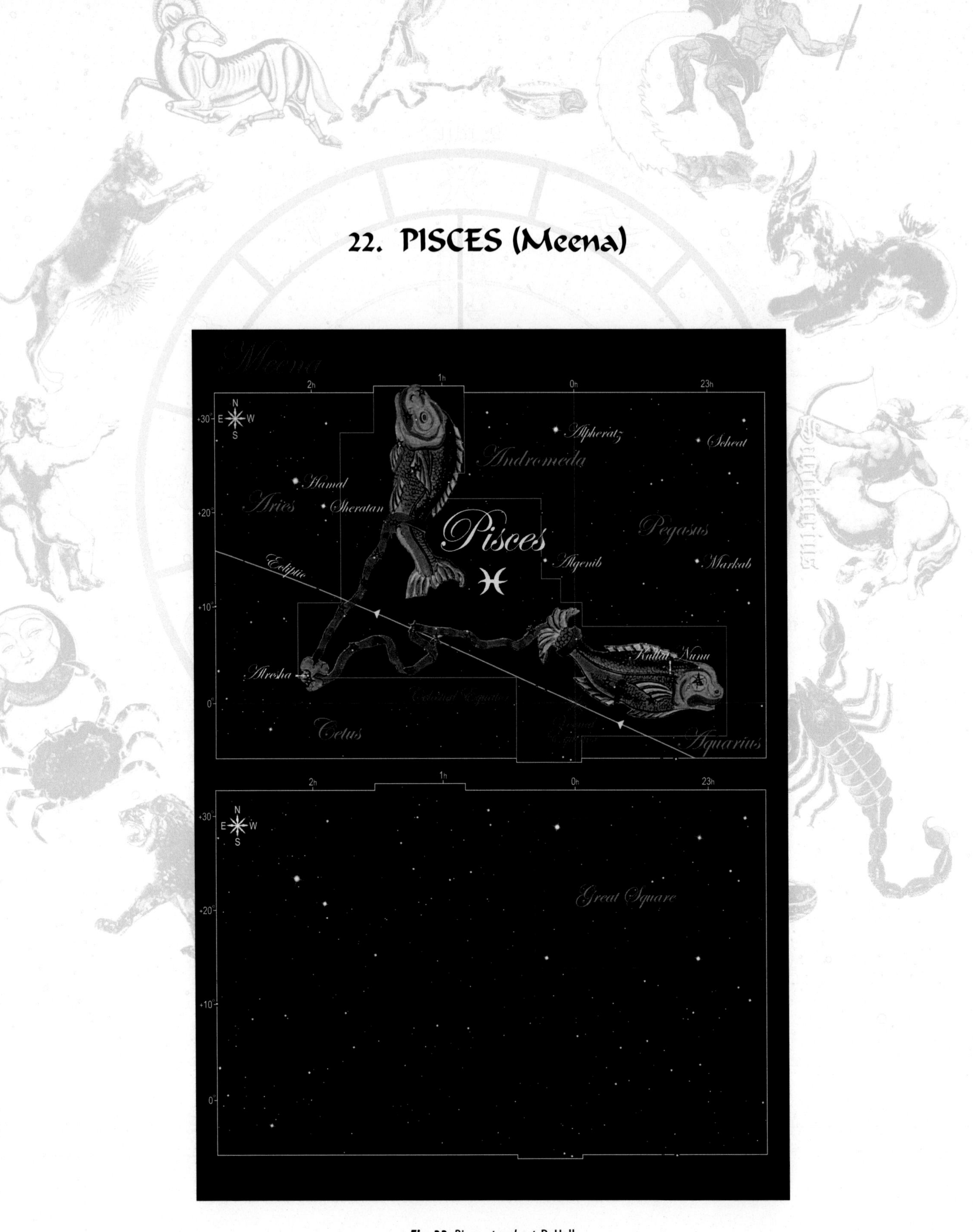

***Fig. 33:** Pisces star chart* R. Hall

And here fantastic fishes duskly float,
Using the calm for waters, while their fires
Throb out quick rhythms along the shallow air.

Elizabeth Browning; A Drama of Exile, c.1836

PISCES – The Fishes **MEENA – The Fish** (Rashis)

Solar Conjunction

Astronomical	March 12th to April 17th
Rashis (sidereal)	March 15th to April 13th
Tropical Calendar	February 20th to March 20th

Opposition: October 3rd **House** of Jupiter **Element:** Water

Pisces is an ancient Sumerian and Babylonian constellation, the origin of which dates back to the time of the Great Flood, 5,500 B.C. The Babylonian story of the Great Flood is associated with the heliacal rising of Aquarius at the time of the winter solstice. At this time the sun was in Pisces.

Pisces was originally the fish of Ea, the god of wisdom. In Babylonian mythology, there was a never ending struggle between the gods of salt and fresh water, which resulted in the Great Flood. Hence the two fishes, one salt and the other fresh water, one pointing north and the other west, tied together by the star Alrescha (Alpha Piscium).

Geological evidence suggests that the Biblical great flood occurred in the region now occupied by the Black Sea (see Figure 14). Originally this was a fertile valley basin the size of Mesopotamia with a large fresh-water lake. This basin was below sea-level but land-locked. As the ice-age ended melting glaciers produced a rising sea-level and eventually the waters of the Mediterranean burst into the Black Sea basin at Bosporus, near Istanbul (Constantinople). Core samples from the floor of the Black Sea show a sudden change to salt-water sediments 7,500 years ago. It is estimated that when the breach occurred water flooding into the area exceeded that flowing over the Niagara Falls a thousand fold. Is this the meaning behind the two fishes – one salt, the other fresh water?

As the salt water poured in, the level of the great lake rose and the shore-line advanced at an estimated rate of four kilometres per day. That's about three metres a minute. For those living in the area it would have seemed that the entire world was being inundated.

Survivors of this deluge carried their story to neighbouring lands where it became part of the folk-lore. The stories of the great flood from different cultures are all similar but the name of the central character differs. In the Bible it is Noah, he was Ziusudra in the Sumerian story, Atrahasis to the Akkadians, Utnapishtim to the Babylonians, and Deucalion to the Greeks.

The Babylonian creation stories centre on the hero/demi-god Gilgamesh. In the 'Epic of Creation' Utnapishtim, one of Gilgamesh's ancestors survived the Great Flood by filling a cup-shaped boat with treasure and plants and creatures of every kind. As the waters rose he set sail with his wife and family. After seven days he sent out a dove to look for dry land, and it came back exhausted. On the eighth day he sent a swallow, with the same result. On the ninth day he sent a raven, and when it failed to come back he knew land was close by. When he found land he beached his boat, released the animals and planted the plants.

It is important to remember when pondering the size of the 'ark' that the Great Flood occurred 7,500 years ago and, the people who built the arc were living in the 'stone age'.

Due to precession Pisces is now the first sign in the Zodiac. Traditionally the beginning of the Zodiac is determined by the vernal equinox, the point at which the sun crosses the celestial equator on its journey north. 2,500 years ago, when the present zodiac was formalized, the vernal equinox was in Aries. Around A.D.300 precession carried it into Pisces, where it will remain until A.D. 2450 when it will move into Aquarius. This equinoctial point is still known as the 'First of Aries' or the 'First Point in Aries', even though it is now in Pisces. It is also known as the 'Greenwich in the Sky'. The Fishes are called 'Leaders of the Celestial Host', for this is the Age Pisces.

In Babylonian astronomical tablets (200-100 B.C.), Pisces appears in the zodiac as the twelfth sign or last month of the year (at this time the vernal equinox was in Aries). Every six years Pisces became two months, the 12th and 13th months of the year. Thus, the one fish became two. Known as the Fishes of Ea, they symbolized the additional month which was inserted every six years into the Babylonian calendar of 360 days.

In Greek mythology the two fishes represent Aphrodite (the Syrian Astarte) and her son Eros. The sign was known as 'Venus et Cupido' (Venus and Cupid). One day the two were walking along a riverbank when they sensed the presence of the monstrous Typhon (who was determined to overthrow the Olympian gods). They plunged into the river where they took the form of fishes and escaped.

The Romans called the constellation Aquilonius, which signified a rain bearing wind that came from the north when the sun was in Pisces.

Pisces was the national constellation of Judea and was known as Dagaiim, the two fishes. Some Jews ascribe the joining of the fishes to the joining of the tribes of Simeon and Levi.

***Fig. 34:** The Megiddo Church mosaic in Israel.*
Photo by Zev Radovan.

Pisces was also a symbol of the early Christians and forms a centre piece to the floor of a church found at Migiddo in Israel.[30] (See Figure 34) They were said to be the two fishes with which (along with five loaves) Christ fed '...*about five thousand men, besides women and children*'.
Matthew 14-21

The fishy or watery nature of this constellation belongs to Europe and western Asia. In the far-east, to the Chinese, this constellation was the 12th sign Tseu Tsze, the Boar.

The Brightest Stars in Pisces
Pisces lies well away from the Milky Way and contains no bright stars. The brightest is Eta Piscium, magnitude 3.6.

Kullat Nunu (Gamma Piscium) means the 'Cord of the Fish'. Magnitude 3.7

Alrescha (Alpha Piscium) is 'the knot' that ties the two fish together. Magnitude 3.8

30 Dated to the 3rd or 4th century, it is the oldest Christian church known.

23. ARIES (Mesha)

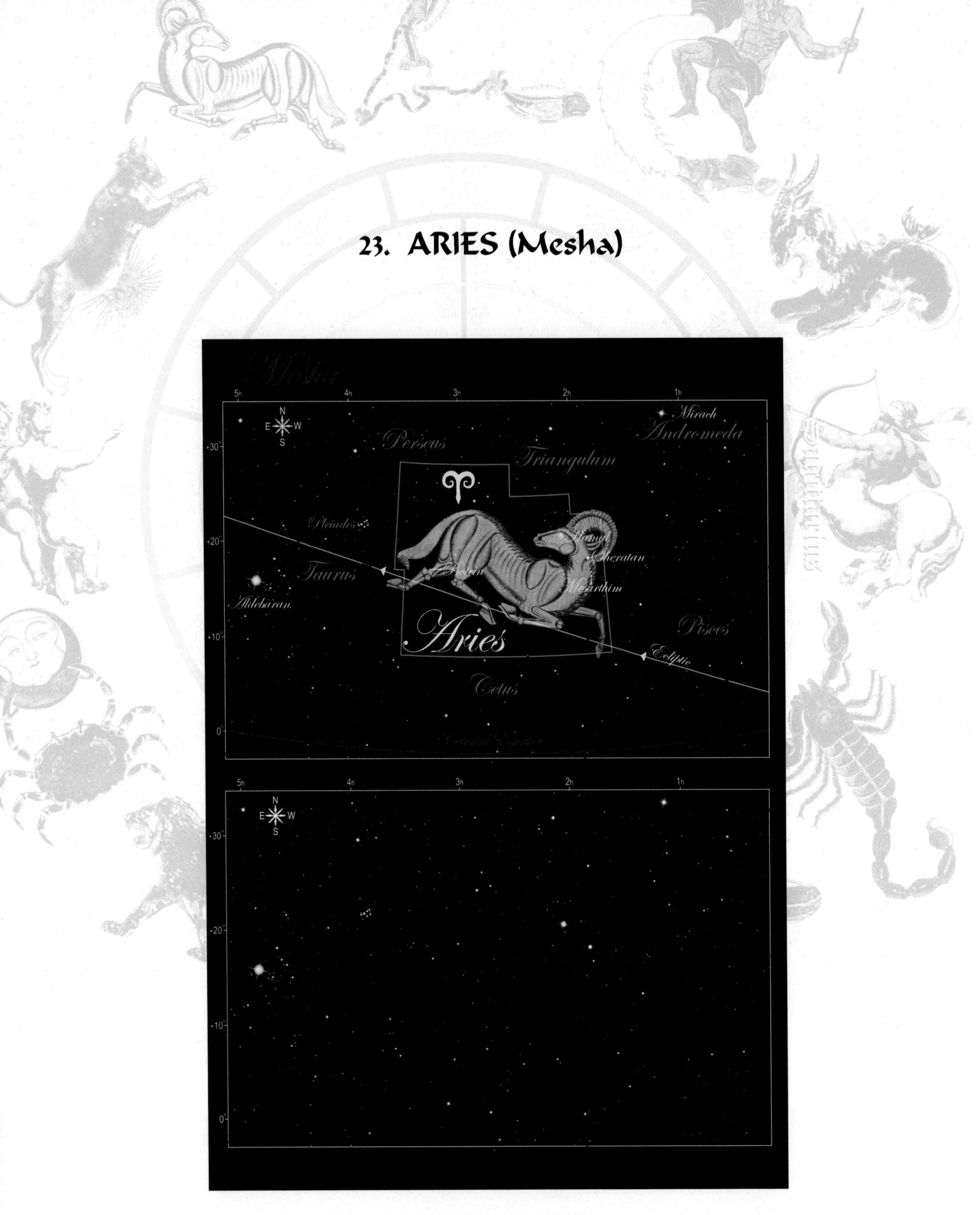

Fig. 35: *Aries star chart* R.Hall

The Ram having pass'd the Sea serenely shines,
And leads the Year, the Prince of all the Signs.

Manilius (first century A.D.)

ARIES – The Ram **MESHA – The Ram** (Rashis)

Solar Conjunction

Astronomical	April 18th to May 13th
Rashis (sidereal)	April 14th to May 14th
Tropical Calendar	March 21st to April 20th

Opposition: November 2nd **House** of Mars **Element:** Fire

Aries was the first sign of the Zodiac from 1845 B.C. to A.D.300 when the vernal equinox lay within its borders, it was the ram that lead the celestial flock across the sky. This equinoctial point, the northern spring equinox, is still known as the 'First Point of Aries', even though it has now moved into Pisces. Today, Aries is the 2nd sign, being the 2nd sign that sun passes through after the vernal equinox.

The year began for the people of Mesopotamia when the Sun moved into Aries; which was described as "the sacrifice of righteousness". This concept survived for thousands of years. To the Assyrians, Aries represented the "Altar and the Sacrifice", where a ram was usually the animal sacrificed. During the plagues of Egypt, at the Passover, the Jews were required to put the blood of a lamb (Aries) over the door to keep the Angel of Death from taking their firstborn sons. In pagan Rome the legions sacrificed a ram before entering battle.

In Babylonia the stars that form Aries was known as 'Mul lu Hun ga' (Hired Man). This constellation was equated with Tammuz the god of spring and marked the beginning of the year. At that time, around 2000 B.C., the 'Hired Man' rose in the dawn at the time of the spring equinox. Around 500 B.C. the Greeks transformed this constellation into Aries the Ram and established it as the first sign of the Zodiac.

In Greek mythology Aries is the magical ram with a golden fleece sent by Hermes to rescue the children of King Athmas of Boeotia from their wicked stepmother, Ino. Helle, one of the children, fell off the ram as it was flying across the straits that divide Europe from Asia, which the Greeks named Hellespont (Helle's Sea). Her brother Phrixes, was carried to safety to the shores of the Black Sea, where he sacrificed the ram. He then placed its golden fleece in the care of a sleepless dragon. Later, the ram's Golden Fleece was sought by Jason of the Argonauts.

In another story, set in the Liberian desert, a ram showed Bacchus and his thirsty companions the way to a well. As a reward Bacchus set the ram under the stars, so that when the Sun wandered through it, it was springtime.

In the Hebrew zodiac, Aries is assigned to Simeon, or by some to Gad. In the Chinese Zodiac it is the 11th sign Kiang Leu, the Dog

The Brightest Stars in Aries

Aries lies in a relatively barren region of the sky but is easily identified by its two brightest stars Hamal and Sheratan which, like Gemini, were also known as "The Twins".

Hamal, (Alpha Arietis), the brightest star in Aries, is a golden-yellow star of magnitude 2.00. Hamal is Arabic and means 'lamb'. Its Euphratean name was Lu-lim, the Ram's Eye. The Babylonians called this star 'Dil-kar' the Proclaimer of the Dawn, the Son of Light, the Sun of God. Hamal is the 13th brightest star in the Zodiac and the 52nd brightest star in the entire sky. It was an important navigational beacon and at least eight Grecian temples are aligned to its rise position.

Sheratan (Beta Arietis), the other twin, is a white star of magnitude 2.6. Its name is derived from Al Sharat, 'the Sign of the opening year'. Sheratan marked the vernal equinox in the days of Hipparchos, the Greek astronomer/mathematician who first charted the heavens.

Mesarthim (Gamma Arietis) is a 4th magnitude star close to Sheratan. Its name may be derived from the Hebrew Mesharetim, which means 'Ministers'.

24. TAURUS (Vrishabha)

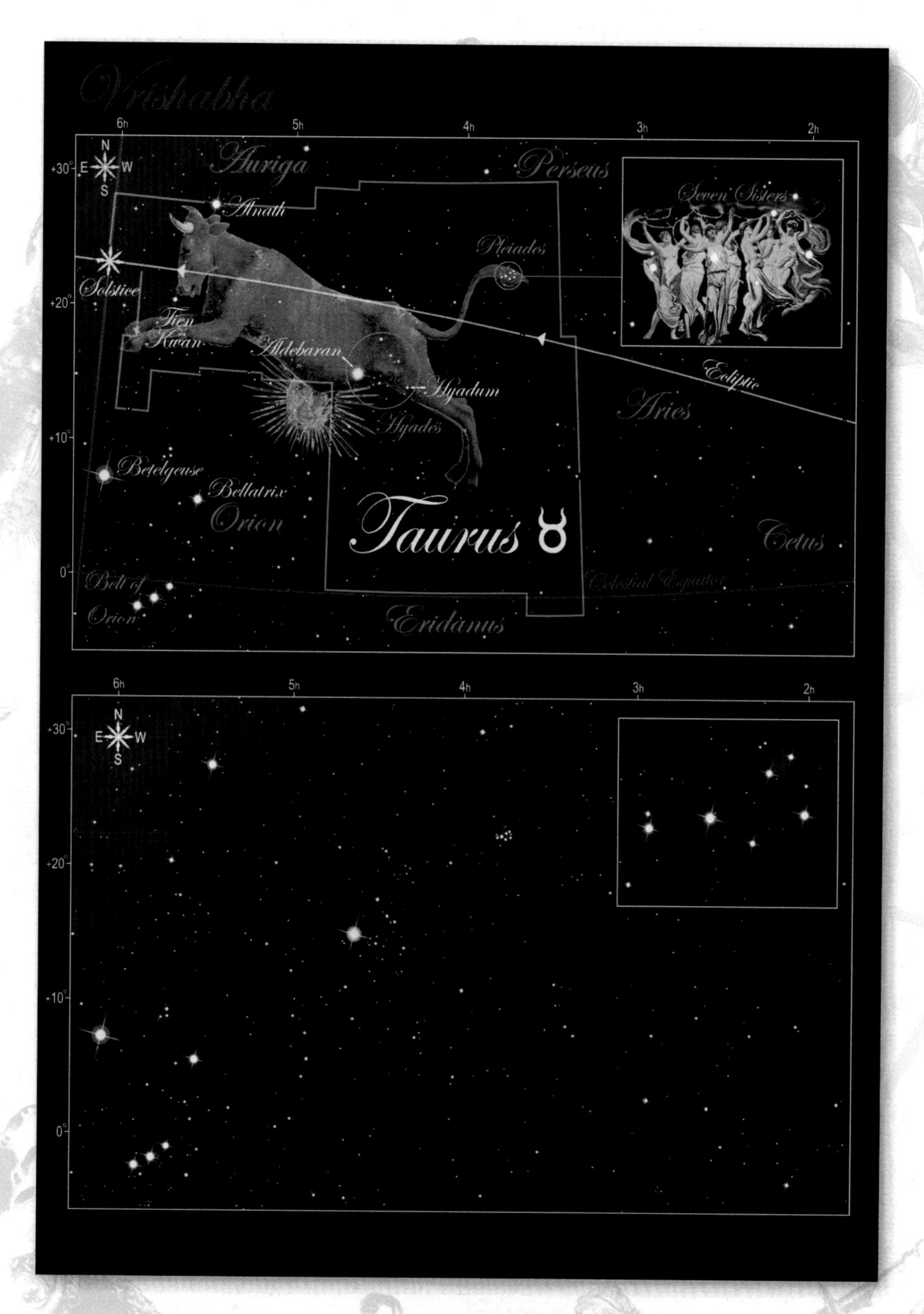

Fig. 36: Taurus star chart R. Hall

Ere the heels of flying Capricorn
Have touched the western mountain's darkening rim.
I mark, stern Taurus, through the twilight gray
The glinting of the horn,
And sullen front, uprising large and dim,
Bent to the starry Hunter's sword at bay.

Bayard Taylor's Hymn to Taurus (c.1855)

TAURUS – The Bull	**VRISHABHA – The Bull** (Rashis)
Solar Conjunction	
Astronomical	May 14th to June 20th
Rashis (sidereal)	May 15th to June 14th
Tropical Calendar	April 21st to May 21st

Opposition: December 4th **House** of Venus **Element:** Earth

Taurus is one of the oldest and most notable of constellations. Indeed, it may have been the very first sign established in the development of the solar Zodiac. In all of the most ancient zodiacs that have survived to this day, Taurus is the first sign and marked the beginning of the year. It contained the vernal equinox from 4000 B.C. to 1850 B.C., a period known as the golden age of archaic astronomy. Since A.D. 300 it has been the 3rd sign in the Zodiac

In the Persian creation story Mithra, the God of Light and driver of the Sun Chariot, captured and killed the 'primeval bull' (Taurus) which released the life force of nature. In Persian temples (now Iran), Mithra is depicted fighting a griffin. A griffin combined the features of a bull (Taurus at the vernal equinox), a lion (Leo at the summer solstice), a scorpion (Scorpius at the autumn equinox) and an eagle (Aquilla representing the winter solstice).

Thousands of years later, in ancient Rome, Mithras became the centre of a mystery-cult. Worshippers believed that the blood and bone-marrow of the sacred bull guaranteed the fertility of the universe. Its marrow turned into bread and its blood into wine. Mithras was the god of resurrection, and redemption.

Fig. 37: *The Mithraic Bull* R. Hall

Images of the Mithraic Bull engraved on gems from around 400 B.C. show only the head and forequarters of the bull, which may be the origin of our modern representation of the figure of Taurus (Figure 37). However, some people believe that it represents the bull that carried off Europa. Only its upper half is shown because the rest of the bull's body is immersed in water.

According to Greek mythology Taurus personified the bull that Zeus changed into to abduct Europa, the beautiful daughter of King Agenor of Phoenicia. She dared to sit upon its back, whereupon the bull ran into the surf and swam to Crete. There Zeus revealed himself and seduced/raped the maiden. The son that resulted became King Minos of Crete...whose wife gave birth to the Minotaur (Minotaurus, the Minos Bull).

In Egypt festivals for the Bull-God Apis were held in summer, when the River Nile gently overflowed its banks and brought life-giving water to the land, readying the land for planting. The worship of Apis, a god of the Nile, predated the building of the pyramids.

In Europe the great Druidic festival, the Tauric, was held when the Sun was in Taurus. Some believe that the 'Tors' of England were sacred sites of the ancient Taurine Druids.

The Chinese figured the stars of Taurus as part of a great constellation known as the White Tiger. But in their Zodiac it was the 10th sign, the Rooster.

In A.D.2450 the summer solstice will move into Taurus.

The Brightest Stars in Taurus

Aldebaran (Alpha Tauri), the "Eye of the Bull", is a brilliant orange-red star. It is the brightest star in the Zodiac and the 13th brightest star in the entire sky. Its magnitude, which is slightly variable, is 0.85. The name comes from Al Dabaran, which means 'the Follower' – because it follows the Pleiades into the sky. Originally this was the name of the Hyades star cluster. Although surrounded by the Hyades, Aldebaran is a foreground star and not a member of the cluster. It was originally known as Na,ir al Dabaran (the Bright One of the Follower).

Aldebaran was very close to the vernal equinox in 3000 B.C. It was one of the Four Pillars of Heaven, one of four first magnitude stars that marked the location of the equinoxes and solstices. The Persians called these 'Pillars' the Royal Stars and Aldebaran was known as 'Ku, I-ku' – The Leading Star of Stars. The Babylonians knew Aldebaran as the god Marduk, the Spring Sun. In Arabia it was 'Oculus Tauri' (Bull's Eye); while the Hebrews knew it as 'Aleph' (God's Eye). The Polynesians knew this star as 'Aumea', the missile hurled at Matariki (see Pleiades). To New Zealand Maori it is known as Taumata-kuku.

Alnath (El Nath, Beta Tauri) is a bright blue-white star of magnitude 1.65. It is the 8th brightest star in the Zodiac and the 29th brightest star in the sky. Its magnitude is 1.65. Alnath means 'the butting one' and represents the tip of the northern horn. In astrology Alnath portended eminence and fortune to all who could claim it as their natal star. To the Hindus it was Agni, God of Fire.

Tien Kwan (Zeta Tauri) is a blue-white star of magnitude 3.0. Its name means the 'Heavenly Gate'.

Hyades and Pleiades

Manilius described Taurus as '*dives puellis*, "rich in maidens", referring to its famous star clusters the Hyades and the Pleiades. The stars of the clusters visible to the unaided eye were known as the fourteen Atlantides. They were fourteen daughters of Atlas, the Seven Hyades and the Seven Pleiades. New Zealand Maori know them as Matakarehu and Matariki.

The Hyades is a cluster of about 200 stars of which about a dozen are visible to the unaided eye (see Figure 37). The brightest members form a distinctive V-pattern which, in more recent sky-representations of Taurus forms the head or face of the bull. Aldebaran, the brightest star in Taurus is part of the V-pattern and is the 'Eye of the Bull'.

In Greek mythology the Hyades were the daughters of Atlas and Aethra, and therefore half-sisters of the Pleiades. The Seven Sisters of the Hyades are Aesula, Ambrosia, Dione, Thyone, Eudora, Koronis, and Phyto.

Hyades means "to rain" and in the ancient world the rains were expected when the Hyades were seen rising in the dawn twilight of May and again with their rising in the evening twilight in November. The Roman country folk called them Suculae, the 'Little Pigs'. According to Pliny this was because the continual rains turned the country roads into mire, which delighted the pigs.

"a violent and troublesome star causing storms and tempests raging both on land and sea"

Pliny (c. A.D. 50)

Thro' scudding drifts the rainy Hyades vext the dim sea;

Tennyson's Ulysses (c. 1850)

The Pleiades, also known as "**The Seven Sisters**", is the brightest and most celebrated star cluster in the sky (see figures 16 and 36). The cluster contains more than 400 stars but only six are easily seen with the unaided eye. The individual cluster stars are not particularly bright, but seen together in a dark sky they are quite prominent and look like a casket of diamonds in the sky.

Many a night I saw the Pleiads, rising thro' the mellow shade,
Glitter like a swarm of fire-flies tangled in a silver braid.

Locksley Hall

Almost certainly derived from their mother. The names of the sisters are Alcyone, Merope, Electra, Celaeno, Maia, Taygeta, and Asterope. In the constellation figures of Taurus they are generally located on the shoulder of the Bull, sometimes its tail.

Five thousand years ago the heliacal rising of these stars occurred close to the time of the vernal equinox. Consequently, their dawn rising marked the beginning of the year for the Greeks and peoples of Asia.

They are among the oldest celestial objects mentioned in literature. They are recorded in Chinese annals dating to 2,357 B.C. where they were worshiped by girls and young women as the 'Seven Sisters of Industry'. In India they were known as Krittika, the first sign of the Hindu lunar zodiac, which dates back to about 2,300 B.C.

Around 800 B.C., when their helical rising occurred in May, their rising signaled the opening of the sea lanes and the beginning of sea trade for that year; their setting in late autumn signaled the closure. Thus, when the Pleiades were visible it was a sign to the ancient mariners that it was safe to undertake a sea voyage. The Pleiades were known to mariners as the 'Sailors Stars'. When the stars rose and the sea lanes opened it was customary to release pigeons, consequently the Pleiades also became known as the 'Hen and Chicks'.

When the Pleiades culminated, reached their highest point in the sky, at sunset on November 17, no petition was presented in vain to the ancient Kings of Persia. In Greece the Seven Sisters became identified with Athene (the Roman Minerva), the Goddess of Civilization. Great temples, such as the Parthenon, were aligned to their rising.

They are mentioned in the Bible (Old Testament c.500BC)
"*Can you bind the chains of the Pleiades, or loose the cords of Orion?*"
Job 38:31

The Hindus pictured these stars as a flame (of Agni, the God of Fire). Around A.D.400, the evening rising of the Pleiades in October / November became known as the 'Pleid month', Kartik. At this time they celebrated the great star-festival Dibali, the Feast of Lamps, which gave origin to the present Feast of Lanterns of Japan. In Japan the cluster is known as Subaru, which means to 'unite'.

In Europe, the rising of the Seven Sisters at sunset and their culmination at midnight gave rise to the great Celtic and Druidic new-year festival of Samhain on October 31st. At this festival deceased friends and relatives were honored. The Christians turned this festival into Halloween.

In Aotearoa, New Zealand, the heliacal rising of the Pleiades, known as Matariki (Little Eyes), heralds the beginning of the Maori new-year. Matariki is usually figured as an old woman and her six daughters. She was known as the 'Foodbringer', because her heliacal rising in early June heralds a time of plenty.

In Polynesian mythology Matariki (Pleiades) was at one time a single brilliant star. Tane, who personified the rising sun, became envious of its brilliance and hurled Aumea (Aldebaran) at the star breaking it into six pieces, the six stars that we now see. Another name for the fragments is Tauono, the six.

The Pleiades were not always seen as a friendly portent. The Arabs said that when they disappeared in the sun's rays it was, for forty days, a time of great harm to mankind. It marked the time of disease and plague. When the stars reappeared in the dawn sky they marked the height of the plague.

Muhammad wrote that "*when the star rises all harm rises from the earth*".

Taurid Meteors

Every year from September 15th to November 25th a shower of meteors (shooting stars) radiate from a point close to Epsilon Tauri in the Hyades. Maximum activity occurs on November 3rd when the shower produces about ten meteors per hour. The meteors are slow moving and some of them are fireballs.

25. GEMINI (Mithuna)

Fig. 38: *Gemini star chart.* R. Hall

Starry Gemini hang like glorious crowns
Over Orion's grave low down in the west.
Tennyson's Maud.

GEMINI – The Twins **MITHUNA – The Pair** (Rashis)

Solar Conjunction

Astronomical	June 21st to July 19th
Rashis (sidereal)	June 15th to July 16th
Tropical Calendar	May 22nd to June 21st

Opposition: January 4th **House** of Mercury **Element:** Air

Gemini is the most northerly constellation of the zodiac and lies partly within the Milky Way. Since A.D.300 it has been the 4th sign and marks the summer solstice. In ancient times it was known as Jauzah (the Centre of the Heavens), probably because in Egypt and Asia Minor it was a zenith constellation – it came directly overhead. Originally the "Twins" referred to only the two bright stars Caster and Pollux, the constellation was built around these stars at a later stage.

The concept of the 'Heavenly Twins' is almost universal. To the early Christians they were Adam and Eve, the 'Twin Sons of Rebecca' or 'David and Jonathan'. In ancient Greece they were Anaces, the 'Two Gods of Sparta'. To the Romans they were Romulus and Remus, the founders of Rome.

In India they were Acvini, the two magnificent Ashwins (horse lords) who galloped across the sky before the appearance of dawn each day, riding a golden chariot pulled by winged horses. A more common name for them in India was Mithuna, the pair – the boy and the girl.

In Egypt the Twins were Horus the Elder and Horus the Younger, while in Arabia they were known as 'Duo Pavonis', the Twin Peacocks. To the Chinese they were Yin and Yang, the 'Two Principles', although as a zodiacal sign they represented the 9th sign, the Ape. New Zealand Maori know the twins as Whakaahu.

About 9,000 years ago, the vernal equinox was in Gemini, and a line joining the two equinoxes (the equinoctial colure) passed between the two bright stars. At this time the twins represented the equal day and night of the equinoxes. Thousands of years after they had ceased to mark the equinox, the Greeks called them Castor and Polydeuces (in Latin, Pollux), brothers who, according to legend, were "possessed of an immortality of existence so divided among them, that as one dies, the other revives" – day and night.

The Twins were believed to be the 'protectors of mariners' and in ancient times the constellation was symbolized by two stars over a ship. This became a sign or figurehead attached to the mast or side of a ship.

The two stars are associated with an electrical phenomenon that can be observed shortly before or during an electrical storm. Electrical discharges can produce a corposant, glowing plasma on the mast of a ship (or the wing tips of an aircraft). Best known as 'Saint Elmo's Fire', it has a vivid blue or violet colour and can sometimes look like a flame. If a double light appeared, it was believed to be Castor and Pollux, and was said to be a favourable sign that the ship would fare well in the storm. However, if only a single light was seen, it was "that dreadful, cursed, and threatening meteor called Helena," the sister of the twins who brought such misfortune to Troy.

Last night I saw Saint Elmo's stars,
With their glittering lanterns all at play
On the tops of the masts and the tips of the spars,
And I knew we should have foul weather to-day.

Longfellow's Golden Legend (c. 1855)

The Brightest Stars in Gemini

Pollux (Beta Geminorum), is a golden yellow star of magnitude 1.16. It is the 4th brightest star in the Zodiac and the 17th brightest star in the entire sky. In Greek mythology Queen Leda of Sparta was seduced by Zeus and, on the same night, slept with her husband King Tyndareus. She bore the Twins, one of whom was mortal, and the other immortal. Pollux' father was Zeus and is therefore the immortal one. Astrologically it portended eminence and renown.

Castor (Alpha Geminorum) is a brilliant white star of magnitude 1.56. It is the 6th brightest star in the Zodiac and the 24th brightest star in the sky. Castor's father was Tyndareus and is therefore the mortal one. In astrology he is the opposite of his brother and is a portent of mischief and violence.

Alhena (Gamma Geminorum) is the third brightest star in Gemini, the 11th brightest star in the Zodiac and the 45th in the entire sky. It is a bright white star of magnitude 1.93. Its Arabic name means the 'Brand' (on the right side of the camel's neck).

Geminid Meteors

Every year between December 4th and the 16th brilliant meteors, many of them fireballs, appear to radiate from a point in the sky close to Castor. Peak activity occurs on December 14th when the shower can produce up to 75 meteors an hour.

26. CANCER (Karka)

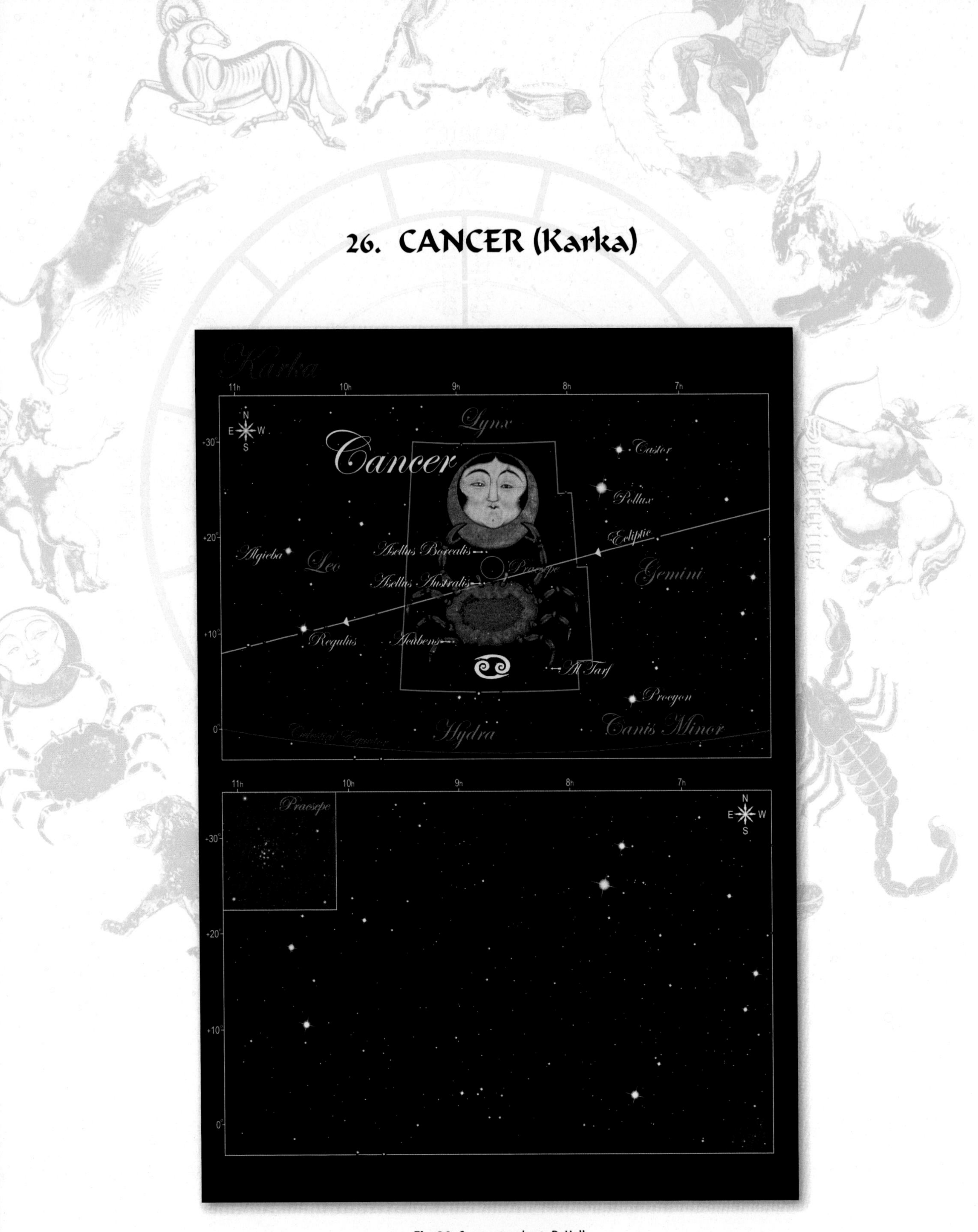

Fig. 39: *Cancer star chart.* R. Hall

. . . and there a crab
Puts coldly out its gradual shadow-claws,
Like a slow blot that spreads, - till all the ground,
Crawled over by it, seems to crawl itself.
Browning's Drama of Exile (c.1850)

CANCER – The Crab **KARKA – The Crab** (Rashis)

Solar Conjunction

Astronomical	July 20th to August 9th
Rashis (sidereal)	July 17th to August 16th
Tropical Calendar	June 22nd to July 22nd

Opposition: January 28th **House** of the Moon **Element:** Water

Cancer is the most inconspicuous constellation in the Zodiac, but symbolically one of the most important. It contained the vernal equinox from 8300 to 6150 B.C. and the Summer Solstice from 1850 B.C. to A.D.300. At noon at the time of the solstice the sun is directly overhead at latitude 23.5° north. Twenty-five centuries ago the Sun was in Cancer at this time and this latitude became known as the Tropic of Cancer. It retains this ancient name till this day despite the fact that precession carried the solstice into Gemini in A.D.300. At present Cancer is the 5th sign in the Zodiac.

According to one legend Cancer is so called because 5,000 years ago, when these stars rose just before dawn, all the crabs in the Nile crawled out on the land. Another tale claims that the Sun was identified as a Celestial Crab. This was because the sideways walk of the crab mimicked the seasonal sideways change is the rise position of the sun. At the summer Solstice (1850 B.C. to A.D.300), the solar crab reversed its path along the horizon.

In ancient times Cancer was the sign that ruled over India and Ethiopia. To the Persians and the Platonists it was the 'Gate of Men', through which souls descended from heaven into human bodies. In Christian times this gate was called the 'Manger', being the portal in the heavens from which the spirit of god descended into the infant Jesus.

Having few stars Cancer was considered to be a 'Dark Sign', black without eyes. In Coptic Egypt, it was known as "The Power of Darkness" and was associated with Anubis, a god of the Underworld. The Egyptian Cancer represented a scarab beetle, a symbol of rebirth. In the Underworld, Anubis supervised the weighing of dead people's hearts before the judgement seat of Osiris. When bodies were embalmed (to prepare them for their journey to the Underworld), the supervising priest wore a jackal-mask to signify the presence of Anubis. He was however, also associated with rebirth and is identified with Sirius, the 'Dog Star'.[31]

In Greek mythology Cancer joined the dreaded Hydra, an enormous water snake with nine heads, in battle against Hercules. At Hera's command, it scuttled out from the swamp to bite at Hercules feet,

31 The heliacal rising of Sirius signaled the inundation of the Nile and the beginning of the Egyptian calendar, the Sothic cycle.

but he crushed it underfoot. The goddess Hera rewarded the Crab by placing it among the stars.

In ancient Babylonia, a tortoise, not a crab, represented this zodiacal constellation. In some Eastern zodiacs it represented two asses. In the Chinese zodiac it was a sheep, the 8th sign. The Hebrews assigned Cancer to *Issachar.*

The Brightest Stars in Cancer

Al Tarf (Beta Cancri) means 'the End' – of the southern foot. At magnitude 3.5 it is the brightest star in Cancer.

Acubens (Alpha Cancri) means 'the claw'. Its magnitude is 4.2.

Asellus Australis (Delta Cancri) and **Asellus Borealis** (Gamma Cancri) are the Arabian southern and northern donkeys respectively. Their magnitudes are 3.9 and 4.7. These are the two asses that aided the gods in their battle with the giants on the peninsula of Macedonia. They were rewarded with a resting place either side of the manger.

The Praesepe (The Beehive) star cluster is the most conspicuous object in Cancer. It is a cluster of some 50 stars of which only the brightest are just visible to the unaided eye. However, the combined light of the cluster stars renders them visible as a misty patch of light. Hipparchos called them the 'Little Cloud'. It is also known as the 'Manger' and the 'Beehive'. The Beehive was one of the emblems of the Eleusinian Mysteries. In these myths (c.1600 B.C.) the goddess Rhea (Mother of the Gods)[32] was represented with a beehive beside her.[33] From the hive arose corn and flowers which denoted the cycle of the seasons and the return of the sun to the summer solstice.

The Praesepe, along with some of the other stars in Cancer were, in ancient times, important in astrological predictions of the weather.

A murky Manger with both stars
Shining unaltered is a sign of rain.
If while the northern Ass is dimmed
By vaporous shroud, he of the south gleam radiant,
Expect a south wind; the vaporous shroud and radiance
Exchanging stars harbinger Boreas.
Prognostica, Aratos (c.270 B.C.)

32 Rhea means 'flow'. She was the wife of Chronos (time) and represented the eternal flow of time and generations. Her flow was the waters of life, menstrual blood and milk.

33 Later she was identified with Cybele and was accompanied by lions.

27. LEO (Simha)

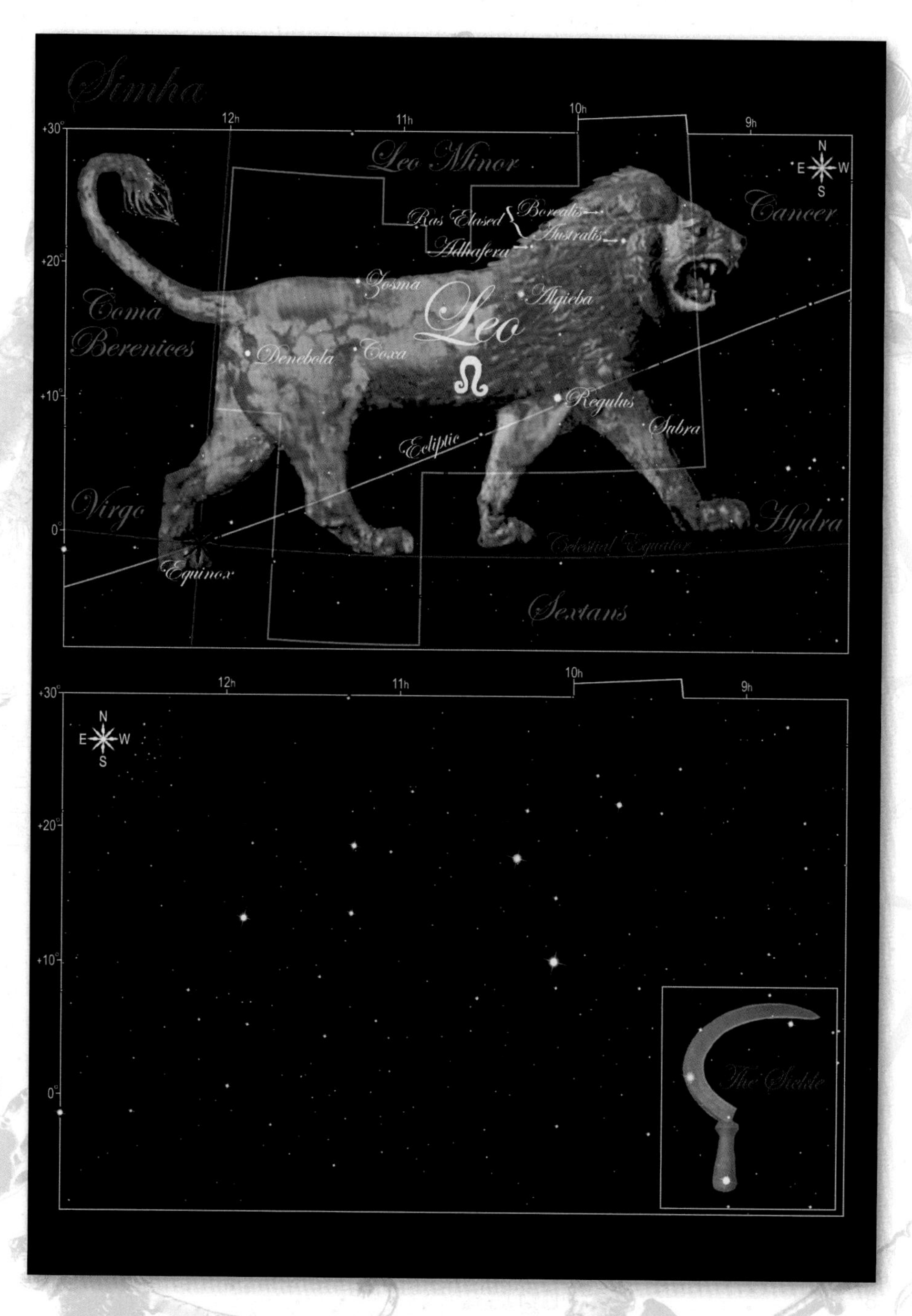

Fig. 40: *Leo star chart.* R. Hall

LEO – The Lion **SIMHA – The Lion** (Rashis)

Solar Conjunction

Astronomical	August 10th to September 15th
Rashis (sidereal)	August 17th to September 16th
Tropical Calendar	July 23rd to August 23rd

Opposition: February 25th **House** of the Sun **Element:** Fire

Leo is one of the most ancient of the Zodiacal signs and was formalized at least 5,000 years ago when it marked the summer solstice (4000 to 1850 B.C.). At this time every spring, rain and melting snow in the Highlands of Ethiopia sent a huge quantity of water down the Nile which reached Egypt about June. The lion-headed water spouts or fountains which originated in ancient Egypt, symbolized the fertile flooding of the Nile when the sun was in the Zodiacal sign Leo. It also signaled the onset of the hottest time of the year and it has been suggested that these conditions, the heat of the desert contrasting with the flood waters and rebirth of vegetation along the banks of the great river, brought the lions in from the wilderness. Leo, it is believed, symbolized the seasonal appearance of the lion, which was also a warning to the villagers along the Nile valley.

***Fig. 41:** Sekmet, the lion-headed goddess.* Wikipedia.

The Egyptians believed that the Sun was born in Leo and, in astrology it is the 'sole house of the sun'. Leo was the Symbol of Sekhmet Figure 41, a lion-headed goddess who carried a fire-spitting cobra. When Leo was at the mid-summer solstice it was a time of intense heat; a lean time before the crops came in. If the floods failed, it was a time of death. The seasonal climate was believed to be a reflection of Sekhmet's mood. When she was at rest, she was a figure of calm, royal dignity. But when she was angry, she transformed herself into the Eye of Ra (Sun), and became a fierce war-goddess. She breathed flames, and searing heat – the parched winds of the desert – radiated from her body. She charred her enemies' bodies and gulped their blood.

Some archeologists believe that the Great Sphinx represents the body of Leo and the head of Virgo because in the past the sun passed through both of these signs when the Nile was in flood. The solar passage through both signs coinciding with the inundation of the Nile occurred around 5000 B.C. To the east in Anatolia, and probably dating back to this epoch, was the growing religion of the Mother Goddess Cybele, who rode a Chariot drawn by lions, Virgo and Leo. The lion was also sacred to the Babylonian Goddess Ishtar, who was known as "The Lioness".

Some people have suggested that the origin of Leo dates back to a time when it contained the vernal equinox (10,450 to 8,300 B.C.). However, there is no evidence for this.

In Greek mythology Leo is the Nemean lion, a monstrous beast which came from the moon and descended to Earth as a meteor. The first labour of Hercules' was to kill the Nemean lion which was terrorizing the people of Nemea. Its claws were razors and its hide was so tough that no weapon could pierce it. After a month-long battle ...Hercules seized the lion and strangled it to death. The angry Hera raised the soul of the lion high into the sky to become Leo, the Lion.

The form the Lion took in the sky changed with time and differed from culture to culture. Around 240 B.C. Leo was robbed of his splendid tail. The astronomer-priests under Ptolemy III chopped off the tail of Leo when they invented a new constellation, Coma Berenices. Berenice was the Queen of Egypt who cut off her hair (coma) in gratitude to the Gods for the safe return of her husband (Ptolemy III) from battle. The Greeks never adopted this constellation which remained part of Leo until it was reintroduced by the Dutch cartographer Mercator in A.D.1551.

The fore-stars of the main of Leo form a well-known and ancient asterism, 'The Sickle'. It was originally a lunar mansion known to the Akkadians as Gis-mes, the Curved Weapon. In medieval Europe the Sickle rose just before dawn at the time to bring in the harvest.

In the ancient world Leo was said to rule over Armenia, Bithynia, Cappadocia, Macedon, and Phrygia. Leo was the tribal sign of Judah, given to him by his father Jacob:

Judah is a lion's whelp;
From the prey, my son, you have gone up.
He stooped down, he couched as a lion,
And as a lioness; who dares rouse him up?

Genesis 49-9

Because of its Royal status the Lion (Leo) was adopted by the British as a national icon.

In pride the Lion lifts his mane
To see his British brothers reign
As stars below.

Edward Young's Imperium Pelagi (c.1720)

The Chinese didn't have lions so this constellation was to them, the figure of 'The Horse'. The Horse was the 7th sign in the Chinese Zodiac.

Leo is currently the 6th sign in the Zodiac and, in A.D.2450, the autumn equinox will move into its domain.

The Brightest Stars in Leo

Regulus (Alpha Leonis) was the 'Leader of the Four Royal Stars', the 'Four Guardians of Heaven' which, 5,000 years ago marked the locations of the equinoxes and solstices. Regulus was the 'Pillar of Summer' and was closest to the solstice position around 2275 B.C. It is a blue-white star, the 5th brightest star in the Zodiac and the 22nd in the entire sky. Its magnitude is 1.36.

Regulus is a name given to the star by Copernicus, which was derived from its earlier name 'Rex' (given by Ptolemy). The Babylonians knew it as 'Sharru' (the King), and in India it was Maghaa (the Mighty). It is also known as the 'Lion's Heart' (Cor Leonis) or the 'Heart of the Royal Lion'. In astrology it was a portent of glory, riches, and power to those born under its influence.

Algieba (Gamma Leonis) is a golden yellow star of magnitude 1.99. It is the 12th brightest star in the Zodiac, 50th in the entire sky. Algieba means 'the forehead'.

Denebola (Beta Leonis) is a white star of magnitude 2.14. Its name is derived from 'Al Dhanab al Asad', which means 'The Lion's tail'. Denebola is the 15th brightest star in the Zodiac, 65th in the entire sky.

The Egyptians believed that, at the beginning of time the Sun was born close to this star. In India it was the 'Star of the Goddess Bahu', the Creator Mother. Astrologically Denebola was an unlucky star, the opposite of Regulus, it was a portent of misfortune and disgrace.

Zosma (Delta Leonis) is a blue-white star of magnitude 2.56. It is the 19th brightest star in the Zodiac, 96th in the entire sky. Its name means the 'Girdle'.

The Leonid Meteors

Each year, around November 17th to 18th, meteors (shooting stars) are observed to radiate from a point close to the star Algieba. Usually only about 10 meteors per hour are observed but, every 33 years a meteor storm occurs which can produce up to 100,000 meteors per hour. The last of these storms occurred in 1999.

28. VIRGO (Kanya)

Fig. 42: Virgo star chart. R. Hall

Soared up to heaven, selecting this abode,
Whence yet at night she shows herself to men.

Aratos (c.270 B.C.)

VIRGO – The Virgin **KANYA – The Girl** (Rashis)

Solar Conjunction

Astronomical	September 16th to October 30th
Rashis (sidereal)	September 17th to October 17th
Tropical Calendar	August 24th to September 23rd

Opposition: April 5th **House** of Mercury **Element:** Earth

Virgo is an ancient constellation and, as has been discussed in earlier chapters, is closely associated with the Earth-Mother Goddess. The constellation contained the vernal equinox from 12,600 B.C. to 10,450 B.C., the summer solstice from 6150 B.C. to 4000 B.C., and has been home to the autumn equinox since A.D. 300. Thus, it is currently the 7th sign of the zodiac.

Virgo was identified with the Babylonian goddess Ishtar who was worshipped all over the Mesopotamian region from Neneveh to Egyptian Thebes, from Cyprus to Babylon. Ishtar was the original of the Aphrodite of Greece and the Venus of Rome. She in turn originated from the Sumerian Innana; who is identified with the Mesopotamian Astarte, the Anatolian Cybele, and Isis of the Nile. Innana, Ishtar and Cebele are also associated with the lion (Leo).

Eratosthenes identified Isis with the Virgo depicted in the zodiacs of Denderah and Thebes. She holds the ears of wheat in her hand that she dropped to form the Milky Way.
She is also depicted as holding the infant sun-god Horus. This same figure reappears around A.D.1200 as the Virgin Mary holding the child Jesus.

To the Greeks Virgo was the 'Queen of the Harvest' because these stars rose just before dawn at the time to bring in the harvest. Virgo was identified with Demeter or her daughter Persephone.

In the ancient world Virgo ruled over Arcadia, Caria, Ionia, Rhodes and the Doric plains. The Romans knew her as Ceres (it is where our word cereal comes from). She was Khosha, or the "the Ear of Wheat" to the Persians, and Bethulah, meaning "Abundance in Harvest" to the Hebrews.

In the Chinese Zodiac the stars of Virgo are the 6th sign and represent a Serpent.

The Brightest Stars of Virgo

Spica (Alpha Virginis) is the 3rd brightest star in the Zodiac and the 16th in the entire sky. It is a blue-white star and its magnitude of 1.0 is slightly variable. Spica was very close to the point of the Autumn equinox in A.D. 350. Spica means 'Ear of Wheat' and she has been known as such in different languages for at least 2,500 years. Shibboleth (Hebrew), Chuushe (Persian), Salkim (Turkish),

Shebbelta (Syrian), all mean 'Ear of Wheat'. The Chinese called Spica 'Kio', which means the Horn or Spike. The commonality of names is of course, due to the association of this star with the harvest. The Babylonians gave this star the title 'Emuku Tin-tir-Ki', the 'Might of the Abode of Life'. The Great Temple of Diana (715 B.C.), along with many others temples from Greece to Egypt, were orientated to the rising of Spica. In New Zealand the Maori know this star as Mariao.

Spica, together with Denebola in Leo, and **Arcturus** in Bootes, forms an equilateral triangle in the northern hemisphere late spring evening sky (southern hemisphere late autumn). Arcturus, the 4th brightest star in the sky, is identified with Mazzaroth and is one of the stars mentioned by Job: "Canst thou bring forth Mazzaroth in his season". Mazzaroth and his sons signified the twelve signs of the Zodiac. Arcturus is also frequently alluded to by Virgil in the first book of the "Georgics." The rising and setting of this star were supposed to portend great tempests. In the time of Virgil it rose about the middle of September.

Porrima (Gamma Virginis) is the name of one of two ancient goddesses of prophecy. The Chinese knew this star as Shang Seang, the High Minister of State. Porrima is a yellow-white star of magnitude 2.8.

Vindemiatrix (Epsilon Virginis) means 'Grape Gatherer', a name that is believed to date back to a time when the heliacal rising of this star coincided with the vintage. This would be around 650 B.C. Vindemiatrix is a yellowish star of magnitude 2.8.

Zavijava (Beta Virginis) means 'Angle' or 'Corner' and refers to one of the corners of an Arabic constellation, The Kennel. With **Zaniah** (Eta Viriginis) the Babylonians knew this star as 'Ninsar', the Lady of Heaven (which is believed to refer to Ishtar). Zavijava is a yellowish star of magnitude 3.6.

29. LIBRA (Tula) & CHELAE

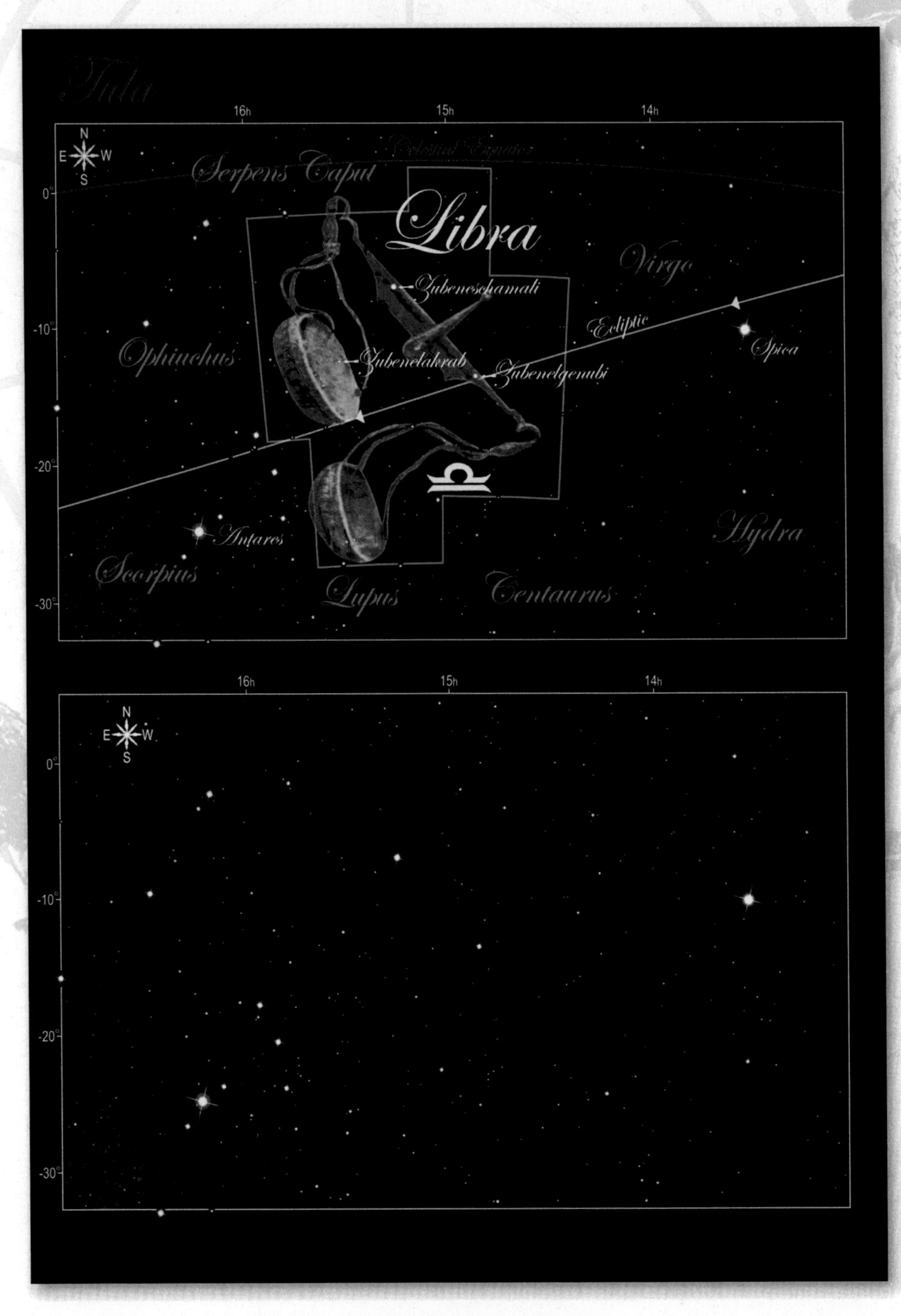

Fig. 43: Libra star chart. R. Hall

Th' Eternal Father hung,
His golden scales aloft,

Homer (c. 850 B.C.)

LIBRA – The Scales
CHELAE – The Claws

TULA – The Balance (Rashis)

Solar Conjunction

Astronomical	October 31st to November 21st
Rashis (sidereal)	October 18th to November 16th
Tropical Calendar	September 24th to October 23rd

Opposition: May 8th **House** of Venus **Element:** Air

Libra, the 8th sign, is a small and faint constellation between Scorpius and Virgo. Libra is the Scales, the 'Balance' to the French, and the 'Wage' to the Germans.

As is often pointed out, Libra is the only inanimate sign in the Zodiac. However, originally Libra was represented by the figure of a man or woman (God or Goddess) holding a pair of scales. In the Zodiac we have today only the scales is depicted. This is probably because this Zodiac came from Arabia when, in the 12th century, astronomy/astrology was reintroduced into Europe following the Dark Ages. The human figure is omitted in all Arabian zodiacs, as it is held unlawful in the Koran to make any representation of God.

The origin of the 'Scales' is little obscure. The Zodiac formulated by the Greeks, which is essentially the one that we have today via Arabia, did not include Libra. The Greeks had eleven constellations of which one was a double sign. There was 'Scorpius' the Scorpion and 'Chelae' the Claws of the Scorpion.

The Romans turned Chelae into Libra the Scales. Legend has it that this occurred in 46 B.C. on the instructions of Julius Caesar, his scribe Flavius, and the Alexandrian astronomer Sosigenes. It was said to represent "the Scales of Justice" held by Julius Caesar. It also symbolised the equality of day and night because at that time it contained the Autumn Equinox. The equinox was in Chelae/Libra from 1850 B.C. to A.D.300

"Libra die somnique pares ubi fecerit horas,
Et medium luci atque umbris jam devidit orbem."

"When Libra makes the hours of day and night equal,
and now divides the globe in the middle,
between light and shades.

Virgil, Georgics, Book I

An alternative story claims that Libra was introduced by/for the Emperor Augustus. It was said that the sun was in Libra when Augustus was born and that this became the time when he dispensed justice. Medals were struck which show Augustus holding the Scales of Justice. However, other authors give Capricornus as the birth sign of Augustus. Later, the Romans associated the scales with Astraeia, the Goddess of Justice. In the heavens Astraeia was represented by Virgo who held the scales of justice aloft. Naturally, in ancient astrology Libra ruled over Italy, especially Rome.

Why did the Greeks have a double sign? Well, what was important was what Chelae, the Claws of the Scorpion, were holding. The claws held the Censor or Lamp on top of the Euphratean Altar (Ara). The lamp was Pharos, the Great Lamp or Lighthouse of Alexandria, one of the Seven Wonders of the World (See figure 44). Ara, which according to Greek mythology was the altar upon which the Olympian gods swore an oath of allegiance before their battle with the Titans, has now been reduced to a small faint and obscure constellation south of Scorpius.

However, the Romans did not invent the celestial scales. The symbol of the celestial balance was common to the Hebrews, the Arabs, the Assyrians, the people of the Indus Valley, and the Egyptians. The Scales are clearly depicted in the Zodiac of Denderah and some people think that they originated in ancient Egypt. The Egyptians saw Libra as a set of scales used for weighing the hearts of the dead. Maat, goddess of truth and balance, was always depicted holding a balance. In the heavens her laws guaranteed the stability of the universe.

However, others argue that the Scales originated in Babylonia. The Syrians called Libra Masathre and the Persians Taraazuuk, both of which mean 'scales'. The Persian zodiacal figure shows a person holding the scales in one hand and a lamb in the other; a lamb being the standard unit of weight in those days. In the Hebrew zodiac Libra is ascribed to *Asher.*

In the earliest Chinese solar zodiac the stars of Libra represented a crocodile or a dragon. They were known as 'Show Sing', the Star of Longevity. Later, they became 'Tien Ching' the Celestial Balance, the 5th sign.

In Medieval times the Scales, or rather Scorpius was associated with the plague. The Scorpion rose in the evening twilight at the beginning of the hottest time of the year. This was the time of disease and the plague. The Scorpion held the Scales in its claws which would weigh who would live and who would die that season. When the Archer rose it drove darkness and death away.

The Brightest Stars of Libra

The Arabian name for Libra was Al Zubana, 'the Claws', and the names of the stars retain this ancient heritage.

Zubeneschamali (Beta Librae) means 'the northern claw'. At magnitude 2.6 it is the brightest star in Libra. The star has a greenish tint.

Zubenelgenubi (Alpha Librae) is the 'the southern claw'. It is a blue-white star of magnitude 2.8.

Zubenelakrab (Gamma Librae) means 'the scorpion's claw'. It is a yellowish star of magnitude 3.9.

30. SCORPIUS (Vrishchika) & ORION

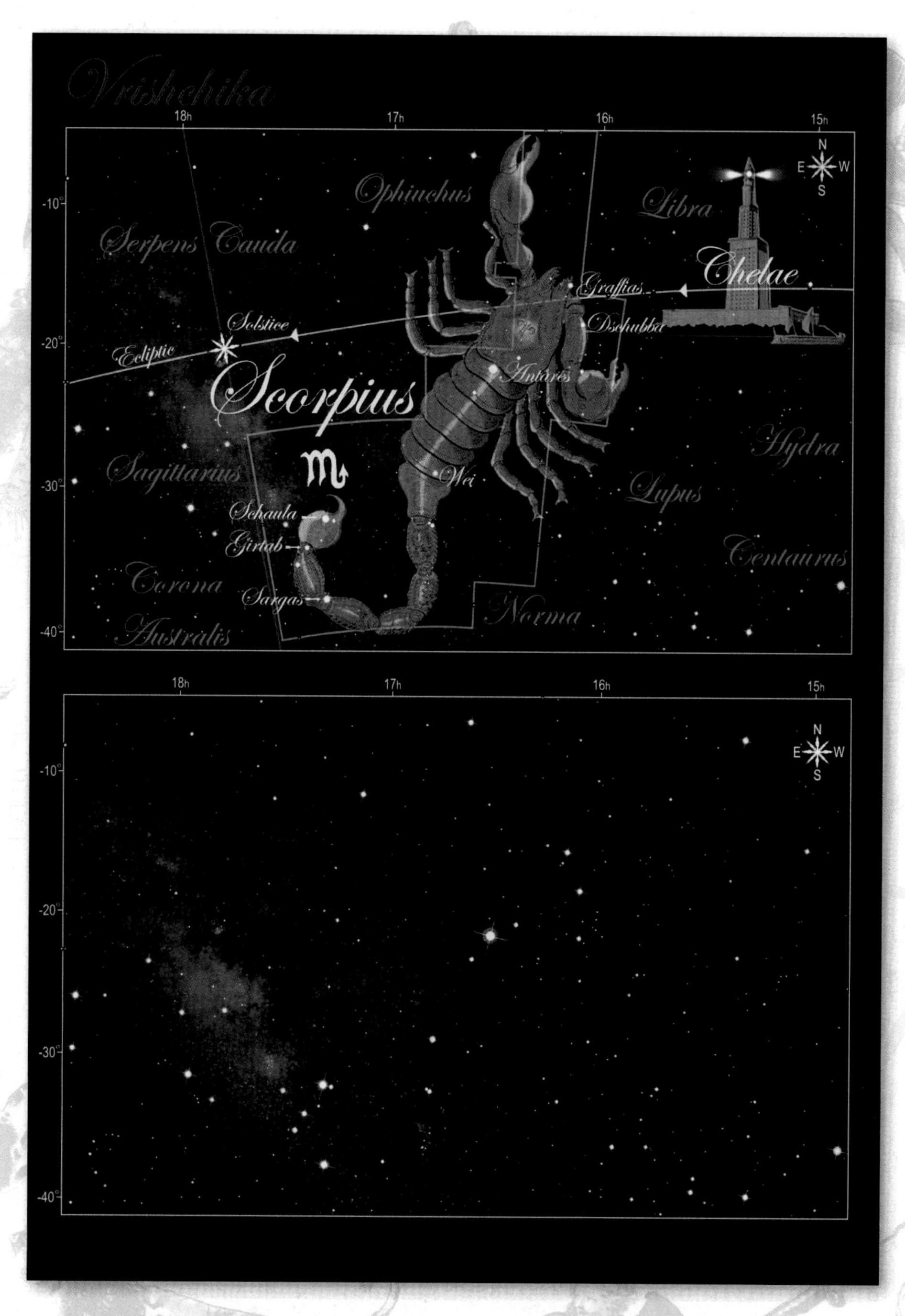

Fig. 44: *Scorpius star chart.* R. Hall

. . . and now in Ocean deepe
Orion flying fast from hissing snake,
His flaming head did hasten for to steepe
Spencer's Faerie Queen (c.1575)

SCORPIUS – The Scorpion **VRISHCHIKA – The Scorpion**

Solar Conjunction

Astronomical	November 22nd to December 16th *
Rashis (sidereal)	November 17th to December 15th

* In modern star charts the Sun passes through the constellation of Ophiuchus from November 30 to December 16.

Tropical Calendar October 24th to November 22nd

Opposition: June 2nd **House** of Mars **Element:** Water

Scorpius, the 9th sign, is embedded in the most brilliant region of the Milky Way. It is one of the most ancient of constellations and was at one time the largest sign in the zodiac. It was originally a double sign - Scorpius (the Scorpion) and 'Chelae' (the Claws of the Scorpion). The Romans transformed the stars of Chelae into the Scales (see Libra). Because of its double nature some people have suggested that the first zodiac consisted of just six (double) signs. But, I find no evidence for this.

The stars of Scorpius represented, for most cultures, a scorpion. Sometimes they were seen as a serpent, a 'Crowned Snake' or a 'Basilisk' – a fabulous reptile hatched by a snake from a hen's egg that had poisonous breath and an evil eye (glance). Indeed, to the Chinese they were part of the mighty Azure Dragon. However, this dragon was a kindly beast. Later, the Chinese transformed these stars into the 'Blue Emperor' and then, in the time of Confucius, it became the double sign of 'Ta Who' – The Great Fire (Antares and the body of the Scorpion) and 'Shing Kung' - Divine Temple (the Tail of the Scorpion). Following the arrival of Jesuit missionaries it was transformed into 'Tien He', the Celestial Scorpion'. However, in the traditional Chinese zodiac it is the 4th sign, the Hare.

Scorpius is believed to have originated when the autumn equinox moved into this constellation. This occurred in 4000 B.C. and Scorpius retained the equinox until 1850 B.C., up until 45 B.C. if you include Chelae. To the people who lived along the banks of the Euphrates 6,000 years ago, it was a symbol of darkness, marking the decline of the sun's power. They pictured it as a 'Scorpion-Man' holding a lamp.

To the Phoenicians, who flourished in what is now Syria, Lebanon and Israel, the sun's entrance into Scorpius marked the 'Darkening of the Sun', the dwindling hours of sunlight as it moved towards midwinter solstice. It was the herald of the end of abundant food supplies. Scorpius is directly opposite the Pleiades (the food-bringers) which at that time marked the spring equinox; when one rises, the other sets.

The Phoenicians were sea-traders and to the ancient mariner the setting of Scorpius in the evening twilight was seen as a malignant influence that would be accompanied by storms. Alchemists however, held Scorpius in high regard, because it was believed that only when the sun was in this sign could iron be transmuted into gold.

In traditional astrology it was a fruitful sign, "active and eminent", that governed the genitals. It was also known as the "accursed constellation", the source of war and discord. This was because Scorpius was 'Martis Sidus', the birthplace and the House of Mars. However, this applied only to the tail and sting. Chelae the Claws (now Libra) were dedicated to Venus. It represented the 'Yoke' under which the goddess united couples. In the ancient world Scorpius ruled over Carthage, Libya, Egypt, Sardinia and islands off the Italian coast. In the Hebrew zodiac Scorpius is referred to Dan.

Constellation boundaries adopted in modern star charts have a 13th constellation along the ecliptic, the path of the sun. Consequently the sun spends only nine days in Scorpius. The rest of the time usually allotted to this sign is spent travelling through the 13th sign, **Ophiuchus the Serpent Bearer**. Ophiuchus was never part of the original zodiac but plays a part in an ancient story of the origin of the Scorpion.

Before I tell that tale I must introduce another constellation closely related to the Zodiac and the Scorpion, **Orion the Hunter (**Figure 45). Orion is the most brilliant constellation in the sky. It straddles the celestial equator and is therefore visible from every part of the world. The constellation is wedged between Taurus and Gemini with its northern most stars close to the ecliptic. Indeed, sometimes it was used as a zodiacal marker in place of Gemini.

Because of its brilliance and location in the sky Orion has been used as a seasonal marker by peoples around the world. Almost universally it has been seen as the figure of a giant man who was both a hunter and a warrior. To New Zealand Maori for example, he was Tautoru, the Bird Snarer. The story that I shall now relate pre-dates the Greek civilization but due to precession can be applied today in the southern hemisphere. If you live in the northern hemisphere just swap seasons.

Orion was a great hunter, a giant of a man who was beloved by the male gods. He had however, one flaw – he was a very boastful man. One day, while he was sitting down socialising with the male gods and drinking wine, he came out with a boast – Such was his prowess as a hunter he could hunt down and kill every animal on earth. Now, it may have been just an idle boast but unfortunately for Orion his words were overheard by a goddess, a very powerful goddess. Her name was Artemis (Diana). She was the Goddess of the Hunt and Protector of the Wild and she wasn't going to have Orion going around slaughtering all those things that she was there to protect. So, she went to Mother-Earth and the two conspired to send a minion to sort Orion out.

If you go out on a mid-winter's evening here in New Zealand, look up on a clear dark sky around midnight, and you will see the most brilliant region of the entire Milky Way directly overhead. You will also notice that there is a dark, almost black rift running along the centre of the brightest region. Today we know that this dark rift is formed by clouds of cosmic dust and gas which blot out the light of the more distant stars in the Milky Way. But in ancient times it was believed to be a hole in the sky where the minion of Artemis ripped open the heavens as it descended to the earth. This minion was the Scorpion. The monstrous creature hunted Orion down and a great battle pursued between the two giants. But eventually the scorpion's sting found its mark and Orion fell in agony to the ground and died. It is the reason why Scorpius is known as 'Slayer of the Giant'.

Asclepius (Ophiuchus) was the first to discover Orion's body. He was the son of Apollo, and the founder of medicine who learned the art of healing from his father and his tutor, the wise Centaur Chiron. Such were his healing powers that he had learnt how raise the dead. He tried to restore the life of Orion but Zeus struck him down, because that was a power reserved only for gods.

The male gods were upset when they came across Orion's body, because they liked Orion and enjoyed his company. They discussed the matter and decided that instead of letting Orion pass on to the Underworld they would keep him with them for ever more. They would give him immortality. This they did and they gently picked his body up and placed him in the heavens.

However, this act infuriated the goddess who exacted a terrible revenge. She placed her scorpion in the sky, on the opposite ends of the heavens so that, for the rest of eternity, it would chase poor old Orion around the sky, which is exactly what it does.

The story of Orion and the Scorpion is an example of how information was encoded in ancient times. Before the time of the written word information was stored in stories poetry and song. We can easily recall stories that were told to us as children, so it makes good sense that if you want someone to remember something put it in a story. Let's see how this story works.

If you go out on a summer's evening here in New Zealand, look up and you will see Orion, the most brilliant constellation in the sky. The Scorpion will be nowhere to be seen. But come late March you will see the Scorpion rise in the south-east. As the Scorpion crawls up into the heavens Orion flees from the sky.

When the Scorpion comes
Orion flies to utmost end of earth

Aratos (c.270 B.C.)

If you go out on a winter's evening the most brilliant constellation in the sky will be the Scorpion. Now it is Orion who is nowhere to be seen. Here in Aotearoa, New Zealand, Scorpius passes directly overhead. The tail of the Scorpion is to the Polynesians the 'Fishhook of Maaui', a navigation beacon and zenith constellation for these lands.

The Scorpion is hunting Orion but he can't find him. Eventually the Scorpion gives up and descends back into the Underworld (he sets). And when the Scorpion is all but gone, just his tail above the south-western horizon, you will see Orion bobbing up in the east.

Tell people facts about solstices and equinoxes most of them will forget. But it is easy to remember the story of Orion and the Scorpion. And, so long as you can recognise these two very distinctive constellations, you too will know the following: While Orion is in our evening sky we will have the lazy hazy days of summer. And while the Scorpion is in our evening sky it will be frigid and frosty. But in both cases the weather will be settled. The only time you will see both constellations in the sky, one in the east, the other in the west, one rising as the other sets, is when the seasons are about to change. So beware my friend, everything in your environment, the winds, weather and food supplies are all about to change. The times at which they are both in the sky together is or course, near the equinoxes. And this is how mythological stories conveyed important information.

In A.D.2450 the winter solstice will move into Scorpius. It will then mark the darkest of times… in the northern hemisphere. Down here in the southern hemisphere it will mark the onset of the hottest time of year and the longest hours of daylight.

The Brightest Stars of Scorpius

Antares (Alpha Scorpii) is a glorious red star and, like all red stars, varies in brightness. Its average magnitude is 0.90. It is the 2nd brightest star in the Zodiac and the 15th brightest in the entire sky. Antares is believed to mean either 'similar to' or 'rival of' Mars. This may be because of its colour and/or that Scorpius is the House of Mars. The Arabians called it Kalb al 'Akrab, the Scorpion's Heart.

Antares is one the Persian Royal Stars and was known as the 'Guardian of the Heavens'. It was one of the four Pillars of Heaven and in 3000 B.C. was close to the point of the autumn equinox.

In ancient Egypt this star was the goddess Selkit who herald the sunrise at the autumn equinox. It was also a symbol of Isis.

In Aotearoa, New Zealand Maori know this star as Rehua. It also represented Hine-Raumati, the Summer Maid, one of the two wives of Te Ra (The Sun), who rose just before the sun at the summer solstice (Southern Hemisphere). The other wife, the Winter Maid, was Hine-Takurua (Sirius).

Shaula (Lambda Scorpii) is a blue-white star and, it too is variable in brightness. Its average magnitude is 1.60. Shaula means the 'sting', and is the brighter of two stars that form the tip of the Scorpion's tail. To the Polynesians it is the barb of the Fishhook of Maaui. Shaula is the 7th brightest star in the Zodiac and the 27th brightest in the sky.

Sargas (Theta Scorpii) is the Euphratean name for this star; the Persians called it 'Vanant', the Smiter. Sargas is a yellow star of magnitude 1.87 and is the 10th brightest star in the Zodiac, 42nd in the entire sky.

Wei (Epsilon Scorpii) means the 'tail'. This golden-yellow star of magnitude 2.29 is the 16th brightest star in the Zodiac, 74th in the sky.

Dschubba (Delta Scorpii) means 'forehead'. This blue-white star of magnitude 2.32 is the 17th brightest star in the Zodiac, 79th in the entire sky.

Girtab (Kappa Scorpii) means 'seizer'. This blue-white star of magnitude 2.40 is the 18th brightest star in the Zodiac, 85th in the entire sky.

Graffias (Beta Scorpii) is a blue-white star of magnitude 2.6. Its name means 'claws'.

Scorpid Meteors

Each year from March 26th to May 12th meteors will be seen radiating from a point close to Antares. This meteor shower is known as the Alpha Scorpids and peaks on May 3rd. On this date up to 10 meteors an hour will be observed.

ORION – The Hunter

***Fig. 45:** Orion star chart.* R. Hall

Solar Conjunction: June 16th **Opposition:** December 17th

As discussed above Orion is the most glorious constellation in the sky and, almost universally has been seen as a figure of a giant man, a hunter and warrior. The Arabians called it Al Jabbar, which means 'the Giant'. This word gradually evolved into Algebra!

In Egypt, in the Zodiac of Denderah Orion is Horus. This zodiac and planisphere was sculptured around 34 B.C. Horus/Orion is in a boat accompanied and followed by Sirius which has the form of a cow. Orion is also sculptured on the temple walls of Sakkara at Thebes, which dates back to 3285 B.C. He represents Sahu, and is the place where the soul of Osiris was laid to rest.

In Greek and Roman times Orion had a stormy character, and was to the mariner a dangerous sign. The Greek historian Polybios (c. 150 B.C.) maintained that the loss of the Roman fleet in the first Punic war was due to the fact that it sailed just after 'the rising of Orion'.

Later Orion became a calendar sign. Its heliacal rising marked the beginning of summer; its midnight rising the season of grape-gathering; and its rising just after sunset forewarned the coming of winter and the season of storms.

The Brightest Stars of Orion

Rigel (Beta Orionis) is a brilliant blue-white star, the brightest in Orion and the 7th brightest in the entire sky. Its magnitude is 0.12. Rigel is derived from the Arabic *Rijl Jauzah al Yusra*, the left leg of Jauzah (the Giant). The ancient astrologers maintained that splendour and honours would fall to those born under this star.

In Norse star-lore Rigel is one of the toes of the giant Orwandil. The other toe was broken off by Thor and hurled into the northern sky where it became the star Alcor of the Great Bear (Ursa Major).

To New Zealand Maori Rigel is known as Puanga. It is the flower or decoy on the bird-snare of Tautoru. For many tribes, particularly those of the South Island, its heliacal rising in June heralds the beginning of the new-year. Puanga rises about 30 minutes before Matariki (Pleiades).

Betelgeuse (Alpha Orionis) has a strong reddish hue making it easy to identify among the other bright stars of Orion which are blue-white. The star slowly varies in brightness with an average magnitude of 0.50. At its faintest it is orange-red in colour, at its brightest it has been described as a celestial topaz. Betelgeuse is the second brightest star in Orion, the 10th brightest in the entire sky. Its name is derived from *Ibt al Jauzah*, Armpit of the Giant, which degenerated into *Bed Elgueze,* and finally into Betelgeuse. Other Arabic names for this star are *Al Mankib*, the Shoulder; *Al Dhira*, the Arm, and *Al Yad al Yamna*, the Right Hand.

While most red or variable stars were believed to be malevolent, ancient astrologers believed Betelgeuse to be a Kingly star that would bring fortune, martial honours and wealth to those born under it.

Bellatrix (Gamma Orionis) is the 'Amazon Star' and its name means the 'Female Warrior'. Bellatrix is a blue-white star and, with a magnitude 1.64, it is the 3rd brightest star in Orion, the 28th in the entire heavens. In ancient astrology those born under this star, particularly women, were said to be destined for great civil or military honours.

"women born under this constellation shall have mighty tongues"

Thomas Hood

Saiph (Kappa Orionis). The Arabic name for this star is *Rijl Jausah al Yamna*, the right leg of Jauzah. Its common name however, is Saiph which is derived from *Al Saif*, the Sword. This name originally belonged to Theta Orionis, which really does mark the sword. In the Orient Saiph is known as the 'Eye of the Tiger'. This blue-white star has a magnitude of 2.06 which ranks it as the 58th brightest star in the sky.

Al Nijad - The Belt of Orion

The three bright stars that form the Belt of Orion are a well known asterism. In the present era they have been important navigational beacons, laying close to the celestial equator the Belt rises due east and sets due west. Other Arabic names are Al Nasak, the Line, and Al Alkat, the Golden Nuts. They are also known as the 'Three Kings' or Magi. To the Norse they were Fiskikallar, the Staff; and in Britain they were known as Jacob's Rod. Down here in Aotearoa, New Zealand Maori know these stars as Tautoru, or more correctly the Bird Perch of Tautoru. All three stars are blue-white in colour and their individual names also apply to the three. They are, from east to west:

Alnitak (Zeta Orionis) means 'the Girdle'. Its magnitude is 1.77 and it is the 34th brightest star in the sky.

Alnilam (Epsilon Orionis) is the brightest and central of the three stars. Its name means 'String of Pearls'. Its magnitude is 1.70 and it is the 31st brightest star in the sky.

Mintaka (Delta Orionis) is the fainter of the three and the one closest to the celestial equator. Its name means 'the Belt'. Mintaka is the 71st brightest star in the sky and its magnitude of 2.23 is slightly variable.

The Orionids is an annual meteor shower active from October 2nd to November 7th. The radiant is close to the border with Gemini. The shower peaks on October 25th when up to 25 meteors per hour may be observed. New Zealand Maori knew these meteors as 'Maui's Darts'. They signalled the time to plant the kumara seed tubers.

31. SAGITTARIUS (Dhanus)

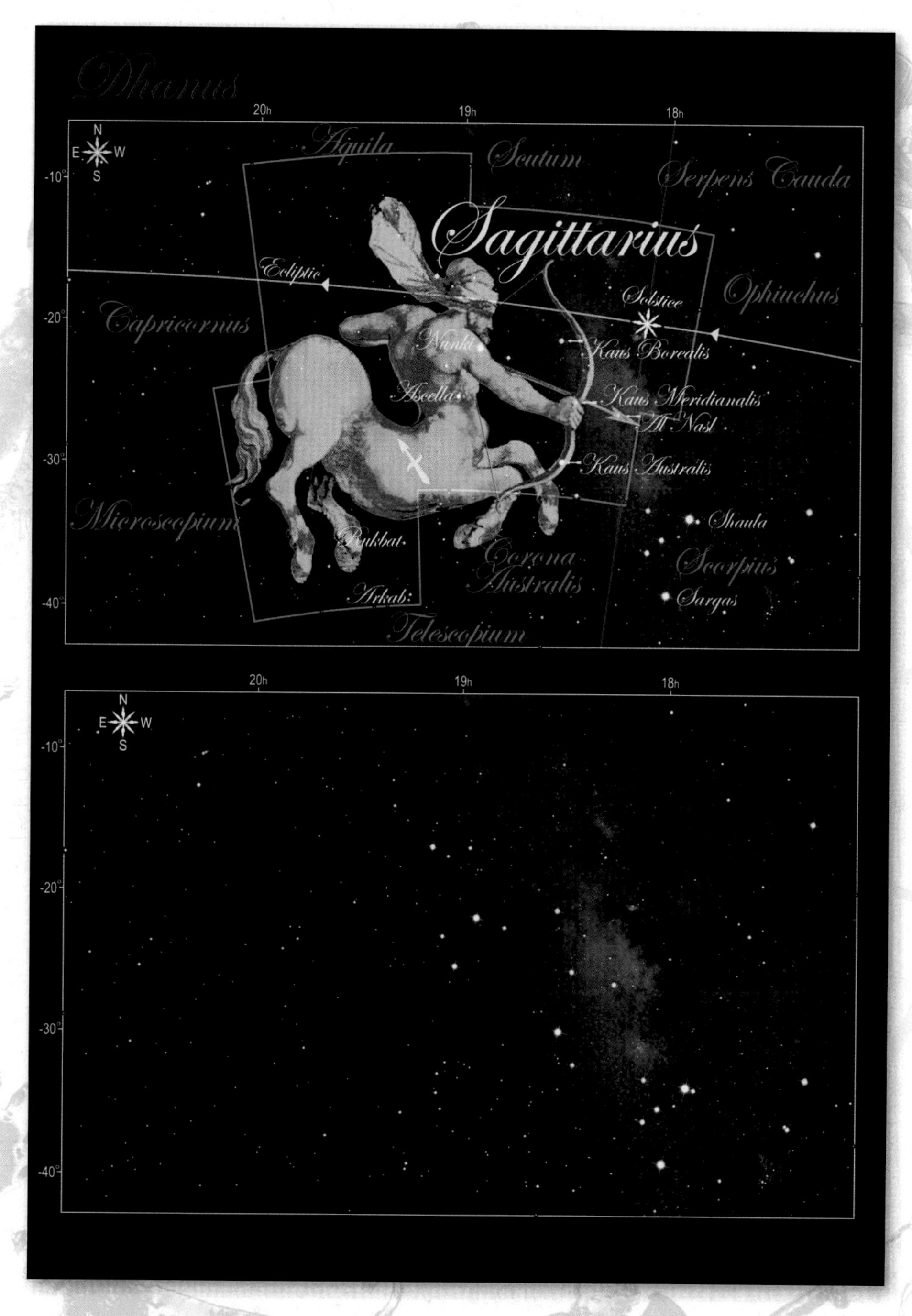

Fig. 46: Sagittarius star chart. R. Hall

. . . glorious in his Cretian Bow,
Centaur follows with an aiming Eye,
His Bow full drawn and ready to let fly.
Creech's Manilius (c. first century A.D.)

SAGITTARIUS – The Archer **DHANUS – The Bow** (Rashis)

Solar Conjunction

Astronomical	December 17th to January 19th
Rashis (sidereal)	December 16th to January 14th
Tropical Calendar	November 23rd to December 21st

Opposition: July 3rd **House** of Jupiter **Element:** Fire

Like adjacent Scorpius the stars of Sagittarius occupy a brilliant region of the Milky Way. The figure of Sagittarius that we know today is that of a centaur firing an arrow. But it has not always been seen as such. Originally, in most cultures, it was just a bow and arrow. In Persia it was known as Kaman, it was Yai in Turkey, Keshta in Syria, it was the Hebrew Kesheth, and the Chaldaean Kertko. All of these names mean bow or bow and arrow. In India, over 3000 years ago, this group of stars represented a horseman. Later it became Dhanus, the Bow. In Egypt they originally represented an Ibis but in the Zodiac of Denderah it is an archer with the face of a lion. To the Chinese it was a Tiger. Since A.D.300 the northern winter solstice has been located in Sagittarius. It is therefore the 10th sign.

Five thousand years ago as the sun set in Taurus at the spring equinox Sagittarius rose in the east. The Assyrian Royal Standard had an archer (Sagittarius) standing on or above a bull (Taurus). In the oldest zodiacs the archer holds a crown - the constellation of the Southern Crown (Corona Australis). The archer represented the god Ahura Mazda, the Ruler of the Universe. He was the twin of Ahriman (Taurus), and was his opposite in every way; light instead of dark, calm instead of fury, giving instead of taking, forgiveness instead of implacability. At the time of the spring equinox, as the Archer rose in the evening twilight he chased the Bull into the underworld.

As a dual creature the constellation was first depicted as 'Minotaurus', half man and half bull. Later it became half man and half horse, a centaur. It has been suggested that the myth of the centaur originated with the arrival of the first horsemen. They were armed warriors that swept down from the north on horseback firing arrows. To the locals they appeared to be some kind of strange half-human creature, combining the upper body of a warrior with the four legs of an animal – a centaur.

According to Greek mythology Sagittarius is the centaur Chiron, a friend of Hercules. In this story Hercules was very thirsty and drank from a jar of wine belonging to the centaurs. The centaurs attacked and Hercules killed many of them. Chiron, a centaur, but Hercules' friend, had not taken part, but was accidentally shot. Zeus gave the good centaur a resting place among the stars as Sagittarius.

Sagittarius is often associated with Crotus, the son of Pan (the goat-god) and Eupheme (the Muses' nurse). Crotus was raised by the Muses. He was a skilled hunter who loved the arts.

The Archer aims his arrow to strike the Scorpion's heart. The Scorpion and the Archer are usually seen as two opposites. Scorpius is a symbol of darkness while Sagittarius represented light. For the last two millennia the winter solstice has been in Sagittarius. As the Archer rose in the winter dawn twilight it drove darkness (the Scorpion) away – the days would begin to grow longer. To the Chaldeans however, the Archer was a God of War. But I guess that whether in this case it represented light or dark depended upon which side you were on.

Sagittarius is the House of Jupiter and has always been seen as a fortunate sign. It was associated with the south-west wind and represented fruitfulness. It was also known as Dianae Sidus, the Domicile of Diana. In the ancient world it ruled over Crete, Latium and Trinacria.

The Brightest Stars in Sagittarius

Kaus Australis (Epsilon Sagittarii) is the brightest star in Sagittarius. Its name means 'southern part of the bow'. Kaus Australis is a blue-white star with a magnitude of 1.79. It is the 9th brightest star in the Zodiac, the 36th in the entire sky.

Nunki (Sigma Sagittarii) is, in the Euphratean 'Tablet of the Thirty Stars', the 'Star of the Proclamation of the Sea'. This sea was the quarter of the sky occupied by the watery signs Aquarius, Capricornus, Pisces, Delphinus, and Pisces Australis. It is a blue-white star of magnitude 2.02. Nunki is the 14th brightest star in the Zodiac, the 53rd in the entire sky.

Ascella (Zeta Sagittarii) or Axilla, means the Armpit of the Archer. At magnitude 2.60 it is the 3rd brightest star in Sagittarius, the 20th brightest star in the Zodiac, and the 100th in the entire sky.

Kaus Borealis (Lambda Sagittarii) means the 'northern part of the bow'. It was also known in Arabia as Rai al Na'aim, the Keeper of the Desert Birds (Ostriches). Kaus Borealis is a yellowish star of magnitude 2.8.

Al Nasl (Gamma Sagittarii), a yellow star of magnitude 3.0, is the 'point of the arrow'.

Arkab (Beta Sagittarii) is a double star that can be seen with the unaided eye. The two stars, magnitude 4.0 and 4.3, one bluish the other white, are not physically related. Its name means the Tendon and it has been dubbed 'Achilles Tendon'.

Rukbat (Alpha Sagittarii). Although listed as Alpha, this blue-white star is, at magnitude 4.0, the 7th brightest star in Sagittarius. Its name means the 'Knee' of the Archer.

32. CAPRICORNUS (Makara)

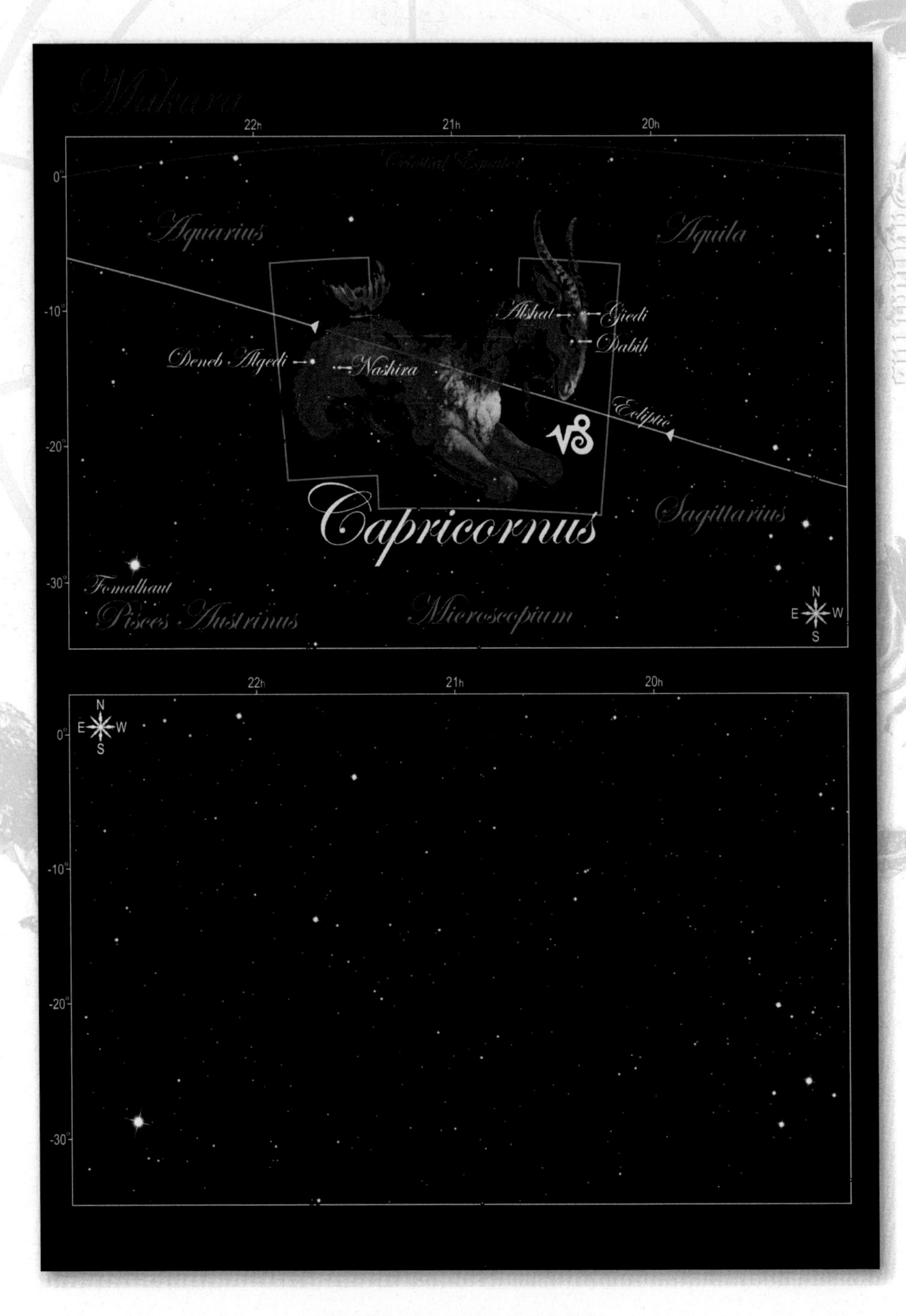

***Fig. 47:** Capricornus star chart.* R. Hall

Then grievous blasts
Break southward on the sea, when coincide
The Goat and sun; and then a heaven-sent cold.
Aratos (c. 270 B.C.)

CAPRICORNUS – The Sea-Goat **MAKARA – The Sea-Monster** (Rashis)

Solar Conjunction

Astronomical	January 20th to February 16th
Rashis (sidereal)	January 15th to February 12th
Tropical Calendar	December 22nd to January 20th

Opposition: August 6th **House** of Saturn **Element:** Earth

Capricorn, the 11th sign, is depicted as a goat with the tail of a fish. One legend links this figure to the Babylonian Ea, an ancient god with a human upper body, and the tail of a fish who emerged from the sea to bring knowledge and culture to humanity. Capricorn was also known as 'Neptuni proles' (Neptune's offspring), 'Pelagi' (the Ocean Storm), and 'Imbrifer' (the Rain-bringing One). All of these titles date back to the time when the winter solstice was in Capricornus, 1850 B.C. to A.D. 300. This solstice marked the onset of the season of rains and storms.

In Greek legends from that time Capricorn represented Amalthoea[34], the divine goat that suckled the infant Zeus. When the Sun entered Capricornus (Zeus was suckled by Amaltheia) the Sun (Zeus) triumphed over darkness and, as it emerged from that sign, its warmth and light grew in strength. Out of gratitude Zeus turned one of the goat's horns into 'Cornucopia' – the 'Horn of Plenty', which was always filled with whatever its possessor needed. Thus, the name Capricorn is derived from *"cornucopia"*, which denotes plenty. The word *"Amalthoea,"* when used figuratively also has the same meaning.

The whole story (Zeus & Amalthoea) is a solar allegory, alluding to the arrival of the sun among the stars of Capricorn, at which time the fruits of the earth—"corn, oil, and wine"—have all been gathered in and stored away, so that, although winter comes to desolate the land, the industrious husbandman is yet blessed with "plenty."

Stellar Theology & Masonic Astronomy
Robert Hewitt Brown, 1882

At noon on the day of this solstice the sun is directly overhead at latitude 23.5°south. This latitude became known as the Tropic of Capricorn because at that time, 2,500 years ago, the sun was in Capricorn on the day of the solstice. It retains this ancient name despite the fact that precession carried the solstice into Sagittarius in A.D. 300. In the Orient Capricornus was known as the 'Southern Gate of the Sun'.

While Capricornus is usually a watery sign, in ancient Egypt, when it culminated opposite the Sun at the summer solstice, it presided over the least honored time of the Egyptian year. This was the arid season preceding the flooding of the Nile. Unlike rivers such as the Tigris and Euphrates that flood in

34 In some traditions Amaltheia was a nymph who nourished Zeus with honey and the milk of a divine goat.

the winter season of rains, the Nile, which flows from the south, floods in the summer. The flow of the Nile increases in the summer due to the heavy rainfall that occurs in the tropical Ethiopian highlands in March. Flooding begins in southern Sudan in April and reaches Egypt in July. The river continues to rise until it peaks in mid-September. After this peak the levels fall quickly during November and December. The Nile is at its lowest ebb between March and May.

Capricornus is the House of Saturn which was associated with the Grim Reaper. It was the 'Gate of the Gods' through which the souls of the deceased ascended into heaven. It was also known as *Vestae Sidus*, the domicile of Vesta, goddess of the home and hearth; and the 'Mansion of Kings' because the Roman Emperors Augustus and Vespasian were said to have been born under this sign. In the ancient world Capricornus was said to rule over France, Germany and Spain.

In one story from antiquity Capricorn is Pan. Pan fled to Egypt from the monstrous Typhon and hid in the river Nile. He was already halfway submerged before deciding to wear the form of a goat. So a goat he became, but only from the waist up. From the waist down he took the form of a fish. In another, almost identical version of this story, Capricorn is Bacchus, the God of the Vine.

Sometimes Capricornus was just a goat and in Germany it was known as 'Steinbock', which means Stone-buck or Ibex. To the Anglo-Saxons it was the 'Buccan Horn'. Indeed, the prominent stars in Capricornus form the distinctive pattern of a horn. To the Chinese however, these stars represented an Ox, which is the 2nd sign in their zodiac.

According to the books of Babylonian astrologer Sargon c.3850 B.C., translated into Greek by Berossos about 260 B.C., the world would be destroyed by a great conflagration when all the planets met in this sign. This grand conjunction would not be visible because it would occur in daylight when the sun (classed as a planet by the ancients) was in Capricornus along with the new moon. I haven't worked out when this will occur but you are welcome to have a go. The Chinese recorded a conjunction of the five planets Mercury, Venus, Mars, Jupiter and Saturn in Capricornus in 2449 B.C.

The Brightest Stars in Capricornus

Deneb Algedi (Delta Capricorni) is, at magnitude 2.9, the brightest star in the constellation. Its name means 'the goat's tail'.

Dabih (Beta Capricorni) along with Giedi (Alpha) has the Arabic title of the 'lucky one of the slaughterers'. This is derived from the sacrifice made celebrating the heliacal rising of Capricornus. Dabih is a blue-white star of magnitude 3.1.

Giedi (Alpha Capricorni) means 'goat' or 'ibex'. Giedi is, seen with the unaided eye, a double star, Prima and Secunda Giedi. The two stars, which are physically unrelated, are both yellowish in colour and have magnitudes of 3.6 and 4.2.

Nashira (Gamma Capricorni) is a white star of magnitude 3.7. Its name means the 'Fortunate One'.

Alpha Capricornids Meteors
Each year between July 3rd and August 25th meteors are observed to radiate from a point close to Giedi. The meteor shower peaks on July 30th when up to 8 meteors per hour may be observed.

33. AQUARIUS (Kumbha) & PISCES AUSTRINUS

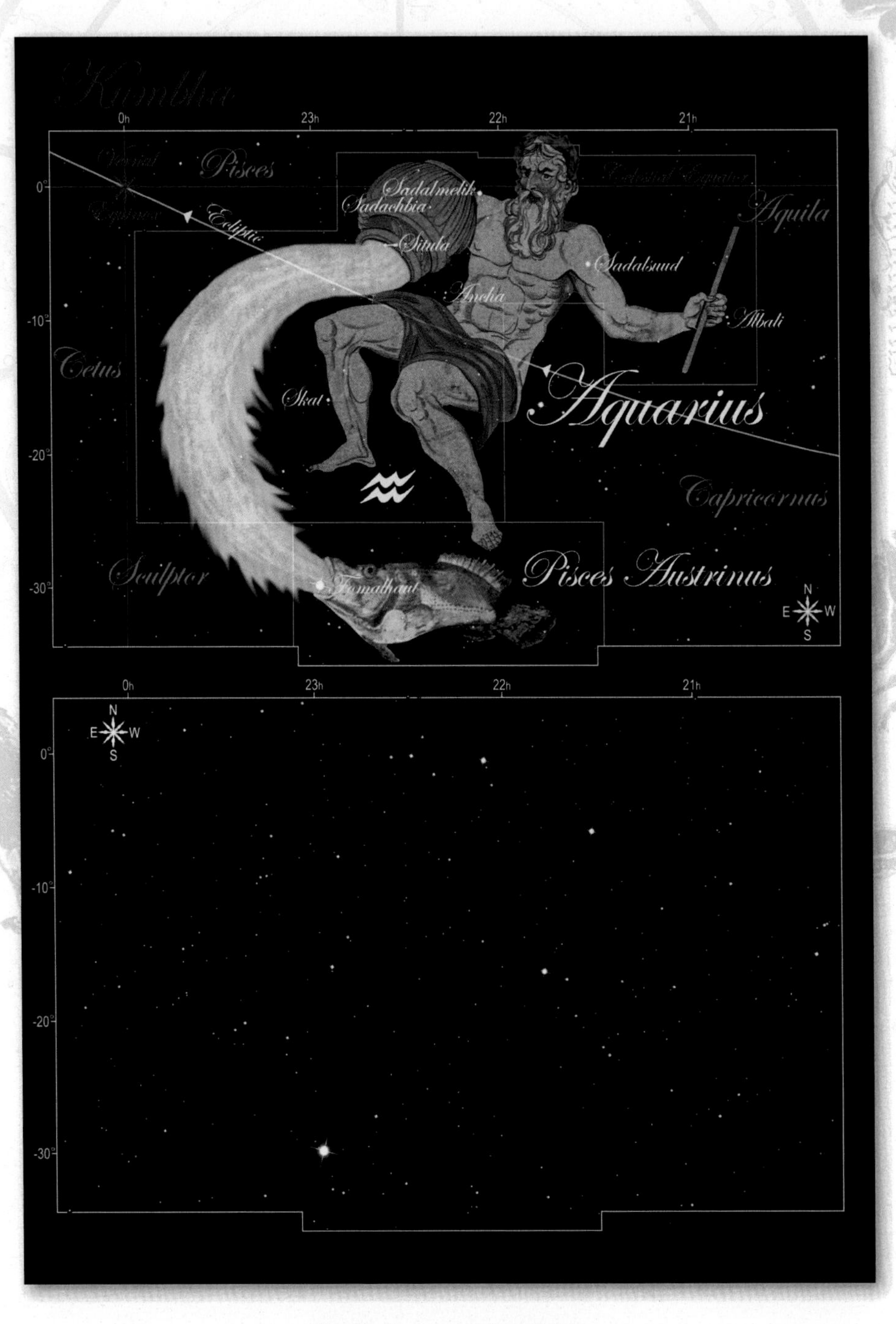

Fig. 48: *Aquarius star chart.* R. Hall

The sun his locks beneath Aquarius tempers,
And now the nights draw near to half the day,
What time the hoar frost copies on the ground
The outward semblance of her sister white,
But little lasts the temper of her pen.

Longfellow's translation of Dante's Inferno (c.1850)

AQUARIUS – The Water Carrier **KUMBHA – The Pitcher** (Rashis)

Solar Conjunction

Astronomical	February 17th to March 11th
Rashis (sidereal)	February 13th to March 14th
Tropical Calendar	January 21st to February 19th

Opposition: September 1st **House** of Saturn **Element:** Air

In ancient times the heliacal rising of Aquarius was a portent of the coming of the rainy season. The Greeks and Romans identified Aquarius as a male figure pouring or bearing water. This figure was the creator Zeus, Jove pouring life-giving water down on the Earth, or Ganymede bearing the libations of the gods. The Romans also saw Aquarius as the source of the celestial river Eridanus that flows from Orion to Achernar.

Aquarius controlled the cosmic 'Sea' of watery signs including Capricornus, Cetus, Delphinus, Pisces and Pisces Austrinus. The Sun passed through or close to these signs during the rainy season which occurred during the winter. Aquarius itself housed the winter solstice from 4000 B.C. to 1850 B.C. Arabian spiritual beliefs forbade their astronomers depicting the living form of the water carrier, so they figured it as 'Al Dalw', the Well-bucket. Arabic names for many of the stars in Aquarius and neighboring watery signs refer to the good luck and bountiful harvest that these rains bring.

...a sign of no small note, since there was no disputing that its stars possessed influence, virtue, and efficacy, whereby they altered the air and seasons in a wonderful, strange, and secret manner.

Longfellow on Dante

The Babylonians knew Aquarius as 'Gu', the water jar overflowing. They associated their story of the Great Flood to the heliacal rising of Aquarius at the winter solstice. This would have been about 5,500 B.C. (See chapter 22, Pisces for additional information on the Great Flood). In the Hebrew zodiac Aquarius represents the tribe of Reuben.

The Akkadians called Aquarius 'Lord of Canals', a title believed by some to be associated with the system of irrigation canals known to have been built along the banks of the Nile. It has been suggested that this dates back to the time when the heliacal rising of Aquarius coincided with the inundation of the Nile. This can't be so because this would place it at about 25,000 B.C. or, if we are talking about the peak of the floods, 10,000 B.C. It is more than probable that similar irrigation systems were built along the Tigris and Euphrates and that the symbolism of Lord of Canals refers to the floodwaters of these rivers in the rainy season when Aquarius marked the winter solstice, about 3,000 B.C.

The Egyptians knew Aquarius as 'Monius'. According to the American astronomer Serviss "The ancient Egyptians imagined that the setting of Aquarius caused the rising of the Nile, as he sank his huge urn in the river to fill it." This too would have a great antiquity, about 19,000 B.C. if it marked the beginning of the inundation, 8,000 B.C. if it refers to the peak of the flood.

In Greek mythology Aquarius represents Ganymede, a shepherd boy carried off by Zeus to Mount Olympus, where he became wine-waiter to the gods. In some Roman Zodiacs it was a Peacock, the symbol of Juno, the Greek Hera, in whose month Gamelion (January/February) the sun was in this sign. It was 'Junonis astrum', the domicile of Juno and Jove. In ancient Europe and western Asia Aquarius ruled over Cilicia and Tyre.

In China, the Emperor Tchuen-Hio, 2510 to 2431 B.C. decreed that the year would begin with the conjunction of the sun and moon close to the star Siou Hiu (in Aquarius). At this time it would have been close to the mid-winter solstice.

The Chinese knew Aquarius as Hiuen Ying, the Dark Warrior. It was a symbol of the Emperor Tehoun Hin, in whose reign there was a great deluge. After the arrival of Jesuit missionaries it became known as Paou Ping, the Precious Vase. In the traditional Chinese Zodiac based upon the 12 year cycle of Jupiter these stars represent The Rat, the Leader of the Zodiac.

Aquarius is currently the 12th sign of the Zodiac but in A.D. 2450, when the vernal equinox moves into this sign, it will become the 1st. This will be the 'Dawning of the Age of Aquarius' (See chapter 8, The Pillars of Heaven).

The Brightest Stars in Aquarius

The stars of Aquarius, although not particularly bright, are noticeable because many of them form pairs or triplets in lines. With a bit of imagination they look like a celestial water spray or rain.

Sadalmelik (Alpha Aquarii) means 'the Lucky One of the King (or Kingdom)'. It is, at magnitude 2.9, the brightest star in Aquarius. The star has a yellow tint.

Sadalsuud (Beta Aquarii) means the 'Luckiest of the Lucky Stars'. It is so called, it is believed, from a time when its heliacal rising coincided with the onset of the season of gentle, continuous rains after winter had passed. Sadalsuud is a yellow star of magnitude 2.9.

Sadachbia (Gamma Aquarii) is a white star of magnitude 3.8. Its name means 'Lucky Star of Hidden Things'. The meaning behind this name is believed to be referring to its heliacal rising that signaled the reappearance of worms and reptiles that had been buried during the preceding cold of winter.

The Aquarid Meteors

Aquarius is rich in seasonal meteor showers. Each shower is named after the closest star where the meteors appear to radiate from.

Eta Aquarids These meteors appear to emerge from a point close to Al Bali, Eta Aquarii. The shower, which is active from April 19th to May 28th, peaks on May 3rd when up to 50 meteors per hour may be observed.

Delta Aquarids South is active from July 8th to August 19th. They peak on July 29th when up to 20 meteors per hour may be observed.

Delta Aquarids North is active from July 15th to August 25th. They peak on August 12th when up to 5 meteors per hour may be observed.

Iota Aquarids peak on August 6th producing up to 8 meteors per hour.

PISCES AUSTRINUS – The Southern Fish

I have included this small ancient constellation with Aquarius because on occasion it formed part of it. In ancient times it was part of the astrological character of Saturn, a property shared with Aquarius and Capricornus. According to Virgil (c.50 B.C.), it marked the time for gathering the honey harvest. It is the fish that drinks the water that flows from the Urn of Aquarius and was also known as **Pisces aquosus**. It was also said to be the parent of the two zodiacal fishes, Pisces. Although small in size the constellation is prominent in the sky because it contains the first magnitude star Fomalhaut, which outshines by far all of the stars in Aquarius.

Fomalhaut (Alpha Pisces Austrini) is a bright white star, magnitude 1.17, located in an otherwise relatively barren region of the sky. It is one of the top twenty, being the 18th brightest star in the entire sky. Its name comes from the Arabic *Fum al Hut*, the 'fish's mouth'.

The Persian's knew it as Hastorang, one of the Four Royal Stars, the Guardians of Heaven which, 5,000 years ago marked the solstices and equinoxes. These stars were spaced approximately 90 degrees apart on or close to the path of the sun. Three of the signs, Taurus, Leo and Scorpius each have a first magnitude Royal Star that marked a solstice or equinox. But Aquarius, which held the winter solstice, has no bright stars so they appointed Fomalhaut which is well south of the Zodiac but at the correct celestial longitude (right ascension) to mark the solstice.

The star was worshiped at sunrise in the Temple of Demeter at Eleusis around 500 B.C. To ancient astrologers it was a portent of eminence, fortune, and power.

The **Pisces Austrinids** is an annual meteor shower that is active from July 9th to August 17th. It peaks on July 29th when up to 8 meteors per hour may be observed radiating from a point near Fomalhaut.

Printed in the United States
By Bookmasters

21st CENTURY ASTRONOMY

Angelo Pettolino

Library of Congress Control Number: 2015915537

ISBN:	Softcover	978-1-5144-4072-8
	Hardcover	978-1-5144-4071-1
	EBook	978-1-5144-4073-5

Print information available on the last page

Rev. date: 12/07/2016

To order additional copies of this book, contact:
Xlibris
1-800-455-039
www.xlibris.com.au
Orders@Xlibris.com.au

THIS BOOK IS DEDICATED TO THE BRAVE ASTRONAUTS WHO ARE THE ONLY PEOPLE TO HAVE WALKED ON THE MOON:

NEIL ARMSTRONG, BUZZ ALDREN, PETE CONRAD, ALAN BEAN, CHARLES M. DUKE, HARRISON "JACK" SCHMITT, PETE CONRAD, EUGENE A. CERNAN, ALAN BEAN, ALAN SHEPARD, EDGAR D. MITCHEL, DAVID RANDOLPH SCOTT, JAMES B. ERWIN, JOHN WATTS YOUNG.

THE MILKY WAY GALAXY

Our Sun is located 30,000 light years from the centre of the Milky Way Galaxy . It's meshed in the spiral arms of Sagittarius in the Orion Spur and contains 10,000 nebulae the zodiac is one of 88 constellations (Plate 2461). Our Sun is 90% brighter than the 200 billion stars comprising our galaxy (7966). The Sun is 1.4 million miles in diameter, is surrounded by Hydrogen which it burns as fuel while exhausting Helium to form the heliosphere. Its (solar) system of planets, asteroids, and comets also formed about 5 billion years ago. It takes the intergalactic winds about 700 million years for our sun to make one orbit around our galaxy. The Milky Way galaxy is spiral and one of the "Poor Clusters" of the 100 billion galaxies in the universe. It belongs to a local group of 35 or 40 galaxies in the Virgo super cluster. The galaxy is predicted to be surrounded by universal gas atoms which are lighter in atomic weight and smaller in size than the galactic hydrogen gas atoms. The Milky Way is 1,000 light years thick and made up of high inclination , halo type, inner stars and rotating outer stars. The outer stars span 100,000 light years (600 quadrillion miles) in diameter, with masses from the 11th power greater than the Sun. The value of one light year is equal to the distance light travels in one year or 5,878,000,000,000 miles. These distances were determined from the spectral lines of Doppler shifts in the 21-centimetre radio emission line (Plate 4668).

Plate 2461

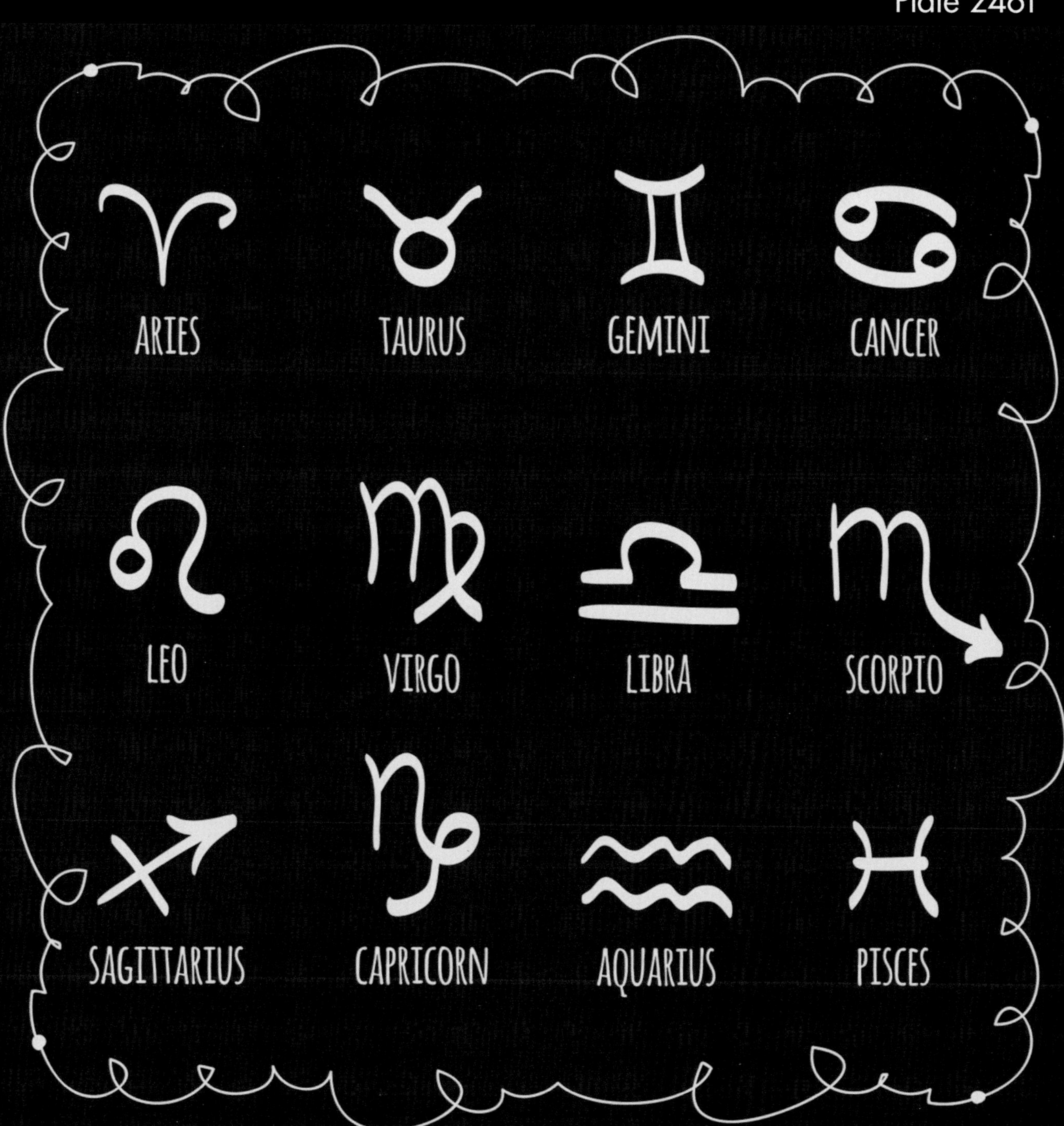
ARIES
TAURUS
GEMINI
CANCER
LEO
VIRGO
LIBRA
SCORPIO
SAGITTARIUS
CAPRICORN
AQUARIUS
PISCES

Plate 7966

Plate 4668

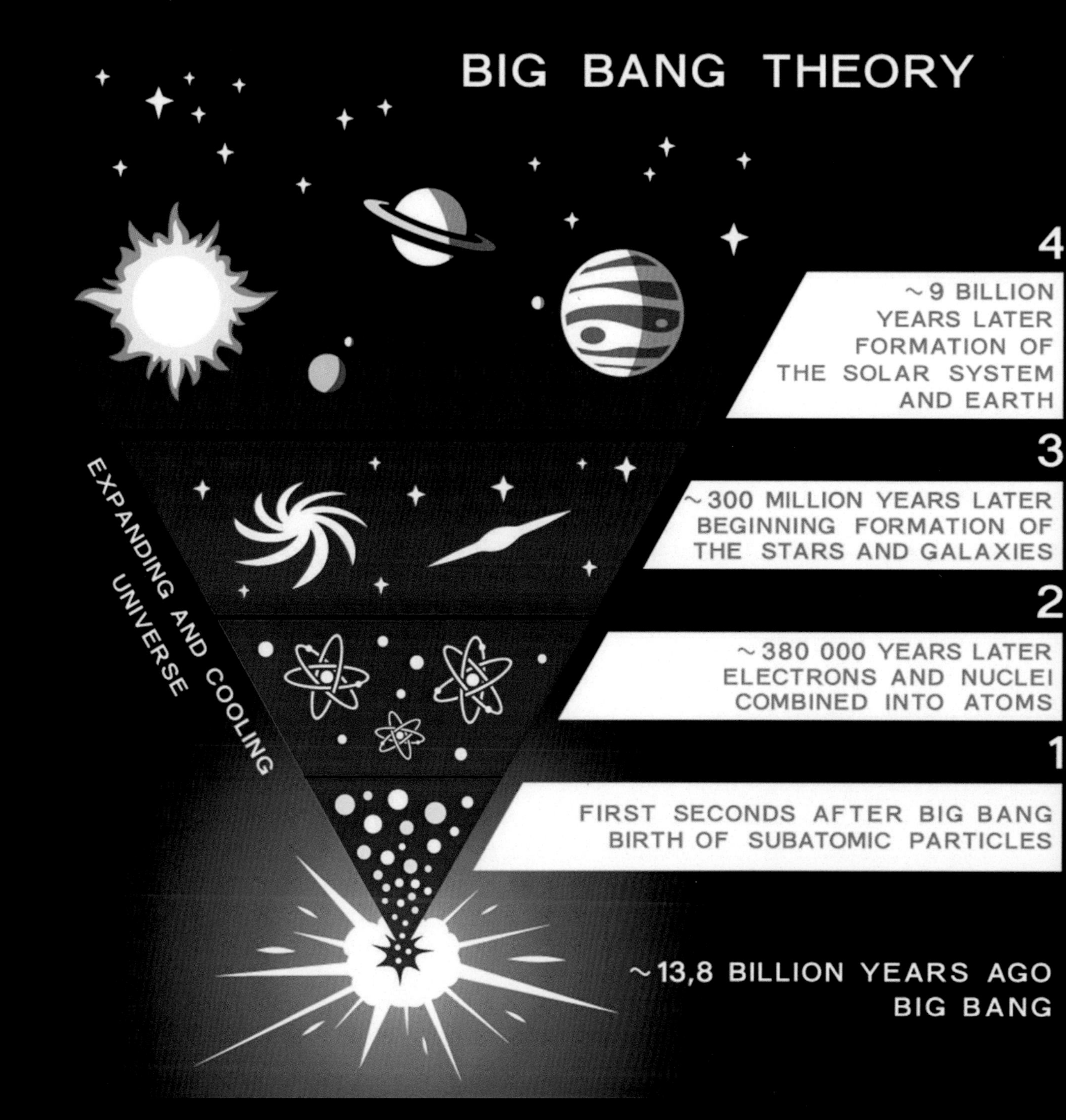
BIG BANG THEORY
4
~9 BILLION YEARS LATER FORMATION OF THE SOLAR SYSTEM AND EARTH
3
~300 MILLION YEARS LATER BEGINNING FORMATION OF THE STARS AND GALAXIES
2
~380 000 YEARS LATER ELECTRONS AND NUCLEI COMBINED INTO ATOMS
1
FIRST SECONDS AFTER BIG BANG BIRTH OF SUBATOMIC PARTICLES
EXPANDING AND COOLING UNIVERSE
~13,8 BILLION YEARS AGO BIG BANG

WATER - H_2O

Chemical Reactions

$$2H_2 + O_2 \longrightarrow 2H_2O$$

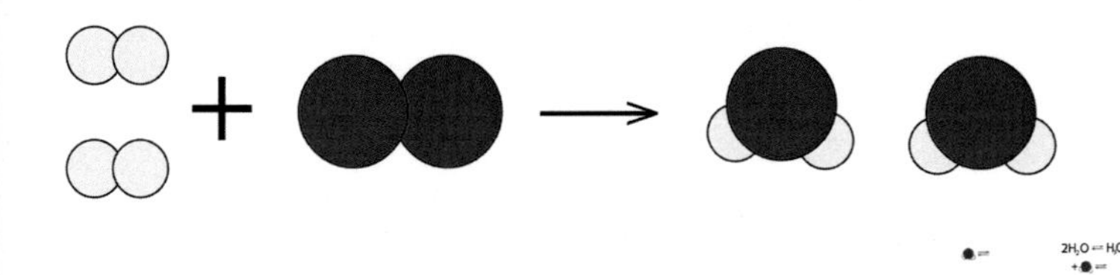

2 hydrogen molekules + 1oxygen molekule yelds 2 water molekules

2 ×(2.02 amu)	+ 32,00 amu	yelds	2 ×(18,02 amu)
4,04 amu	+ 32,00 amu	yelds	36,04 amu

36,04 amu reactans

Water Molecule

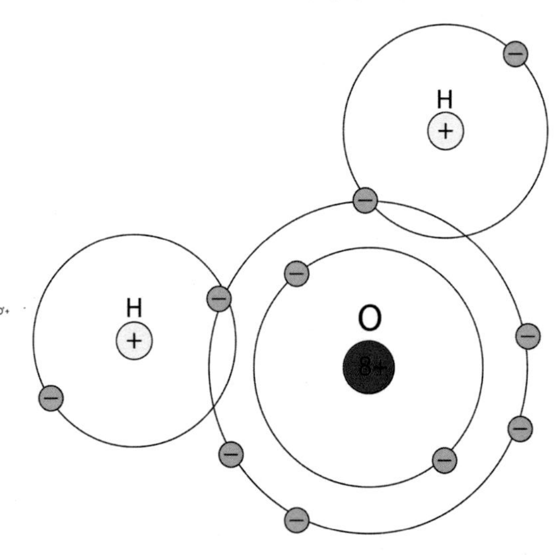

4 step formation of a molecule of water

1\.

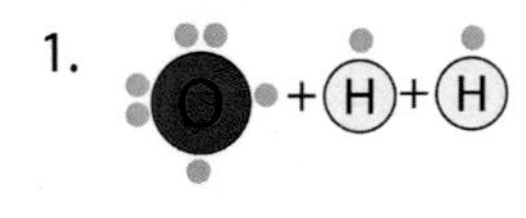

2\.

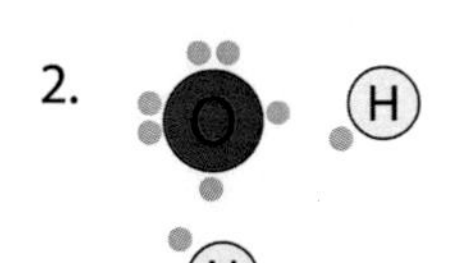

3\.

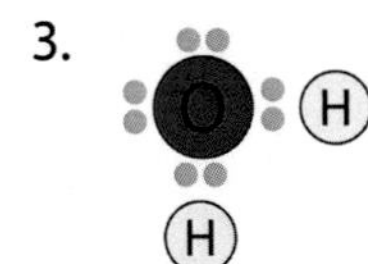

4\.

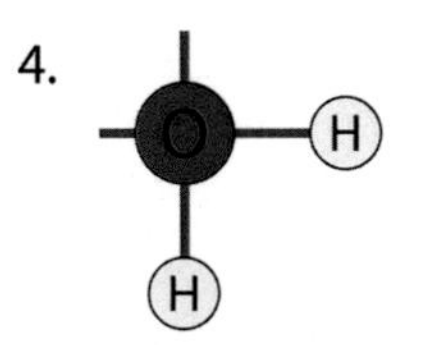

Hydrogen peroxide is thermodynamically unstable and decomposes to form water and oxygen

$$2H_2O_2 \longrightarrow 2H_2O + O_2\uparrow$$

Under the influence of the very high temperatures or electric current water decomposes into molecular oxygen and molecular hydrogen:

$$t\uparrow \quad 2H_2O \longrightarrow 2H_2\uparrow + O_2\uparrow$$

PREFACE

Science can be defined as an ever evolving , never-ending, and relentless pursuit of the best-fit model of reality by calculable and repeatable means. New discoveries lead to mankind's greatest achievements. They have also been responsible for some of our world's most infamous feats as well. This progress is as human as it is to err. It is also human to be persuaded by some over years of misinformation and "deliberate untruths" without much resistance in order to 'own the game' and maintain a sense of superiority over educators, known scientists, and the communities that support them. However, we, as scientists and pursuers of truth, must assume the unpopular role of the sceptic. We must find tangible proof that what is being taught and taken for truth is indeed so. Never should man stop defying the norm to seek the provable truth. New discoveries soon teach that what's true today may be fiction tomorrow. Time has a way of telling all truths.

Some ideas are more popular than others and take precedence over fact and modern knowledge like scientific myths. This happens today as well as it happened hundreds of years ago. For example, a well-known fact such as the Earth being round rather than flat was once a concept punishable by law. Though the enforcement of these false ideals is less barbaric today and refusal to consider alternative theories as solutions based on fact is. Science is math and math is a process of deduction.

Modern scientists have claimed to have unlocked the secrets of the cosmos but have little more than circumstantial evidence and a snapshot of the entire timeline which our existence occupies. It is the beginning of our existence, the beginning of our solar system, galaxy, and universe, which has remained mankind's greatest mystery. Through science's attempts to reveal our origins, one theory spun into another and another, while new ones were created to fit what scientists wanted to make work. An alleged solution known as the 'cosmological constant' was once described by A. Einstein, an infamous scientist as a 'repulsive force' and 'deliberate untruth' and was used as the basis of his own world-renowned flawed theories of Time and Space, Relativity and Special Relativity (gravity isn't everywhere in the cosmos) and his equally unsuccessful formulas e.g. E=mc2. The speed of light "c" cannot be squared because it's the fastest speed that there is and there was no matter (m) because in the beginning there was nothing.

Scientists and laymen alike want truth and seek it on every level. We demand it, but know we often get a slanted version of it. Yet what if someone tells you that what you think you know about the fundamentals of our universe, our galaxy and even our own solar system is wrong? Will you want to believe it? What if the elegant and convenient deliberate untruths were exposed with basic science? Will you change your mind? What if by watching one programme, learning to see from another perspective, the things you know about the solar system and how it was formed will go up in smoke? Will you still want to learn? Here is your chance to decide. It is your chance to rediscover and demand the truth about the beginnings of our mysterious solar system and how its water formed. It is your turn to ask what if, why, and how and our turn to answer in an impartial manner. A true scientist is not concerned with who's right, just what's right.

FOREWORD

Dear colleague: The first second of the formation of our Universe when it went from nothing to something is too deep for us to comprehend. The formation of our solar system is a different case. Since the turn of the last century (1915) there has been a struggle between the Sir Isaac Newton scientists and the Albert Einstein scientists. Newton claimed anything with mass can be attracted by gravity and that there was no gravity in the cosmos. Einstein claimed there was gravity in the cosmos and it connected everything. Newton has been found to be correct while Einstein's work has been discredited. The Accretion theory and all its misconceived variations like Relativity and Special Relativity, positing that gas and dust gravitationally coalesced and became our solar system, is having serious difficulties proving its basic concepts and are flagrantly violating the laws of physics by trying to explain the improvable. Physics tells us that there is no gravity in a vacuum. We now know that although gas has mass it also has an escape velocity greater than Earth's gravitational attraction. Any dictionary or encyclopaedia will define gas as a substance that expands indefinitely until it reaches its natural ceiling of lighter gas. Our atmosphere's natural ceiling is the Helium filled Heliosphere. The heliosphere which completely surrounds the solar system and Earth's magnetic field protects the Earth from radiation. The cosmos is a vacuum and there is no gravity in a vacuum, just inertia. In 1978 NASA's Gravity Probe A confirmed that clocks move faster in orbit than on Earth's surface. This was all finally confirmed by the Laser Interferometer Gravitational-Wave Detector. The GWD Observatory which went into operation in 2002 and collected data has operated for 10 years without detecting any gravitational waves. Now that there is proof that the LIGO actually works the engineers are trying to upgrade it to an Advanced LIGO to try to accomplish their goal of trying to prove Relativity works. In 2015 the European Space Agency launched the LISA Pathfinder to find low-frequency gradational waves. None have yet to be detected. So far no gravity has been found in the cosmos as Sir Isaac Newton predicted.

Accretion is a failed and rehashed theory of the Cartesian hypothesis, which was conceived by René Descartes (1644) ('I think therefore I am') and was successively mimicked by Swedenborg, Kant, Laplace, Chamberlin and Moulton, Carl von Weizsäcker, Whipple, Kuiper, Safronoff, and the astronomy establishment. The 'new' rewrite of the same old theory was named the Nebula hypothesis and was finally uncritically and blindly accepted by the International Astronomical Union and the Royal Astronomical Society for unknown reasons. The same unanswerable questions and improbable scenarios remain as they have for the past 350 years. The absence of explanations or logical answers makes their illogical reasoning open to questions and implies that a more fundamental theory exists.

The best-fit model of reality and the most logical formation explanation is namely 'thermal reaction", the process and results of freezing and thawing gases under extreme conditions'. This monistic concept of the formation of our solar system was first introduced by George Gamow and then Alpher, Herman, Sir Oliver Lodge, Sir James Jeans, Fred Hoyle, Hermann Bondi, Thomas Gold, and Hannes Alfven. They all believed the solar system started from a series of galvanic explosions from within our newly forming photostar 5 billion years ago and was termed monistic. The AP Theory is the most current monistic theory that was put foreword.

Here are some of the unanswered accretion questions:

1. The laws of physics clearly state gas has a built in escape velocity that is stronger than gravity's attraction, is molecularly structured to electrical attraction and expands indefinitely. Q: Why are there no successful experiments to prove that gas can be attracted by gravity or condensed into solid particles anywhere in nature or anywhere on Earth? A: Because none exist.

2. Why weren't the 'rotating discs' thrown out into space instead of accreting?

3. How could Venus's and the Moon's surfaces have been formed all at the same time 4.5 billion years ago?
4. Why do Venus, Uranus, and one-third of the moons in the solar system spin in a direction opposite to the others?
5. What useful information is available on the specific solid-state processes at work in the accretion phase?
6. How was the planetary matter separated from the solar gases?
7. How did the collapsing cloud cross the thermal, angular, and momentum barriers?
8. What were the mechanics of condensation and coalescence?
9. Where did our water and atmosphere come from?
10. Where did the gravity originate in the first place to start the accretion process?
11. Why hasn't the Kuiper belt, asteroid belt, or Saturn's' rings coalesced into a planet after 4 billion years when it's claimed the entire solar system formed in less than 500,000 years?
12. Why couldn't Uranus, Neptune and beyond be coalesced or formed in any way by using the latest computer simulations for accretion?
13. Explain how gravity selects and keeps in suspension, Oxygen and Nitrogen only and attracts none of the other heavy gases like Neon, Fluorine, Radon and Chlorine?
14. Can an example of where (on Earth) Earth's gravity is attracting and holding gas be given?

These questions and many other quantum gravity and Accretion issues have never successfully been proven or addressed. The main reason was that it could never be proven that gravity could attract, hold and compress gas in nature and still can't. Isaac Newton was the first to say "There is no gravity in the cosmos". Information gathered from our 9 journeys to the moon proved Newton to be right and Einstein's Relativity Theory to be wrong.

Jupiter
Saturn
Uranus
Neptune
Earth
Pluto

1
Mercury < Mars < Venus < Earth
3
Jupiter < Wolf 359 < Sun < Sirius
5
Aldebaran < Rigel < Antares < Betelgeuse

CONTENTS

②
Earth < Neptune < Uranus < Saturn < Jupiter
④
Sirius < Pollux < Arcturus < Aldebaran
⑥
Betelgeuse < Mu Cephei < VV Cephei A < VY Canis Majoris

Plate 9533

INTRODUCTION

Our solar system is 3 light years across and is comprised of 0.15 per cent of the Sun's total mass including planets, asteroids, comets, and minor planets (Plate 9533). The Sun formed the solar system 300 million years before it matured into a hot yellow dwarf star. Our solar system is contained in the Oort cloud which extends 150 billion miles out from the Sun. It is also referred to as the Heliosphere most recently discovered in 1952 by Jon Oort a Dutch astronomer. It has a 12 trillion mile diameter surrounding our solar system and is comprised of 90% Helium. The heliosphere is beyond the Oort Cloud and protects Earth from cosmic rays. The Interplanetary Medium, Interplanetary Space and the Oort cloud which reaches the termination shock and is one third the distance of Proximal Centauri or 4.3 light years from Earth all lie within the Heliosphere. These terms are interchangeable and all refer to the same Helium discovered by Pierre Jansen in 1886 and are defined from the galactic gases by a separating border 12.3 trillion miles (2 light years) in diameter, out from the Sun's Astrosphere's Heliosheath and is also referred to as the Heliopause which stops the solar wind.

The Sun rotates once every 25 days. Its gravity and the 1,800,000-mile-per-hour solar wind which extends 12 trillion miles out from the Sun to the edge of the solar system, propels our planets around the celestial plane of the Sun. With its explosive coronial mass ejections,(CME) caused by early galvanic solar eruptions have enough energy to spin the planets and keep them in their respective angular orbits above the celestial poles of the Sun. These explosive beginnings and the solar wind which showers our solar system with Helium were the energy sources responsible for the orbiting of the planets millions of miles above their 200,000-cubic-mile planetary exit point of the Sun (plate 1). All the planets are orbiting above the Sun's celestial pole energized by the solar wind. If the Sun were the face of a clock the position of the planetary and solar wind exit point would be in between 2 and 3 o'clock.

A great scientist, Henry Cavendish was the discoverer of Hydrogen. Hydrogen has one proton and one neutron with a positive charge. Hydrogen is highly reactive and is the

lightest of gases (0.0695 specific gravity). It owes its name to a scientist named Lavoisier, who combined the Greek hydro (water) and gene. Hydrogen has an atomic weight of 1 atomic mass unit and is the lightest gas with the smallest atom yet discovered. Although it is now believed even lighter gases with smaller atoms exist beyond the galaxy. Hydrogen is so light it cannot be held by Earth's gravity, which is why there is none in the heliosphere or our atmosphere. Hydrogen is only found in the atmosphere at trace levels; it is synthesized from hydrocarbons (petroleum and petroleum by-products) and from water, where it constitutes the lightest fraction of the H2O molecule. Hydrogen gas is colourless, highly flammable, cannot sustain life, and reacts easily with other chemical substances. When Hydrogen burns it exhausts Helium and when Helium fuses under pressure, it decomposes and becomes Oxygen the 5th most abundant element in our universe.

An atom of Hydrogen has a stronger escape velocity than Earth's gravitational attraction like all gases and is the smallest atom discovered so far. Hydrogen is the only atom that has one proton and one electron but no neutron. It and all the other atoms absorb their energy from light waves radiated out from the Sun's surface. The propelling force that holds the hydrogen electron in orbit has nothing to do with gravity. It's called electrical attraction (escape velocity) and is stronger than Earth's gravitational attraction according to the world-famous astrophysicist Neil deGrass Tyson. All gases are molecularly structured to expand indefinitely, occupying all available space. That's why this is called escape velocity. Gases expand until they reach their natural ceiling of gases which have no volume and are lighter in atomic weight with smaller atoms. The escape velocity of gas is what causes the vacuum, inertia and kinetic energy in the cosmos.

The escape velocity and size of Hydrogen makes it greater than the attraction of gravity. If gravity attracted gas, Hydrogen's small atom would be the first to be attracted through the heavier Oxygen and Nitrogen atoms. Hydrogen's location in the cosmos is further proof that gravity has no affect on gases.

The planets are not orbiting the Sun's equator at unrealistic speeds as taught by the now discredited Special Relativity and Relativity hypothesis. The solar wind is causing the planets to orbit above the Sun's celestial pole at slower speeds and over shorter distances than previously supposed. The solar wind energy is what keeps the planets orbiting above the celestial poles of the Sun (Bernoulli's Law). The planets are neither orbiting the Sun at its equator nor are they being propelled from the force of gravity. The solar wind propels the planets in their orbits above the celestial poles of the Sun, not around the equator and not propelled by gravity as was previously supposed.

The AP theory states that gravity is not holding our atmosphere to Earth and that the Helium filled Heliosphere is Earth's atmospheric ceiling. The theory takes us one step closer to the truth by predicting that universal gases (could this be dark matter?) with a predicted atomic weight of 0.989 or less will be discovered in-between galaxies and beyond the intergalactic medium of Hydrogen. These universal gases (Plate0597), which are light than Hydrogen in atomic weight and have lesser air pressure than the Hydrogen of the Galactic gases (Plate 4986), are compressing and preventing the galactic gas atoms, which are greater in air pressure and heavier in atomic weight, from expanding (Plate 0204). This in turn keeps the sun-produced interplanetary gases of Helium from expanding into the galaxy. These Heliospheric gases are comprised of mostly Helium and their atoms are larger than Hydrogen atoms. These Heliospheric gases of mostly Helium with 0.139 of specific gravity are keeping the interplanetary gases from rising into the galactic medium of Atomic Hydrogen. These interplanetary gases of Helium, which are low in atomic weight and have a lesser air pressure than Galactic gases (Plate 3489), encompass and compress our Earth's atmosphere, which contains 98% of Oxygen and Nitrogen gases that are heavy in atomic weight and have larger atoms than Hydrogen and Helium (Plate 3596). Thus these atmospheric gases cannot expand further into the Helium dominated Heliosphere. The simple explanation is that yet undiscovered universal gases with a low atomic weight are compressing Galactic gases of Hydrogen which compress the Heliospheric gases of Helium with the smaller in size and lighter in atomic weight atoms than our atmosphere. The Helium gas filled Heliosphere extends about 11.3 trillion miles out from the Sun giving it a much greater overall weight than our atmosphere. Helium is prevented from expanding because it has more atomic weight and decreased air pressure than Hydrogen. The Heliosphere's 11.3 trillion mile diameter gives the Helium mass and is exerting pressure that is completely surrounding Earth's atmospheric gases of Oxygen and Nitrogen with a compressing pressure of one ton per square foot. Gravity is NOT holding our atmosphere as we can now see it's the natural Helium gas ceiling of the Heliosphere which cannot be penetrated by Earth's atmosphere, hemispheric pressure of 14.7 ppsi., Earth's spin and its 0.3 degree tilt per day or 90 degree travel from North to South.

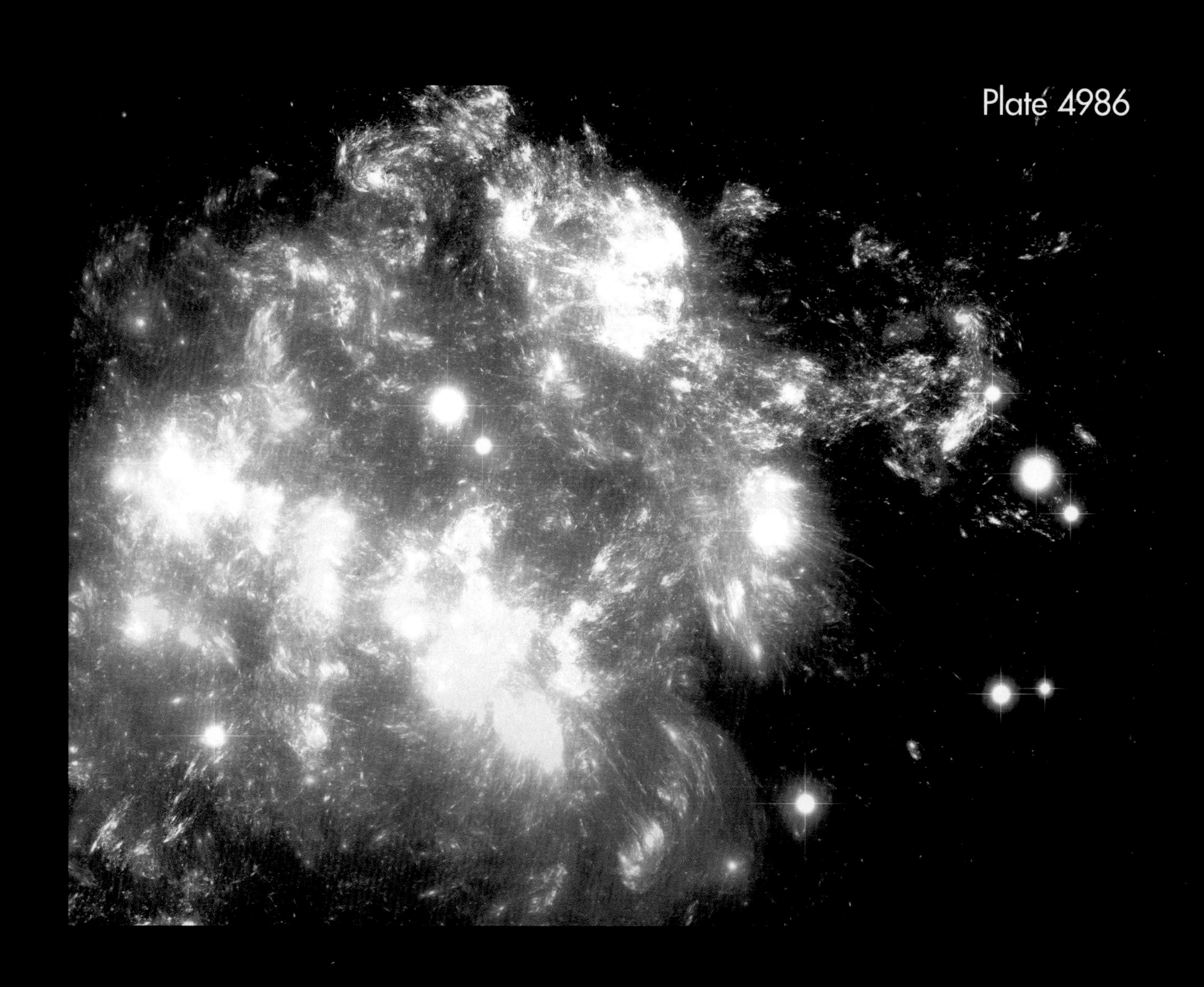

Plate 4986

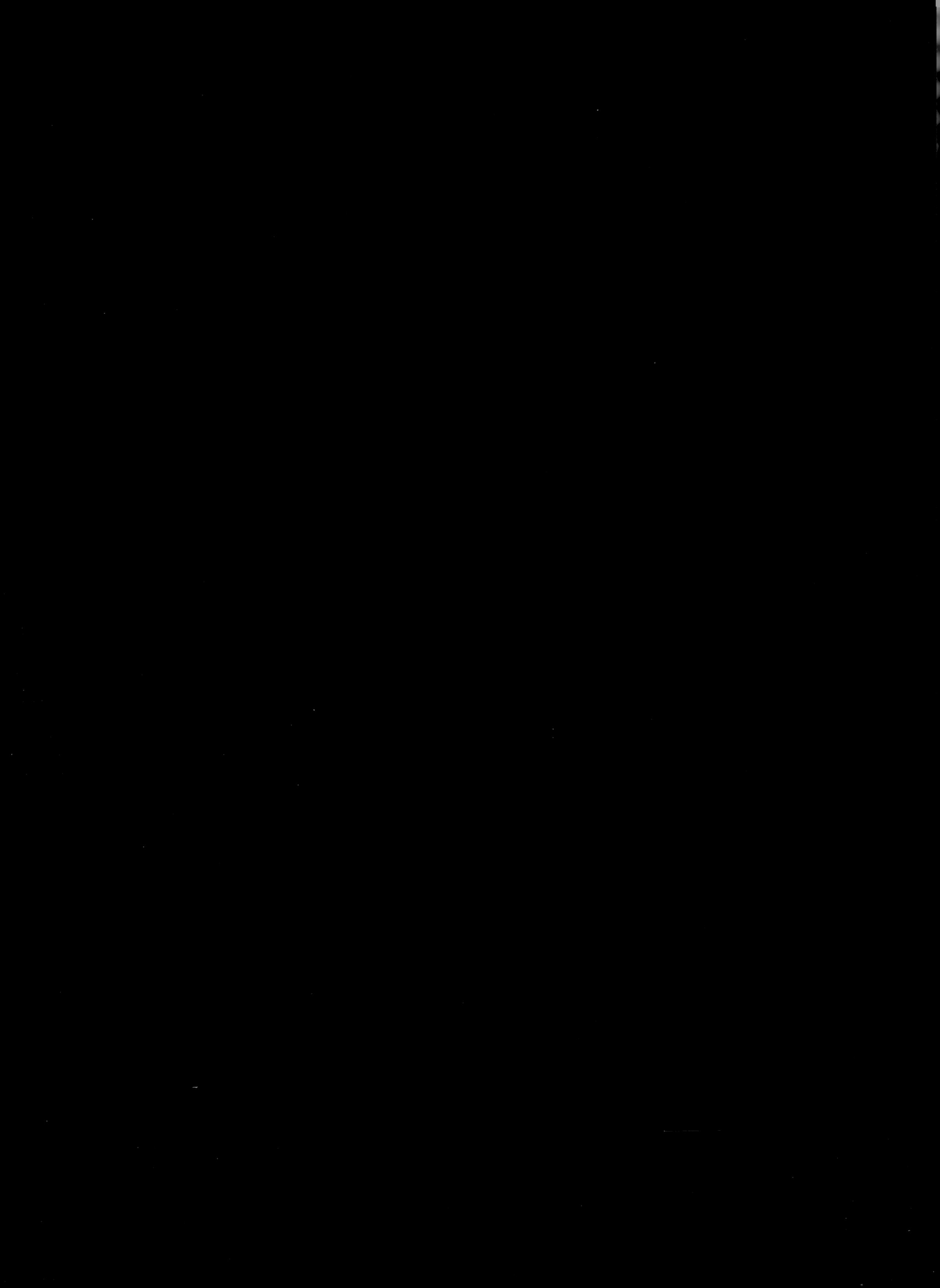

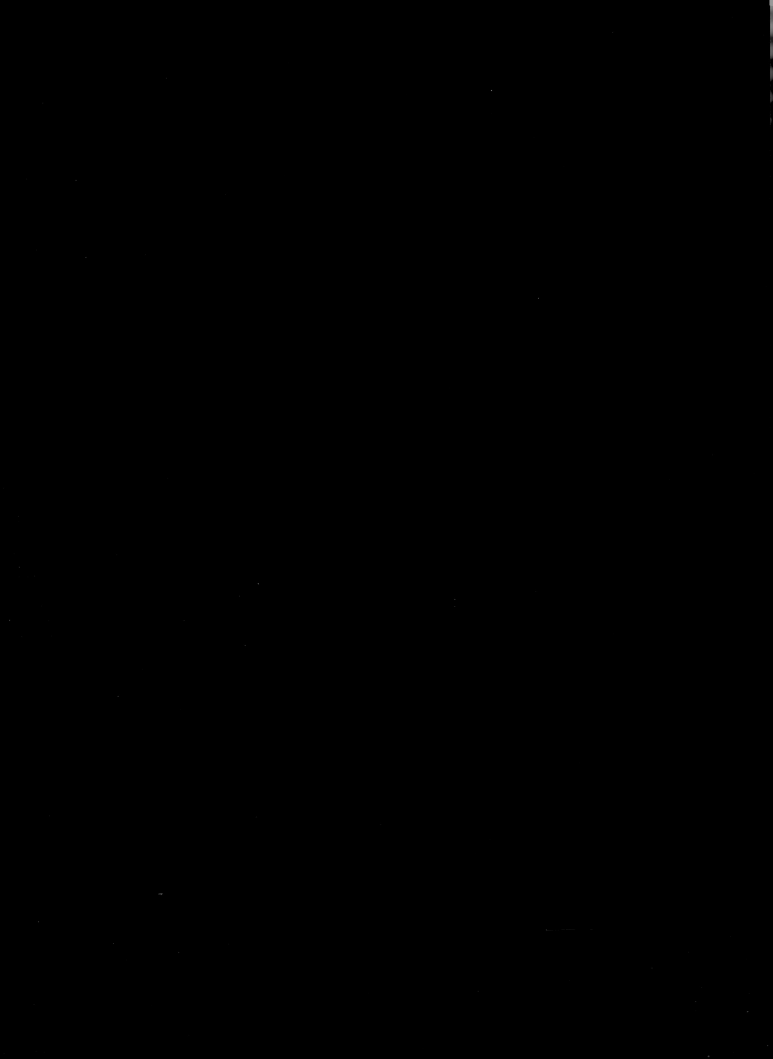

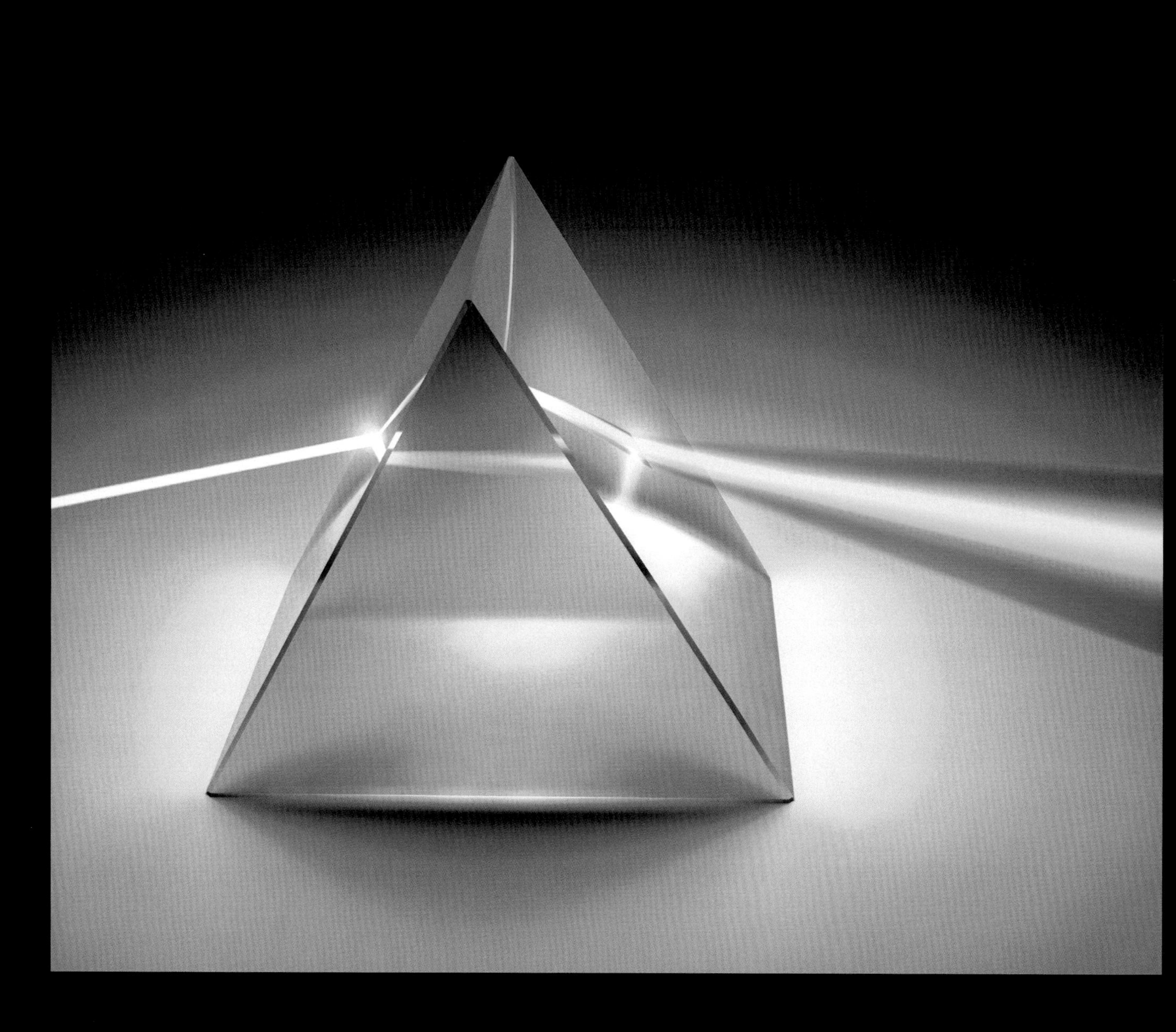

Plate 0597

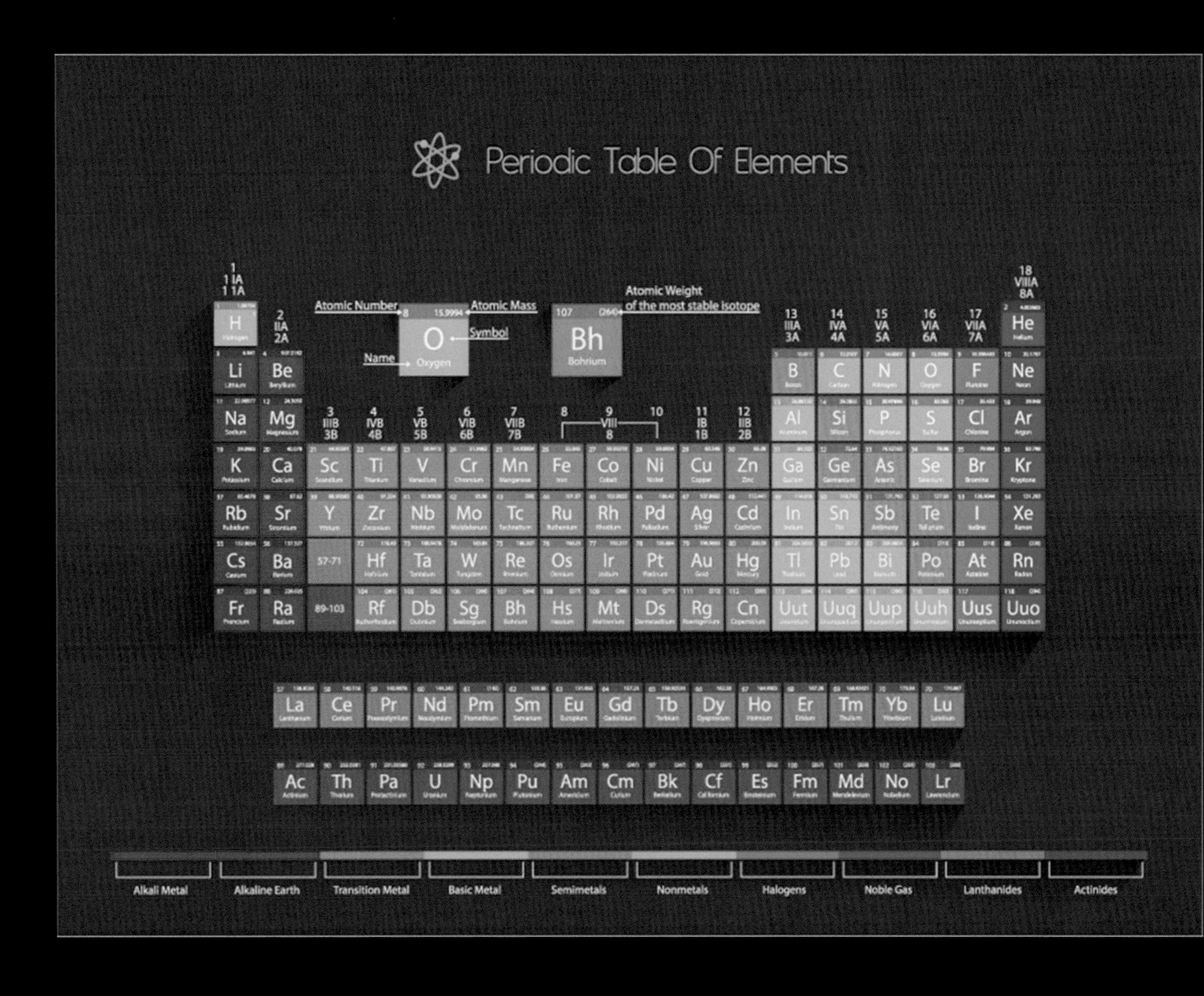
Periodic Table Of Elements
Atomic Number
Atomic Mass
Symbol
Name
Oxygen
Atomic Weight of the most stable isotope
Bohrium
H
He
Li Be B C N O F Ne
Na Mg Al Si P S Cl Ar
K Ca Sc Ti V Cr Mn Fe Co Ni Cu Zn Ga Ge As Se Br Kr
Rb Sr Y Zr Nb Mo Tc Ru Rh Pd Ag Cd In Sn Sb Te I Xe
Cs Ba 57-71 Hf Ta W Re Os Ir Pt Au Hg Tl Pb Bi Po At Rn
Fr Ra 89-103 Rf Db Sg Bh Hs Mt Ds Rg Cn Uut Uuq Uup Uuh Uus Uuo
La Ce Pr Nd Pm Sm Eu Gd Tb Dy Ho Er Tm Yb Lu
Ac Th Pa U Np Pu Am Cm Bk Cf Es Fm Md No Lr
Alkali Metal
Alkaline Earth
Transition Metal
Basic Metal
Semimetals
Nonmetals
Halogens
Noble Gas
Lanthanides
Actinides

Plate 9694

Low-pressure (1.45 pounds per square inch) interplanetary gases of Hydrogen (1 atomic mass unit), Helium (4 atomic mass units), etc. are compressing and keeping Earth's high atmospheric air pressure of 14.7 pounds per square inch (760mmHg) of oxygen and nitrogen down not gravity. Earth produced atmospheric gases of Oxygen and Nitrogen are kept from rising through the Sun produced Helium gas pockets of the Heliosphere. A cubic foot of compressed atmospheric gases weigh 1 ounce . When dissipated, they would cover a vast area and could only be measured by atomic weight. Interplanetary gases with an air pressure less than Earth's are keeping the Earth's atmospheric gases from rising through the smaller, lighter ceiling of Helium atoms contained within the Heliosphere. The atmosphere contains breathable air of 19% Oxygen and 79% Nitrogen. The Helium-filled Heliosphere completely surrounds it and exerts pressure from all sides at 14.7 lb. per sq. in. giving the Earth its gravity. Gases with close atomic numbers will mix, but they will never compound. All these gases are completely surrounded by galactic gases of Hydrogen which is completely surrounded by even lighter yet to be analyzed universal gases, each of which keeps the larger-atom oxygen and nitrogen atmospheric gases from expanding beyond their natural ceiling of Helium. Gravity is not "gas selective". It can't differentiate between one gas and another and so should attract all gases alike but it doesn't. In Earth's atmosphere gravity doesn't attract Hydrogen, Methane, Helium, Carbon Dioxide, Carbon Monoxide, Chlorine, Fluorine, Neon or any of the Noble gases because gravity does not attract gas. Why? Gases of similar atomic weights like Oxygen, discovered by Joseph Priestly and Nitrogen will mix together but not combine and originate from Earth. Hydrogen and Helium gases with great differences in atomic weight than Oxygen and Nitrogen will become a natural ceiling for the heavier gases. Oxygen (Phlogiston) and Hydrogen are a perfect example of gases not mixing. They will not mix together but will settle in layers with the lightest atomic weight Hydrogen expanding to the top of a column as is in nature. We're taught that 75% of the universe beyond the solar systems is Hydrogen and 25% is Helium. The AP Theory disagrees and predicts a gas lighter in atomic weight and smaller in size will be discovered beyond the galaxies. This is what will be described as "Universal Gas", dark energy, dark matter or gas X.

Plate A

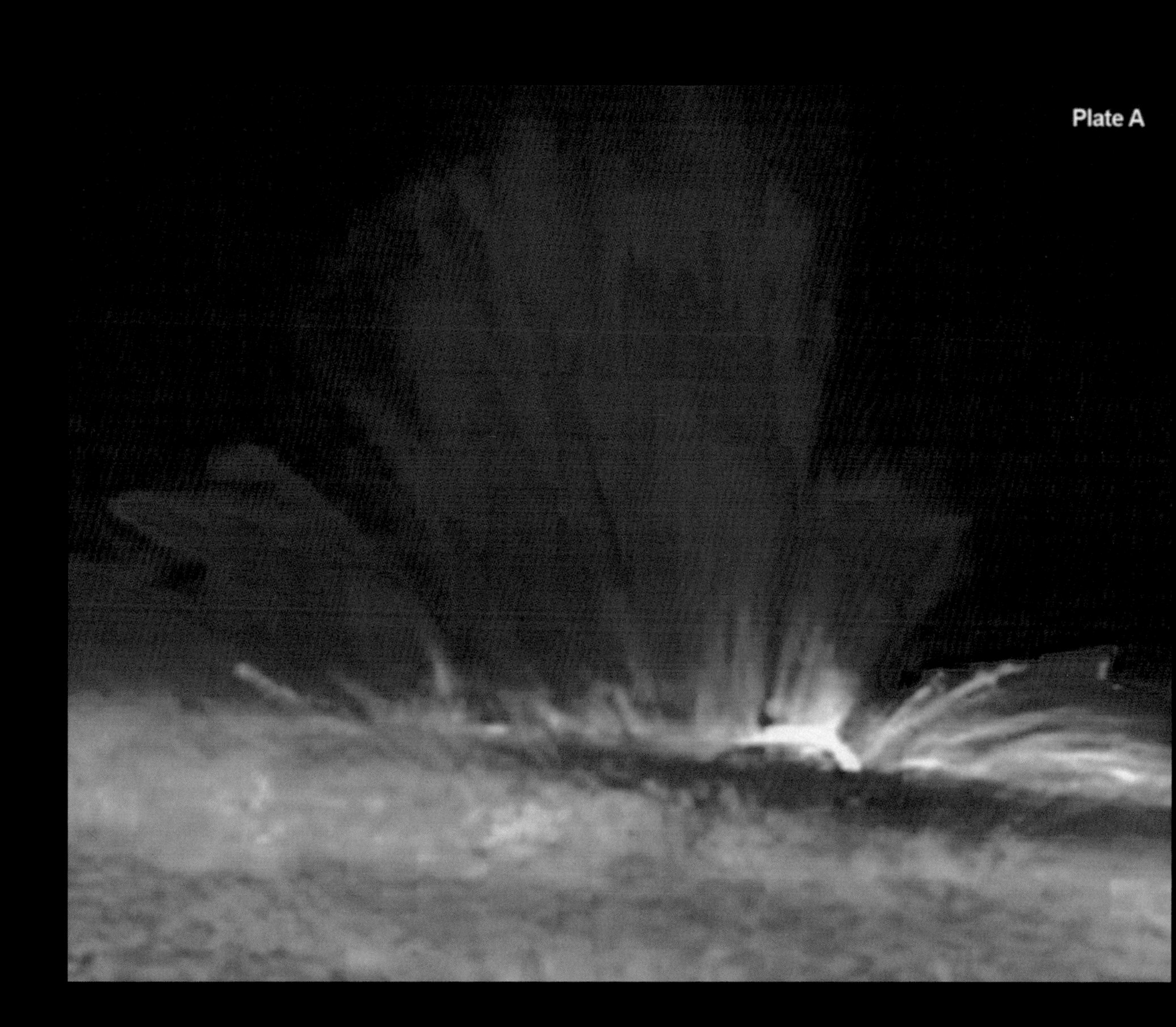

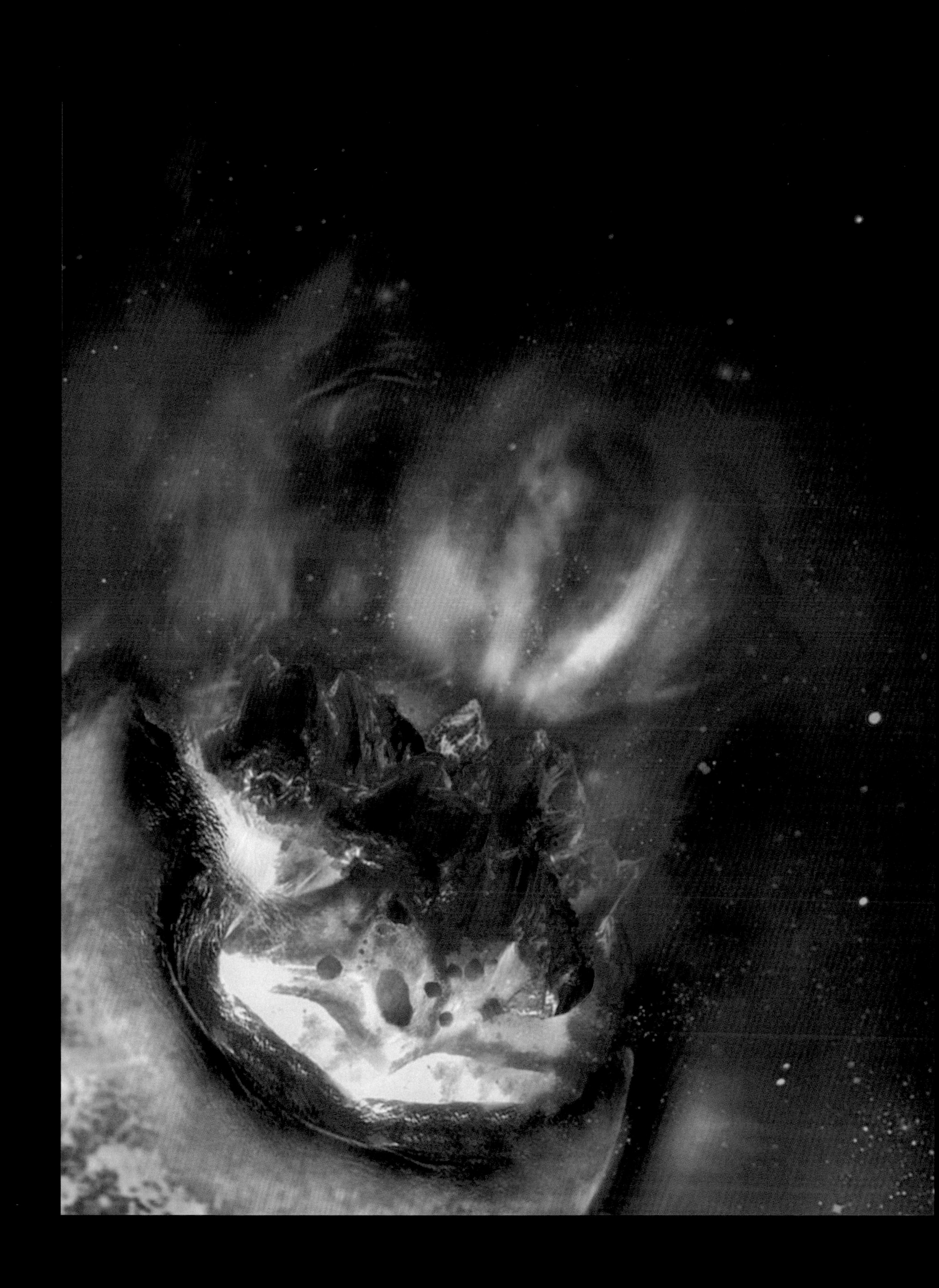

Plate 1A (side view).

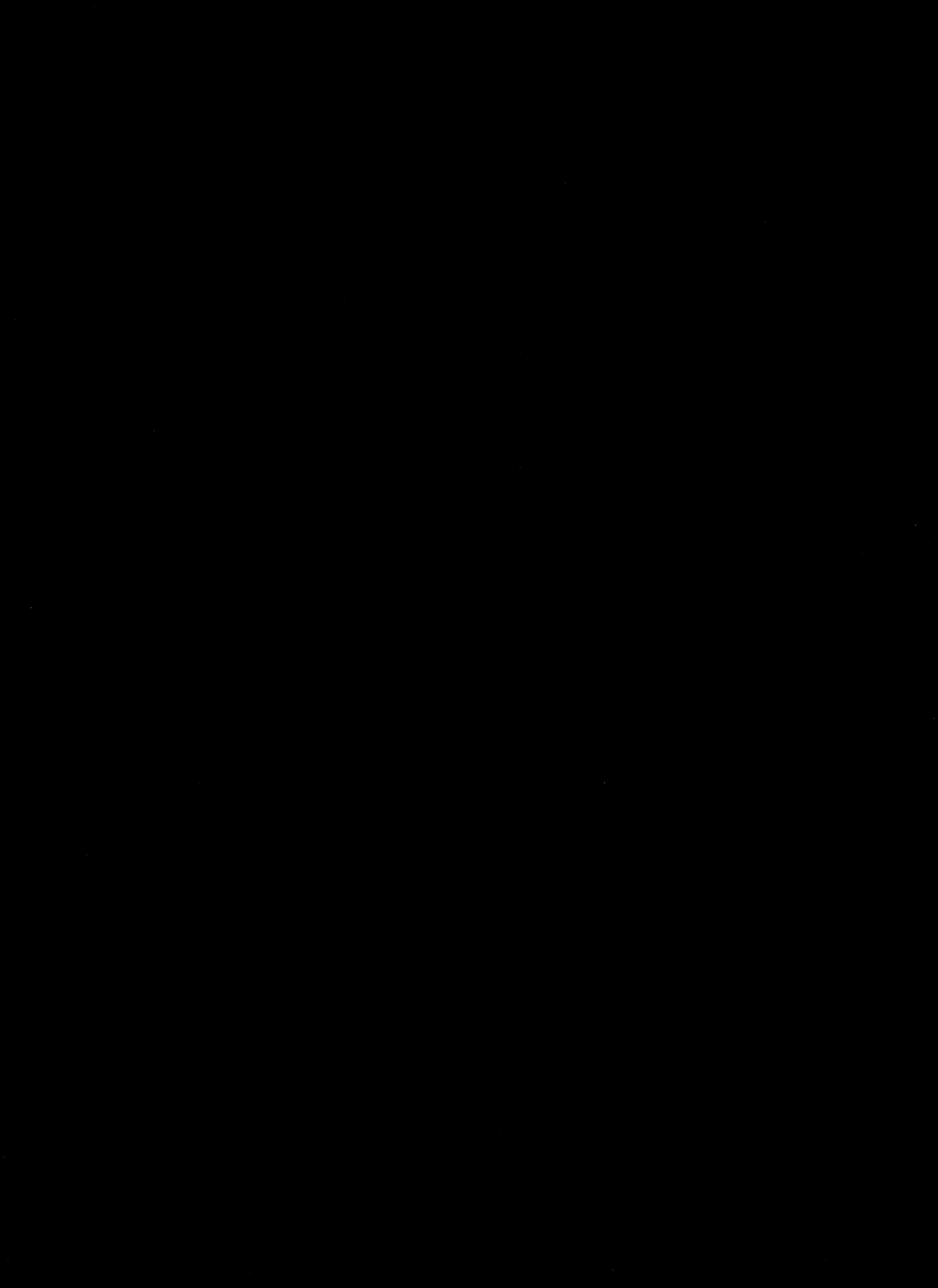

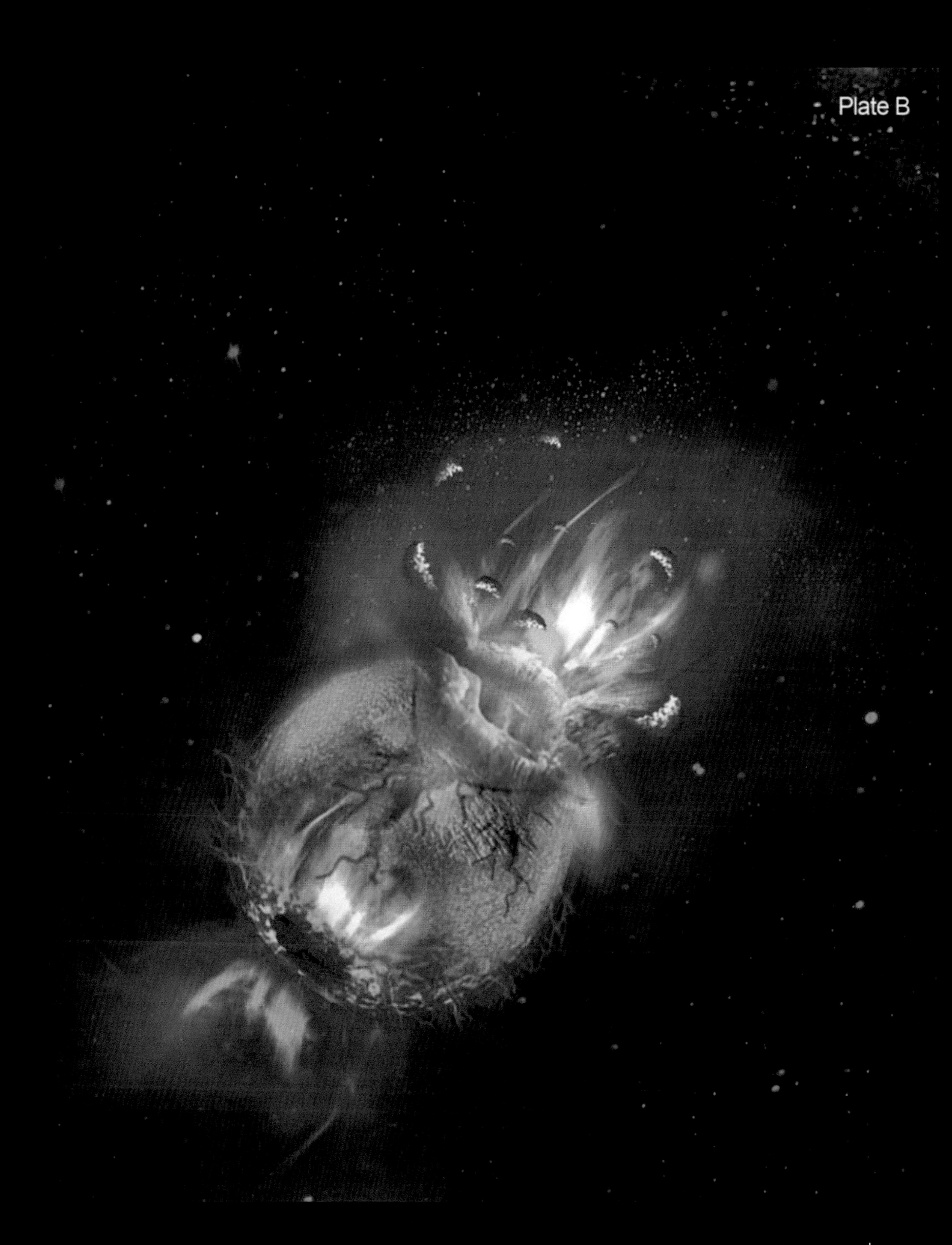
Plate B

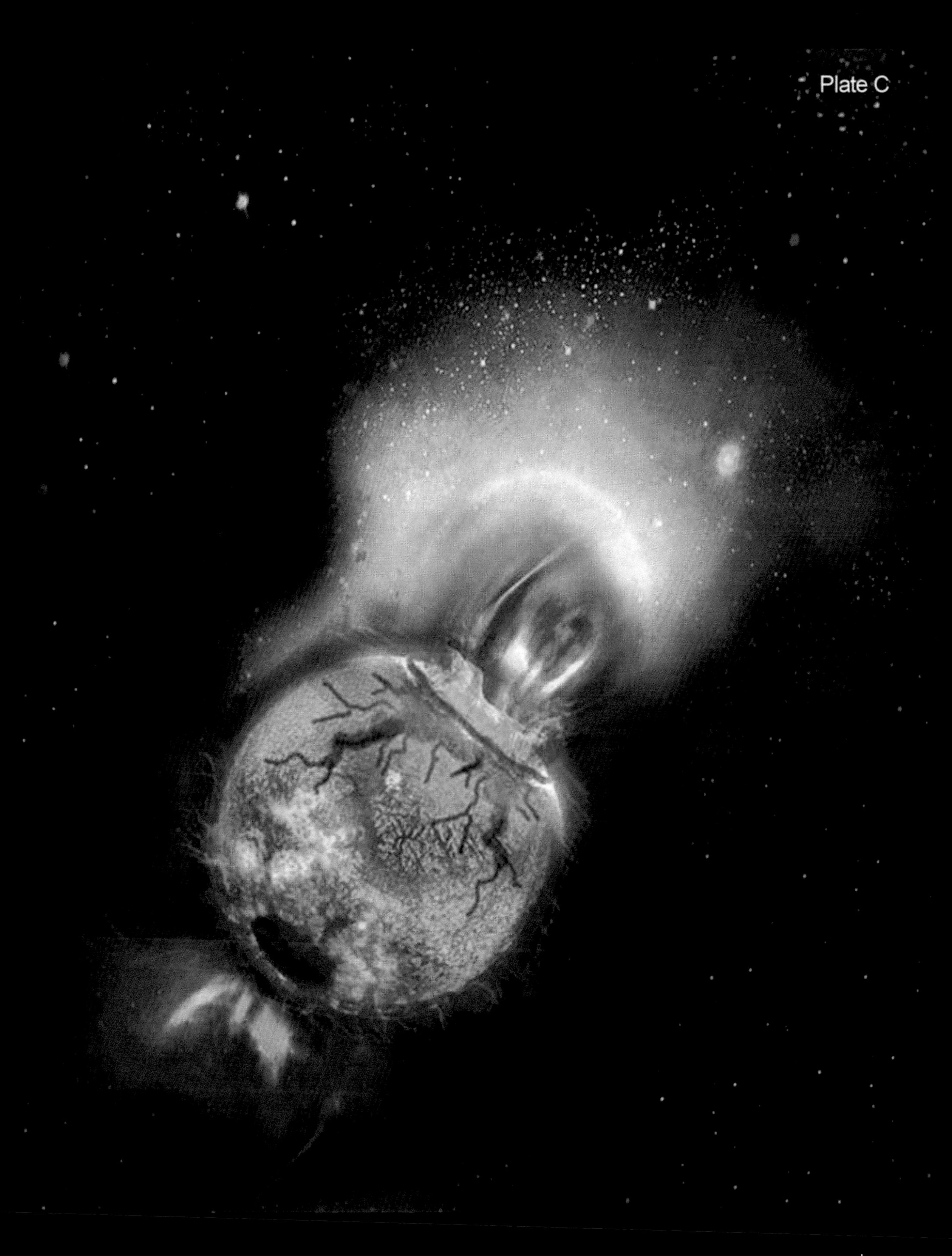
Plate C

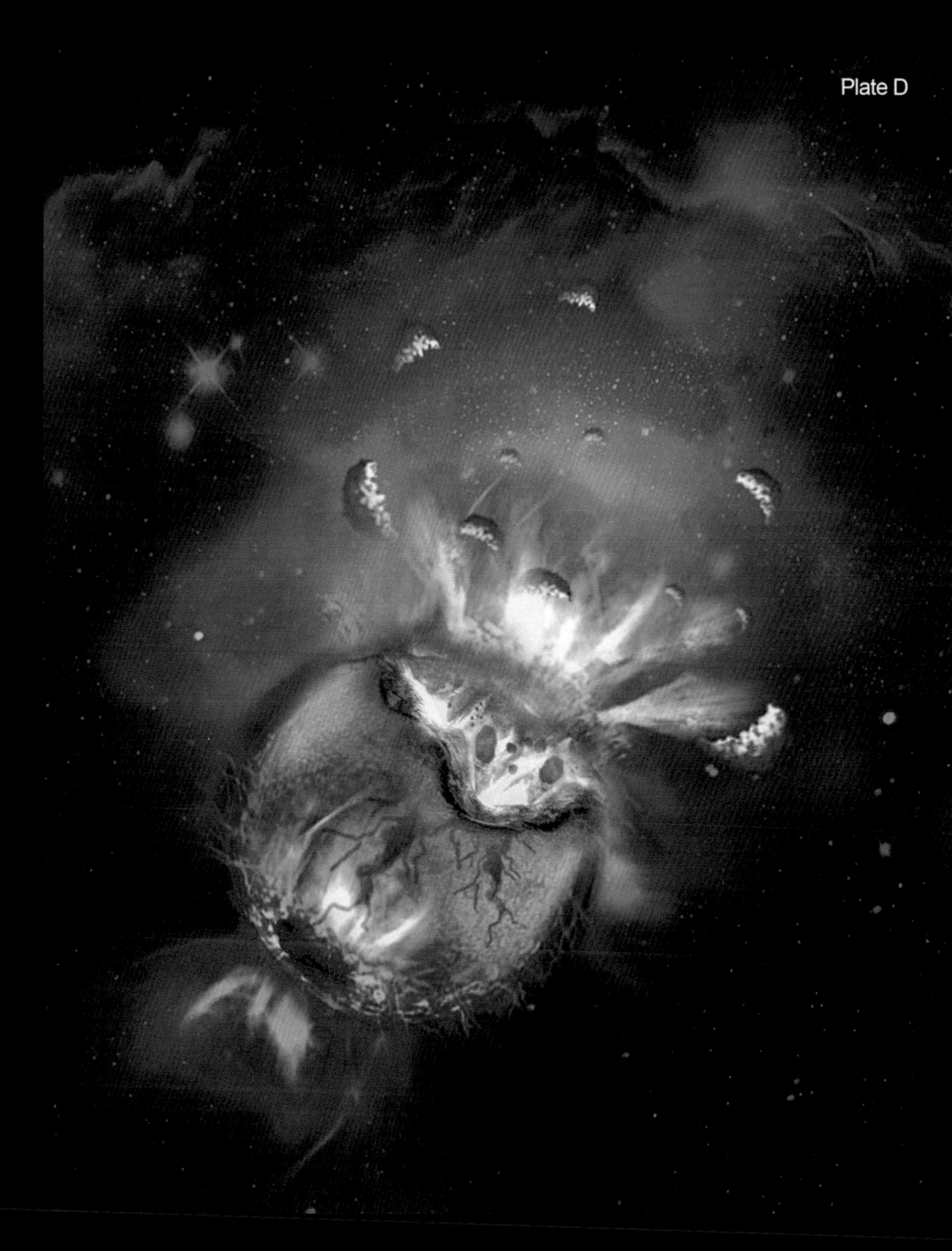

Plate D

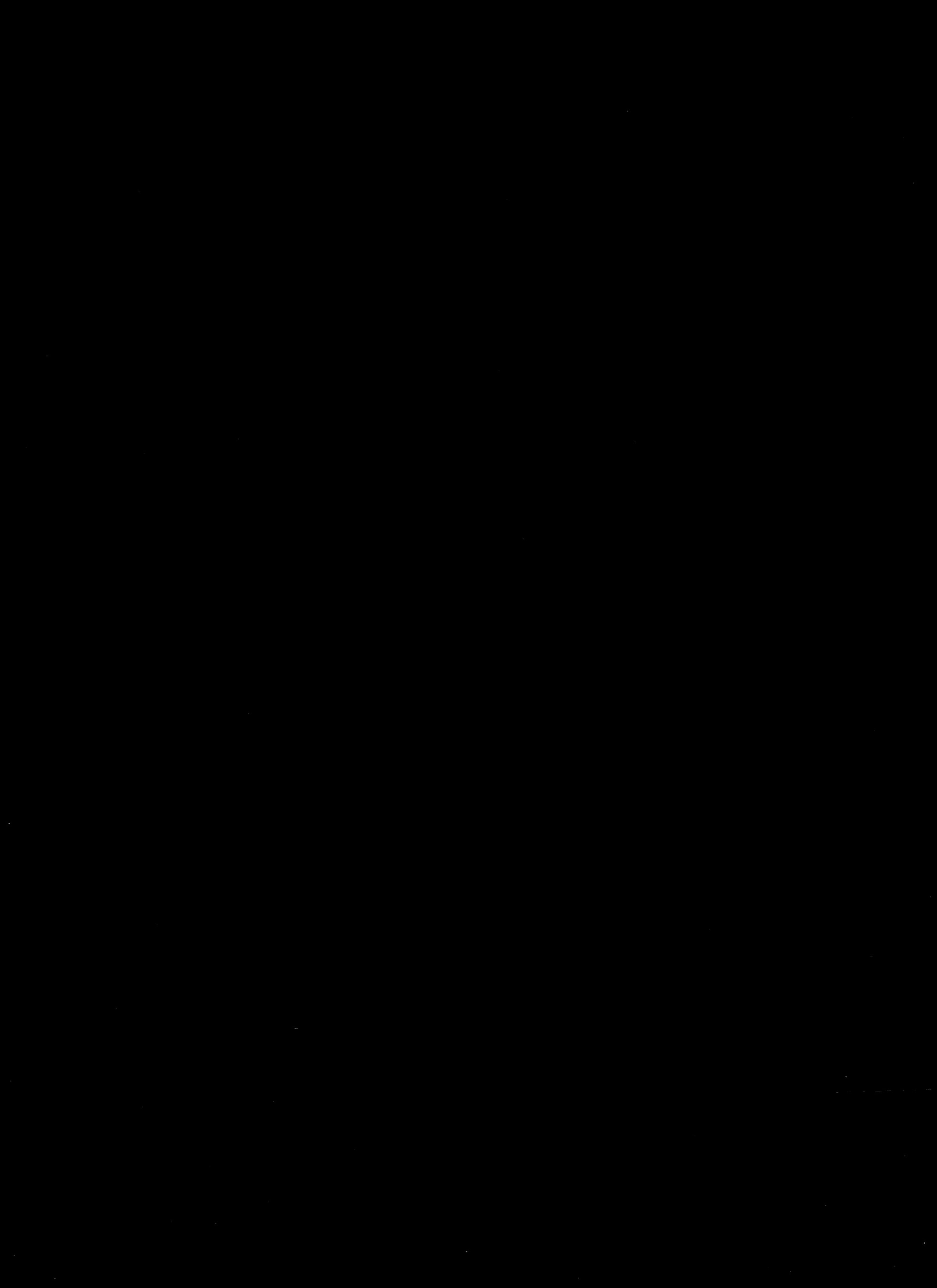

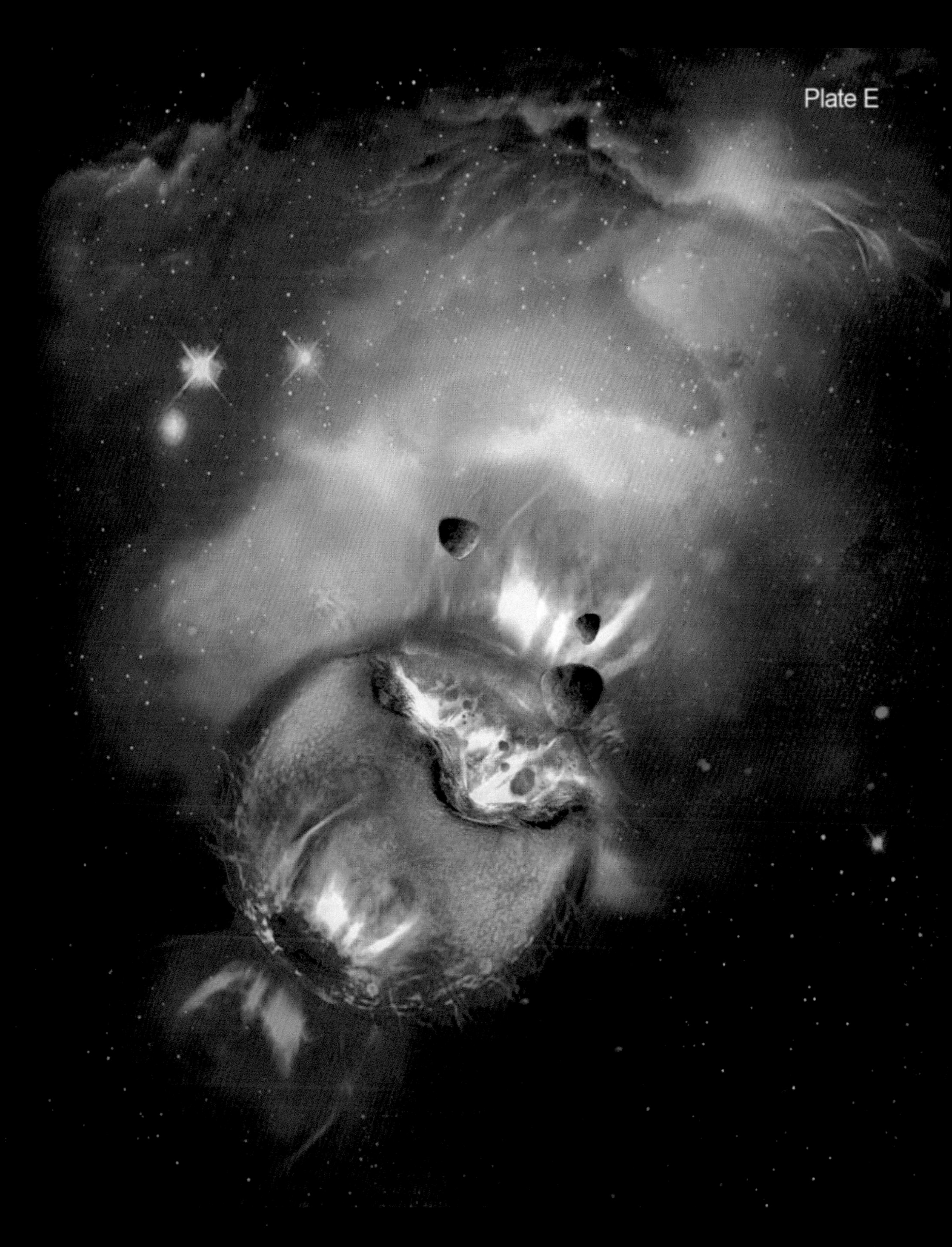
Plate E

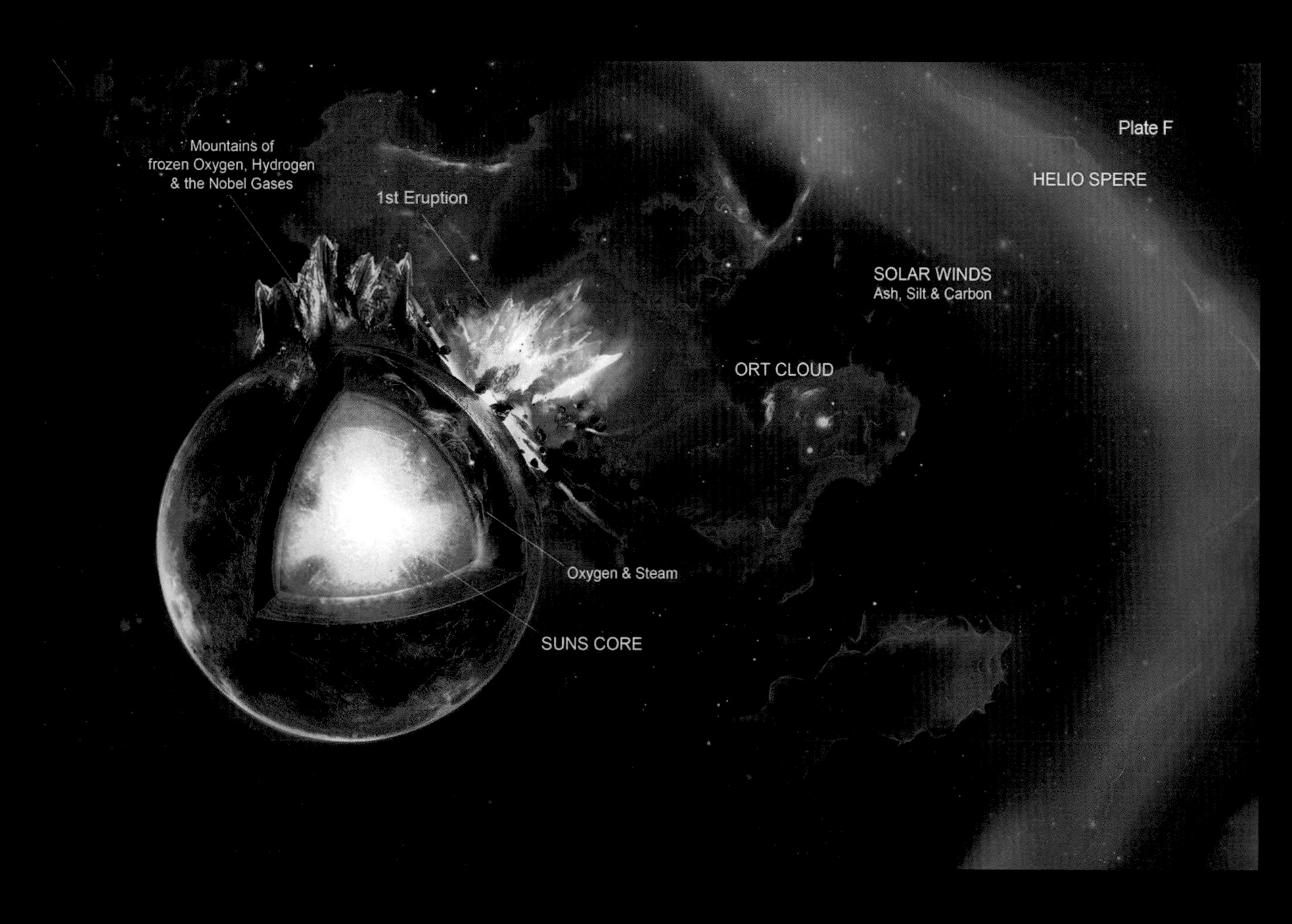
Mountains of
frozen Oxygen, Hydrogen
& the Nobel Gases
1st Eruption
Plate F
HELIO SPERE
SOLAR WINDS
Ash, Silt & Carbon
ORT CLOUD
Oxygen & Steam
SUNS CORE

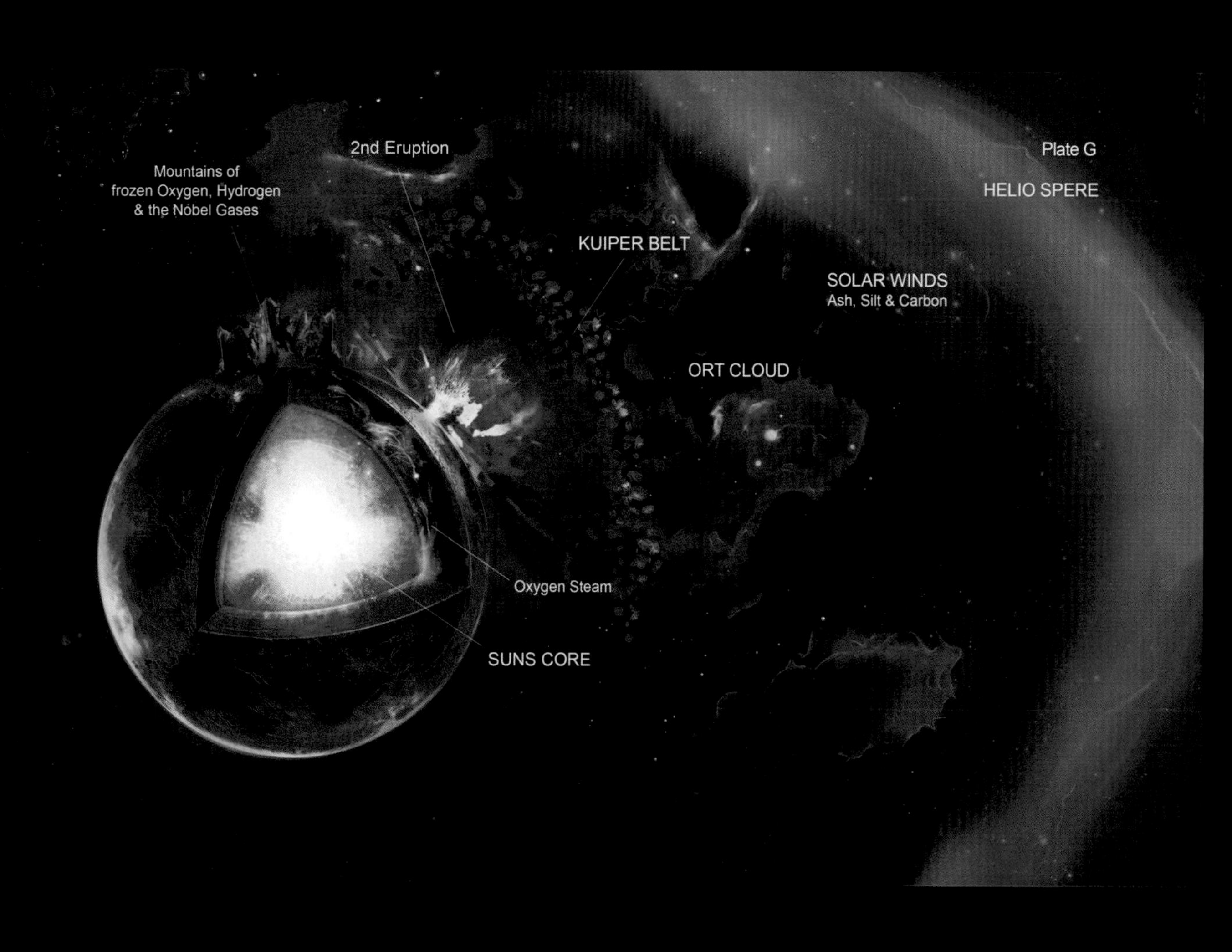
2nd Eruption
Mountains of
frozen Oxygen, Hydrogen
& the Nobel Gases
Plate G
HELIO SPERE
KUIPER BELT
SOLAR WINDS
Ash, Silt & Carbon
ORT CLOUD
Oxygen Steam
SUNS CORE

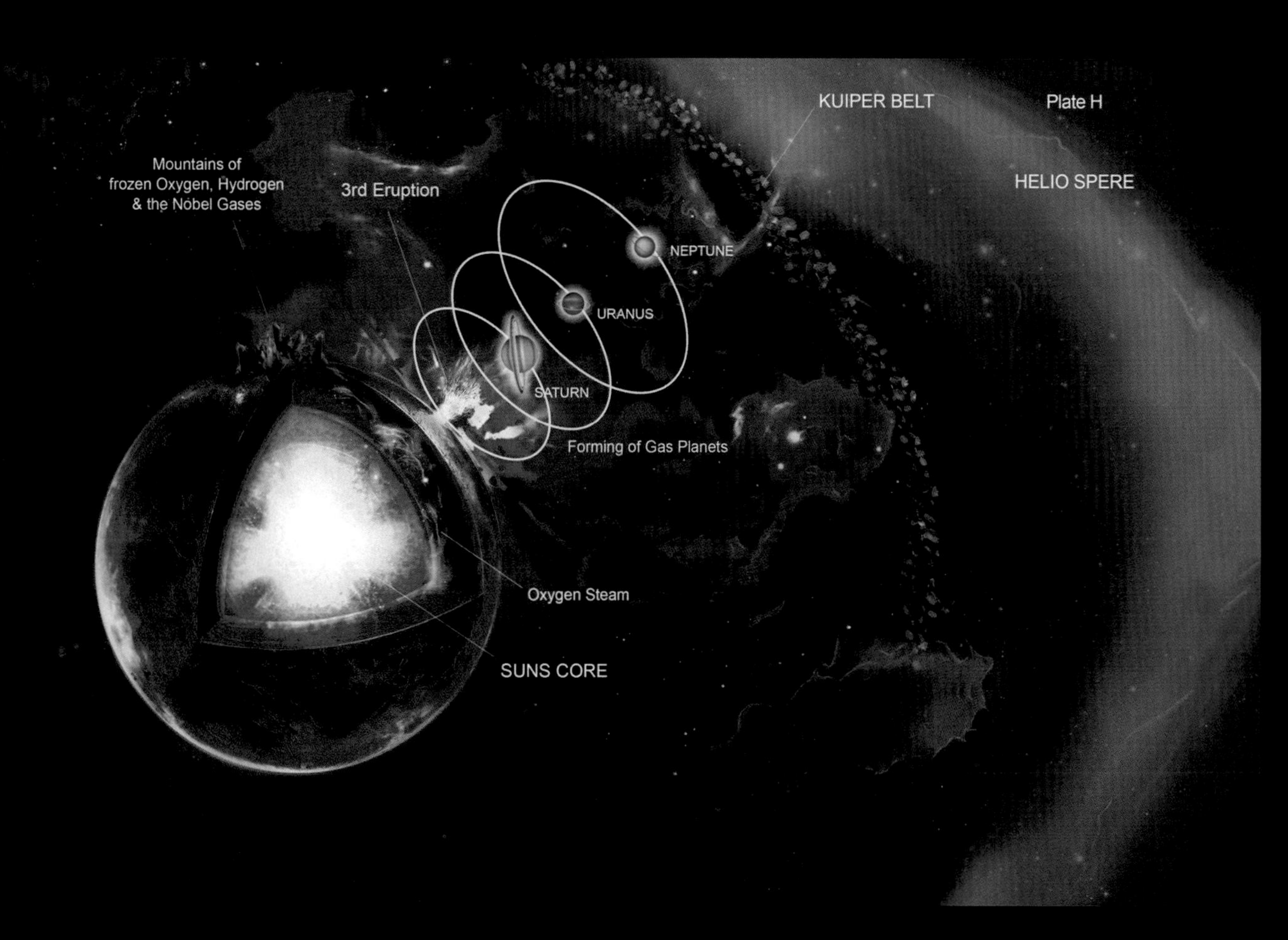
KUIPER BELT
Plate H
HELIO SPERE
Mountains of
frozen Oxygen, Hydrogen
& the Nobel Gases
3rd Eruption
NEPTUNE
URANUS
SATURN
Forming of Gas Planets
Oxygen Steam
SUNS CORE

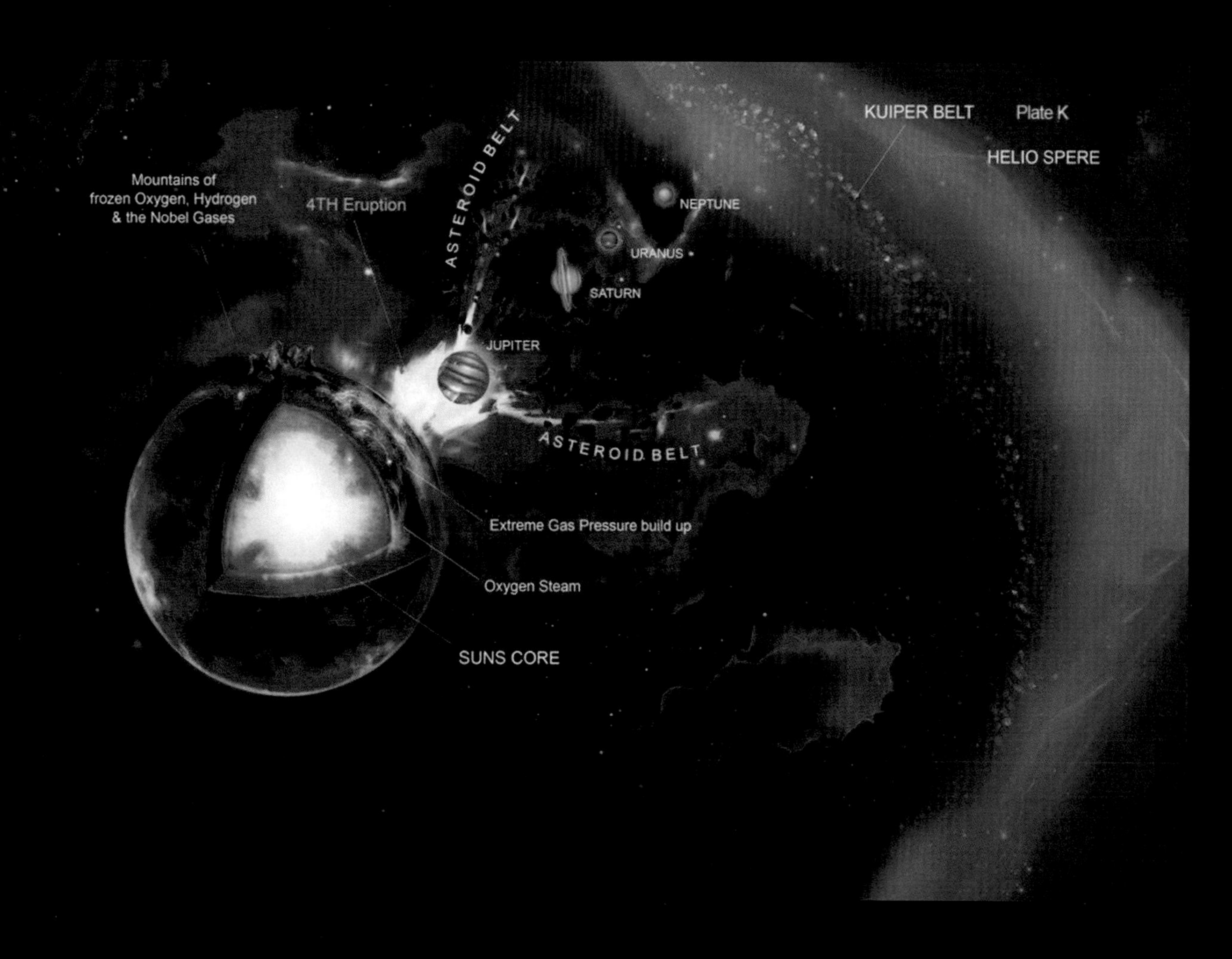
KUIPER BELT
Plate K
HELIO SPERE
Mountains of
frozen Oxygen, Hydrogen
& the Nobel Gases
4TH Eruption
ASTEROID BELT
NEPTUNE
URANUS
SATURN
JUPITER
ASTEROID BELT
Extreme Gas Pressure build up
Oxygen Steam
SUNS CORE

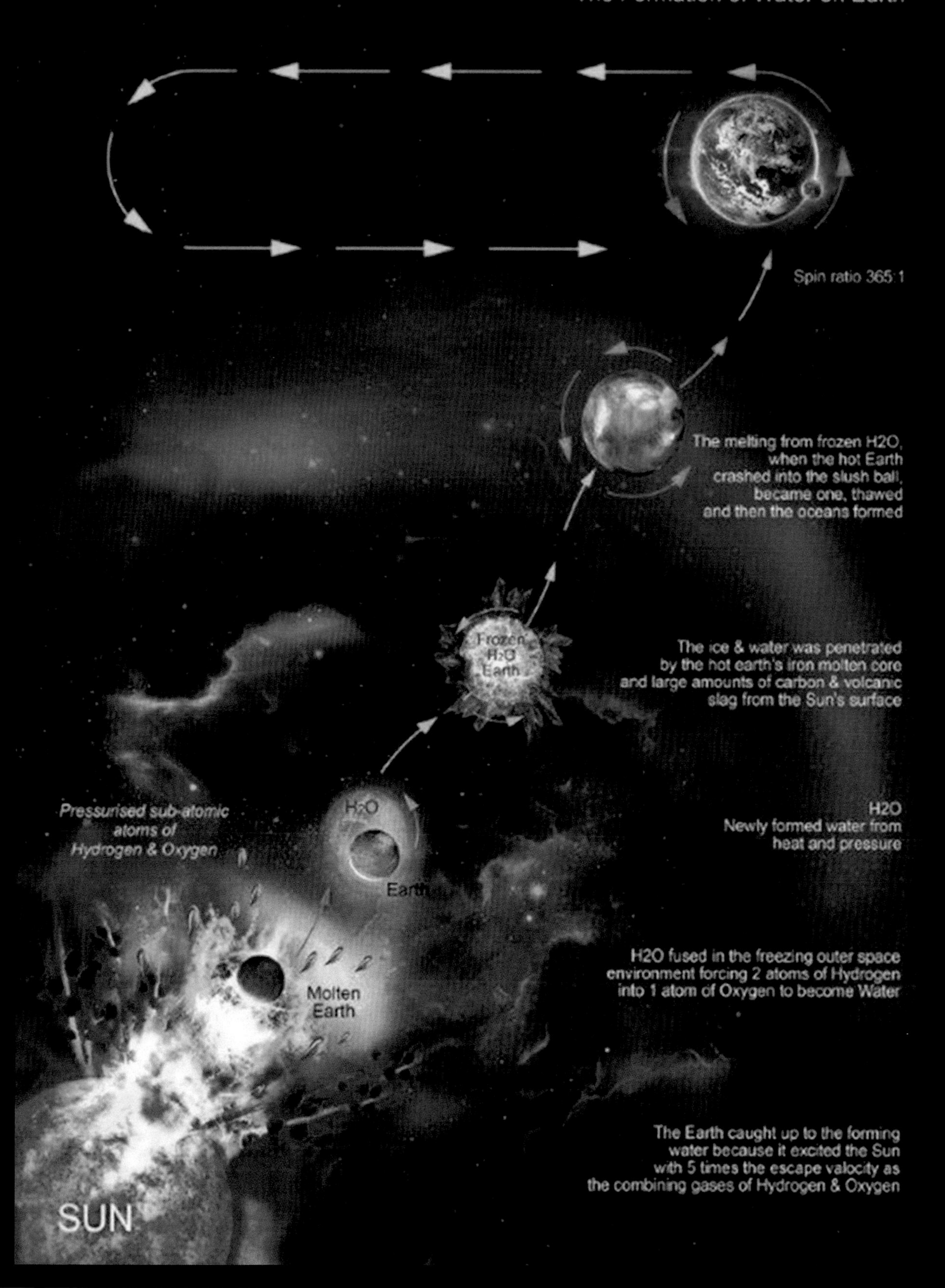
Plate L
The Formation of Water on Earth
Spin ratio 365:1
The melting from frozen H2O, when the hot Earth crashed into the slush ball, became one, thawed and then the oceans formed
Frozen H2O Earth
The ice & water was penetrated by the hot earth's iron molten core and large amounts of carbon & volcanic slag from the Sun's surface
Pressurised sub-atomic atoms of Hydrogen & Oxygen
H2O
H2O
Newly formed water from heat and pressure
Earth
H2O fused in the freezing outer space environment forcing 2 atoms of Hydrogen into 1 atom of Oxygen to become Water
Molten Earth
The Earth caught up to the forming water because it excited the Sun with 5 times the escape valocity as the combining gases of Hydrogen & Oxygen
SUN

CHAPTER 1

TWENTY FIRST CENTURY ASTRONOMY embraces The AP theory's ideology which proposes that gravity is not holding down our atmosphere; the planets are orbiting above the celestial poles of the Sun not around its equator; the planets are animated by the solar wind, and that our entire solar system erupted from a frozen condensed cloud of noble gases and galactic nebula of subatomic particles, which eventually formed into our Sun and the rest of the solar system all at about the same time 4 to 5 billion years ago. The AP theory also logically explains how Oxygen formed within the Sun from extreme heat which passively split its atoms of Oxygen gases into their subatomic forms of Quarks Neutrinos and Leptons. When the split atoms of Oxygen were exploded out from the cauldron of our newly forming Sun into the cosmos, they impregnated the surrounding atoms of Hydrogen and they were compounded and formed into a water frozen cloud of H2O (Theia) for the first time ever. The giant impact of our newly formed Earth with Thiea is what put water on Earth.

Water Formation in a Nutshell

When the fissile atoms' neutrons melted in the presence of extreme heat on a scale beyond comprehension and fervor in the solar cauldron caused the atoms' protons and neutrons to be stripped bare. The extreme environment and heat stripped and exposed the negatively charged electrons and positively charged protons. The nuclei with the bare positively charged protons and negatively charged electrons then collided and produced a tremendous amount of atomic energy similar to an exploding star. The collision of the bare nuclei caused from the extreme heat of the solar cauldron, coupled with the Sun's tremendous internal gas steam pressure, resulted in the unstable atoms of Hydrogen and Oxygen to passively separate. The condensed (frozen) gases then changed from solid into liquid vapour.

The first time frozen Oxygen gas was ever seen in nature was in 1970 by the crew of Apollo 13 as it vented from the command module which was caused by an explosion.

The extreme heat in the stellar oven separated the agitated atoms into their subatomic form of protons, neutrons, and electrons and then split them again into their molecular form of quarks, neutrinos, and leptons and then again. The Galvani process was the final result of the molten 'splash' coupled with steam pressure. This event proved that a galvanic explosion could naturally happen in the cosmos by applying Galvani's law. This cataclysmic galvanic explosion culminated in the compounding of unstable subatomic unionized molecules for the first time ever. These subatomic molecules were then blasted from the solar cauldron of the Sun into the frozen environment—absolute zero (– 273.15 degrees Celsius)—of interstellar space along with the solar mantle which we now call comets. The largest comet found to date is Ceres at 590 miles in diameter which is located near the asteroid belt.

When the superheated subatomic molecules entered the inhospitable and extreme intemperate environment of space, a thermal transformation of metals (Robert Boyle's law) occurred. Upon entering the absolute-zero environments of the cosmos, the split subatomic particles of Oxygen condensed and fused together with the surrounding Hydrogen. (Plate 8588) This resulted in the merging of H2 and O to compound together for the very first time. The explosion formed a large condensed water vapour cloud known as the Theia Cloud in the cosmos. The Theia ice planet, about the size of Mars, was one of the bodies that the super hot Earth collided with, passively melting its way through the massive ice cloud 4.5 billion years ago ("Great Impact") after it exited the Sun. The Earth's atmosphere and the water in our solar system were formed from material produced and provided by a new star, our Sun 300 million years before it matured into a 27-million-degree-Fahrenheit yellow dwarf star.

THE 16 LEVELS OF THE EXISTENCE OF GASES FROM ABOVE EARTH'S SEA LEVEL

THE UNIVERSE completely surrounds everything and is comprised of Hydrogen and the lightest of yet to be analyzed gases. Could this be dark matter/energy?

THE GALAXY is comprised of Atomic Hydrogen gas mixed with Helium which is used as the Sun's fuel source and has a 100, 000 Light year diameter.

THE OORT CLOUD , a mix of Hydrogen and Helium. Discovered in 2012 it extends trillions of miles from the sun, its diameter is light years across and it is home to more than one trillion objects, mostly comets ejected from the Sun's mantle during the first solar ejection.

THE HELIOSHOCK is the border separating the Oort Cloud from the Heliosphere.

KUIPER BELT discovered in 1992 is 3 billion miles from Earth and contains thousands of comets formed from the Sun's mantle.

THE HELIOSPHERE extends 11 billion miles from Earth and contains the comets and its mass completely surrounds and holds down Earth's atmosphere. The Heliosphere (Oort Cloud) is comprised of Helium exhausted from the Sun. It exerts 14.7 Lb. psi (760mm Hg) on Earth's atmosphere and has a 12 trillion mile diameter.

THE HELIOPAUSE is the area of the Sun's magnetic field and stops the solar wind from entering interstellar space. This Heliosheath (termination shock) is about 113 AU's from the Sun.

(AKA. Lagrangian Point, Gravitational equilibrium)

Ozone Region

EARTH'S ATMOSPHERE of Oxygen and Nitrogen gases is divided into 7 levels, all originating from Earth.

Exosphere	340 to 620	miles above sea level	
Ionosphere	50 to 340	miles above sea level	5.5 Lb. psi (87mm Hg)
Mesosphere	19 to 50	miles above sea level	
Stratosphere	8 to 19	miles above sea level	
Troposphere	5 to 8	miles above sea level	
Ecosphere	0 to 8	miles above sea level 14.7 Lb. psi (approximately 250mm Hg)	
Sea Level	0		

Felix Baumgartner set the altitude record in 2012 soaring 135,890 over Roswell, N.M.

It took 15 minutes for him to land from that height.

The International Space Station is 220 miles above sea level and travelling at 17,500 mph. If the Earth is allegedly travelling at 67,000 mph around the Sun, the ISS could never keep up with it further proving The AP Theory. .

IMPROVED HELIOCENTRIC MODEL

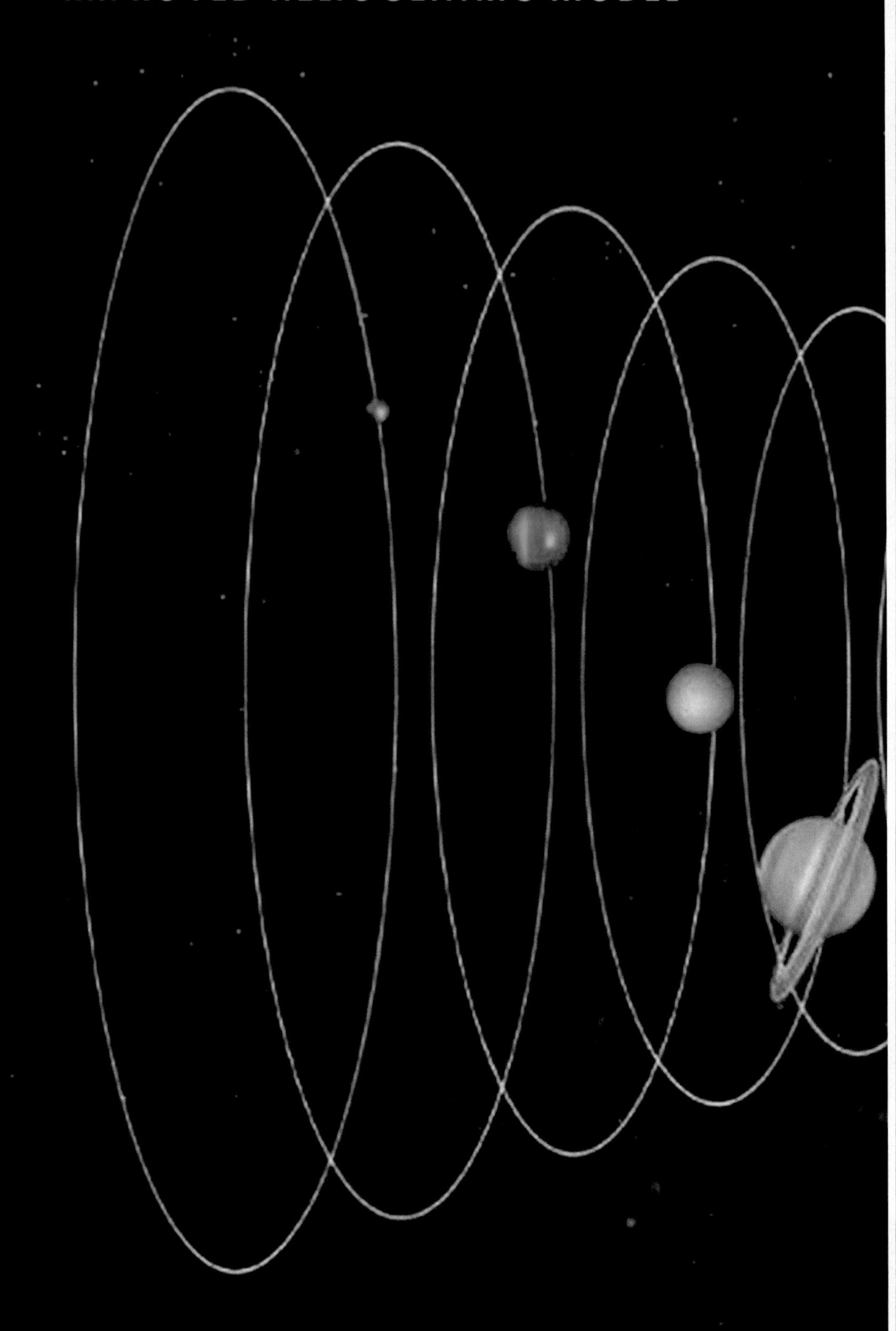

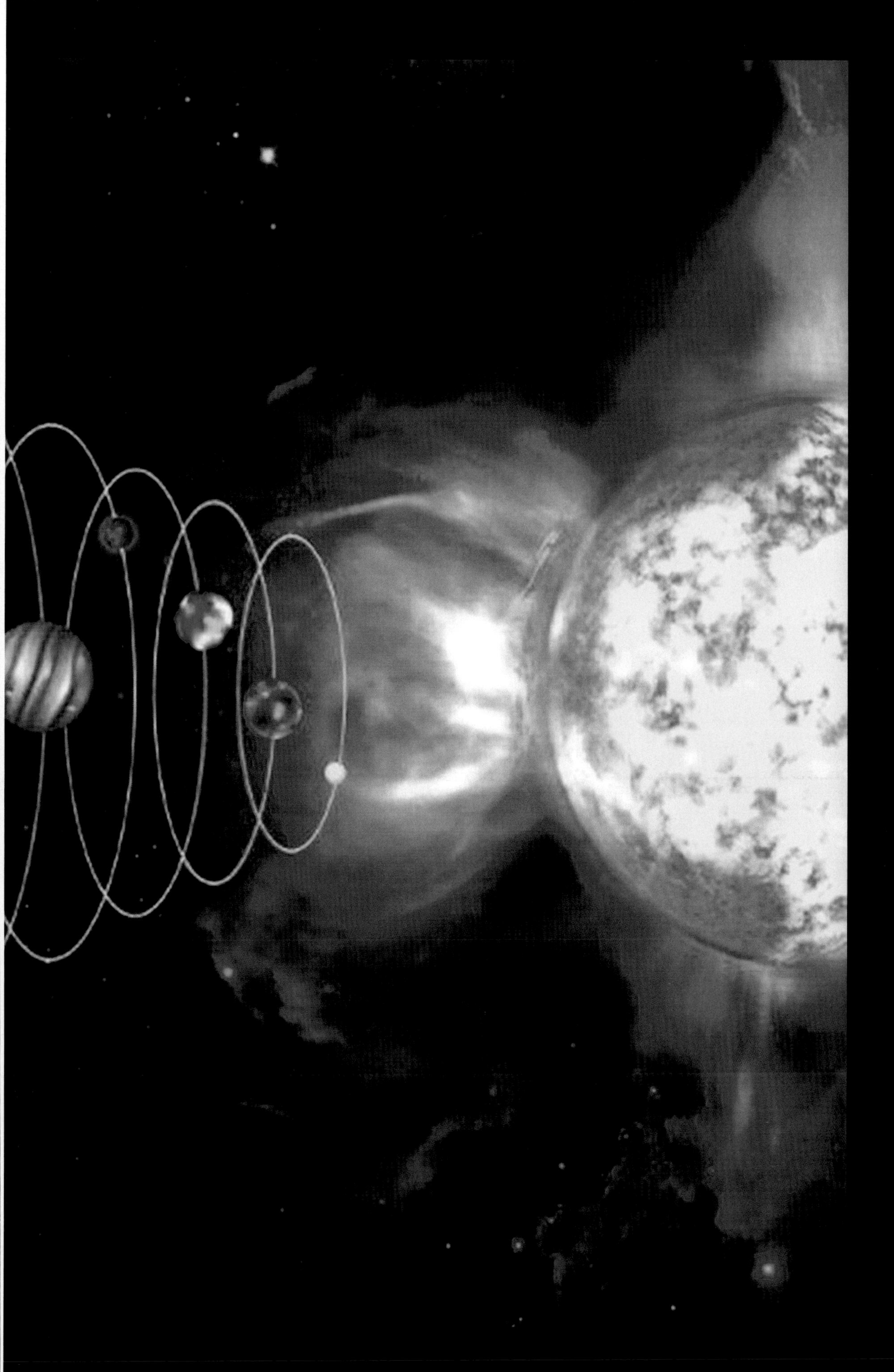

CHAPTER 2

Edwin Hubble was born in Missouri in 1889. Hubble's experience allowed him to lead a research team in the field of astrophysics at Mount Wilson Observatory in California. One of his accomplishments was discovering an expanding universe with Milton by using the Doppler shift and single headedly disproved the theories of relativity and special relativity at the same time. He was a dedicated scientist and also created a classification system for galaxies known as the Hubble sequence. In 1920 Harlo Shapley thought the Milky Way Galaxy contained every star in the universe while Huber D. Curtis claimed it was one of many galaxies. Hubble finally proved Curtis to be right. In Oct. of 1923 Hubble discovered the Andromeda galaxy which proved that the Milky Way Galaxy was not the only galaxy in the Universe. Hubble was a successful lawyer, was a world war ll veteran , and had a Doctorate in astronomy. He went on to win a Nobel Prize in physics by proving other galaxies existed beyond the Milky Way galaxy by comparing degrees of luminosity among variable stars. Hubble died in 1953; he was 63. He and his wife, Grace, had no children.

The Hubble Telescope Launched in 1990 now has a new helper the IBEX telescope, The NASA Kepler orbiting observatory (launched 2009) and Spitzer telescopes designed to find distant planets revolving around a star and objects in the cosmos. The largest telescope is the Keck located in Hawaii. The James Webb Telescope with its 111 sq. metre mirror is to be launched in 2020. It has a large binocular telescope, is 10 xs's sharper and can see 12 billion light years away. The "LIGO" instrument, located at the Hanford Nuclear Reservation in Washington is an improved gravity wave detector and is replacing the "GWD" gravity wave detector. As of 2015, no gravity waves have been detected.

CHAPTER 3

Newton, Hubble, Galileo, and other prominent scientists were intrigued by and worked with a prism (Plate 0597) to separate and define the colours of light in the lines of a spectrum. A Bavarian master glassmaker named Joseph von Fraunhofer finally got the credit for discovering infra-red spectral lines.

Einstein's theorys of General Relativity and Special Relativity* were disproved by Hubble and M. L. Humason in Oct. 1923 by using Mount Wilson Observatory's 250-centimetre telescope. By means of spectroscopy, Hubble's velocity–distance law (velocity of recession = Hubble's constant × distance) shows how the velocity of recession of a galaxy increases with its distance. By using the Doppler Effect, Hubble observed a systematic red shift towards the longer wavelength in the spectra of light and discovered receding galaxies, proving an expanding universe.

This single groundbreaking discovery disproved and made null, and voided the theories of General Relativity, String, Worm Holes, Black Holes, Standard Model, Steady State, Warped Space and Time, Hydrostatic Equilibrium and the Nebular Hypothesis (Accretion) once and for all and forever and ever.

Now that we know it's not quantum gravity or quantum mechanics as allegedly explained in Einstein's infamous relativity theory equation that is propelling our solar system, what then can comprehensively explain it? What causes gravity? Why is global warming a regularly recurring predictable cycle? Why are some of the comets, Earth, and some celestial bodies in the Kuiper belt recently discovered in 1992 are the only objects in our solar system to have water? How did the gaseous and terrestrial planets, their atmospheres, the asteroid belt, and water form? These questions are truly adventures of the mind, and all of them will be addressed in 21st. Century Astronomy. This bold truth book is based on the most current discoveries. It takes us one step closer to the truth by answering the unanswered questions and dispelling misinformation and misconceptions..

*The theories of *Special and General Relativity and Quantum Gravity* stated falsely that gravity had stabilizing, propelling, suspending, and repelling capabilities. The theory has since been proven to be null and void. For example, it allegedly proposed gravity was the force that causes four daily tides on Earth, planetary circumnavigation around the Sun, twenty-four-hour rotation of Earth, and keeping Earth equidistant from other planets *as well* as holding our atmosphere to Earth. Einstein 'fudged' his formula by introducing a cosmological constant and was caught out by Hubble. None of the theory's facts were ever proven, and most of them have now been disproved and discredited. Einstein called this his 'biggest blunder'. It was also revealed that the infamous $E = mc2$ cannot be reversed because E (energy) is an invisible essence impossible to measure and m (matter) is a solid. Nothing can travel faster than the speed of light, which is expressed as c; therefore, c2 is impossible to achieve thus disproving "The Big Bang" theory and recently confirming that it was disproved at CERN in Switzerland.

Plate 8741

CHAPTER 4

Luigi Galvani 1737–1798 was a pioneer in the study of bioelectricity. He earned degrees in both medicine and philosophy. In 1775 Galvani was appointed as a professor and lecturer, taking Guzman Galeazzi's chair. Galvani was a Benedictine member of the academy of sciences, teaching electrophysiology. He showed that when liquid comes in contact with molten materials, a galvanic explosion occurs. The AP theory is based on such explosions, which are now believed to have formed our solar system, and other discoveries. Galvani's prolific work earned him a well-deserved place in history. Galvani made many discoveries and was a pioneer in his time. He was a prolific scientist.

Daniel Bernoulli (1700–1782) came from a prominent and famous family of scientists. Daniel was a physician, PhD, and mathematician. He is best known in the field of fluid mechanics. He published Hydrodynamica , describing the flowing behaviour of liquids and gases as similar in behaviour. Bernoulli's principle was the first to distinguish between hydrostatic and hydrodynamic pressure. The AP theory is based on these principles when describing the motions of the planets. Air is gas, and individual gases have differently sized atoms which act like water. The large atmospheric atoms of Oxygen and Nitrogen gas cannot pass through the small atoms of Helium gas as they rise making the Helium gas a natural ceiling.

Bernoulli is one of the first writers who made an attempt to devise the kinetic theory of gases and used the idea to explain Boyle's law. Bernoulli's principle is of significant use in aerodynamics. Daniel Bernoulli died in 1782 in Basel, Switzerland. During his life, he won or shared ten prizes at the Paris Academy of Sciences. His laws are treated as the standard even today in laboratories around the world. Einstein tried to carry on with Bernoulli's work when he spoke of space time and the fabric of the universe. Nothing ever came of it.

Earth's surrounding Helium-filled Heliosphere is the cause of Earth's atmospheric air pressure of 14.7 lb. per sq. in., which is the force of the unit area exerted on a surface by the weight of air above that surface in the atmosphere of Earth (or that of another planet). In most circumstances, atmospheric pressure is closely approximated by the hydrostatic pressure caused by the weight of air above the measurement point, which is equal to 1 ounce per cubic foot. On a given plane, low-pressure areas have smaller atoms and less atmospheric mass above their location, forming a natural 'ceiling', whereas high-pressure areas have larger atoms and more atmospheric mass above their location.

Likewise, as elevation increases, there is less overlying atmospheric volume and weight so that atmospheric pressure decreases with increasing elevation. On average, a compressed column of air 1 square centimetre in cross section and measured from sea level to the top of the atmosphere has a volume of about 1.03 kilograms and weight of about 2.28 pounds. A column 1-square inch in cross section would have a weight of about 14.7 pounds or 1 ton per square foot. The light atomic weight Helium gases of the 11 trillion mile diameter (about 2 light years) Heliosphere, cause the 14.7 lb per sq in compression of gases of Helium's weight on Earth's atmosphere. This compression combined with Earth's 1,040-mile-per-hour rotations and tilts are the causes and origin of gravity on Earth.

BERNOULLI LAW. A relationship that expresses the conservation of momentum in fluid flow. The usual form applies to the steady inviscid flow of an incompressible fluid and can be obtained by integrating the Navier-Stokes equations along a streamline,

$$gz + \frac{p}{\rho} + \tfrac{1}{2}u^2 = \text{constant on a streamline}$$

If the flow is irrotational so that the fluid velocity is the gradient of a scalar potential ϕ, the restriction to steady flow may be removed and the law is

$$\frac{\partial \phi}{\partial t} + \frac{p}{\rho} + \tfrac{1}{2}u^2 + gz = \text{constant anywhere in the fluid}$$

Lastly, if the fluid is barotropic, i.e., the density is a function of pressure alone, it takes the form,

$$\int \frac{dx}{\rho} + \tfrac{1}{2}u^2 = \text{constant}$$

valid along a streamline for steady flow. For a perfect gas, with p proportional to ρ^{γ}.

$$gz + \frac{\gamma}{\gamma - 1}\frac{p}{\rho} + \tfrac{1}{2}u^2 = \text{constant}$$

The quantity,

$$p\left(1 + \tfrac{1}{2}\frac{\gamma - 1}{\gamma}\frac{\rho u^2}{p}\right)^{\gamma/(\gamma - 1)},$$

is known as the total head or stagnation pressure. For flow at small Mach numbers, i.e., $u^2 \ll \gamma p/\rho = \gamma RT$, it is $p + \frac{1}{2}\rho u^2$.

BERNOULLI METHOD. Given the algebraic equation

$$(1) \qquad x^n + a_1 x^{n-1} + \cdots + a_n = 0,$$

let $h_0, h_1, \ldots, h_{n-1}$ be arbitrary numbers, not all zero, and form h_n, $h_{n+1}, \ldots$, by

$$h_{n+\nu} + a_1 h_{n+\nu-1} + \cdots + a_n h_\nu = 0.$$

If (1) has a unique root of largest modulus, then in general the quotients $h_{\mu+1}/h_\mu$ approach that root. The method can be extended to transcendental equations. Let

$$(2) \qquad f(z) \equiv 1 + c_1 z + c_2 z^2 + \cdots$$

converge in some circle about the origin in the complex plane, and let

$$(3) \qquad g(z) \equiv g_0 + g_1 z + g_2 z^2 + \cdots$$

represent any function analytic in the same circle, and having no zero in common with $f(z)$. Let

$$h_0 = g_0,$$

$$c_1 h_0 + h_1 = g_1,$$

$$c_2 h_0 + c_1 h_1 + h_2 = g_2$$

Then if $f(z)$ has a unique zero of smallest modulus lying within that circle, then $h_\mu/h_{\mu+1}$ approaches that zero. If there are two zeros whose moduli are less than those of all others, then the roots of

$$\begin{vmatrix} z^2 & h_\mu & h_{\mu+1} \\ z & h_{\mu+1} & h_{\mu+2} \\ 1 & h_{\mu+2} & h_{\mu+3} \end{vmatrix} = 0$$

approach those zeros of $f(z)$. Likewise one can form cubics whose roots approach the three smallest roots. (See Alston S. Householder, *Principles of Numerical Analysis*, McGraw-Hill Book Company,

1953.) The QD algorithm improves upon this principle. (See National Bureau of Standards, *Further contributions to the solution of simultaneous linear equations and the determination of eigenvalues*, NBS Appl. Math. Series 49, 1958.)

BERNOULLI NUMBER AND POLYNOMIAL. A Bernoulli number is a coefficient in the power series

$$\frac{x}{e^x - 1} = \sum_{n=0}^{\infty} \frac{B_n x^n}{n!}$$

The first few numbers are $B_0 = 1$; $B_1 = -\frac{1}{2}$; $B_3 = B_5 = B_7 = \cdots = 0$; $B_2 = \frac{1}{6}$; $B_4 = -\frac{1}{30}$; $B_6 = \frac{1}{42}$; $B_8 = -\frac{1}{30}$; $B_{10} = \frac{5}{66}$. Successive values of the numbers may be found from the equation $(B + 1)^n = B^n$ by setting $B^k = B_k$. The numbers occur in the Euler-Maclaurin formula and in the Stirling formula.

The Bernoulli polynomial is a coefficient in the power series

$$\frac{t(e^{xt} - 1)}{(e^t - 1)} = \sum_{n=0}^{\infty} \frac{\phi_n(x)}{n!} t^n$$

The first few polynomials are $\phi_2 = x(x - 1)$; $\phi_3 = x(x^2 - 3x/2 + \frac{1}{2})$; $\phi_4 = x^2(x^2 - 2x + 1)$; $\phi_5 = x(x^4 - 5x^3/2 + 5x^2/3 - \frac{1}{6})$. If the n-th polynomial is expanded in a Maclaurin series, the coefficients are related to the Bernoulli numbers.

Both the numbers and the polynomials are often defined in other ways so that equations containing them should be carefully checked. They are used in numerical integration formulas and in the calculus of finite differences.

The Bernoulli who discovered these relations was Jacob (1654–1705), a member of a distinguished family of mathematicians and physicists. See also **Bernoulli Equation; Logarithmic Spiral.**

Plate 7966

Plate 4560

Plate 4560

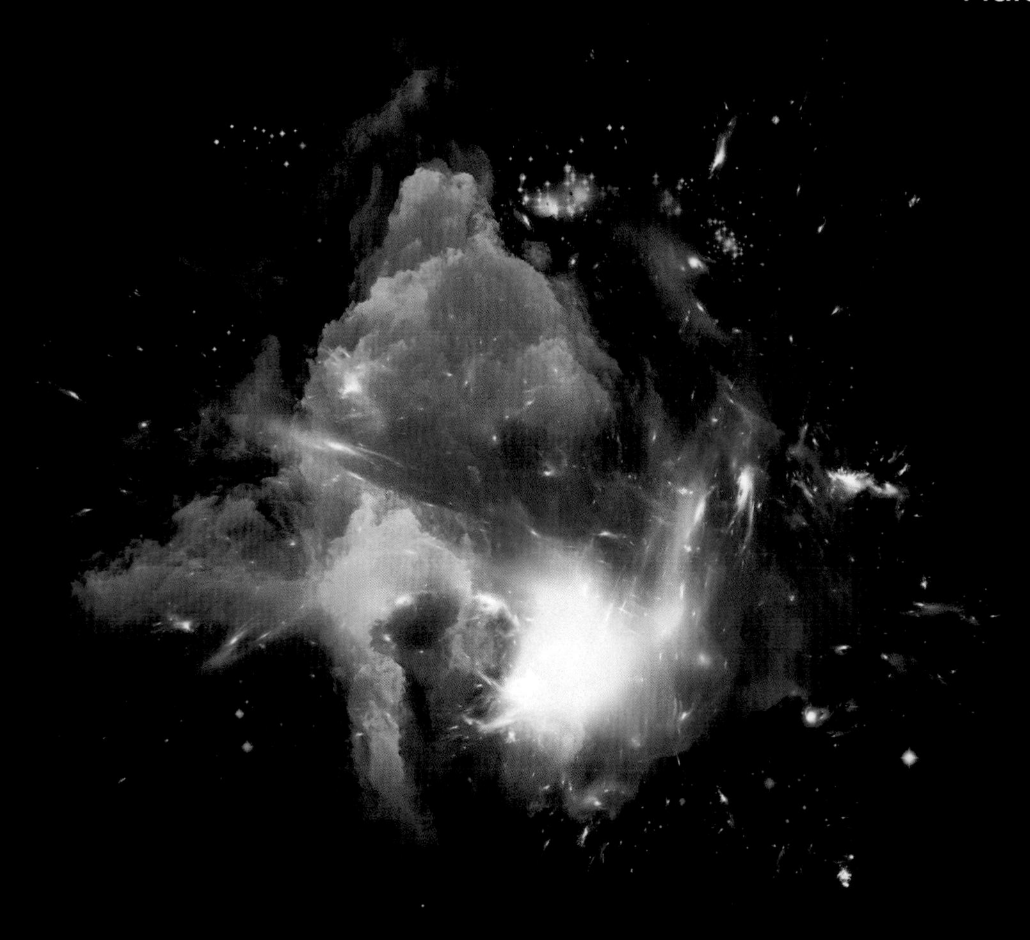

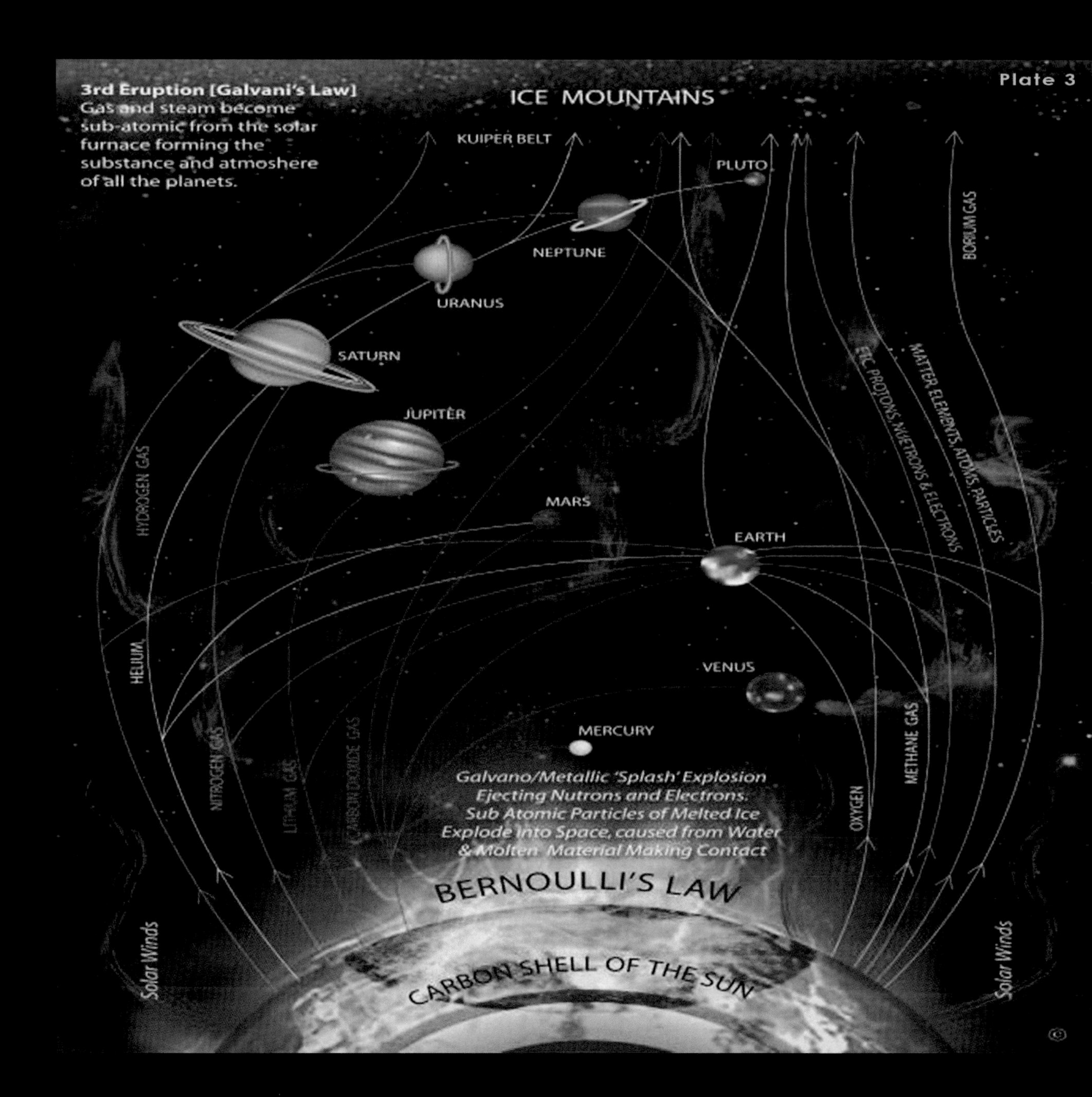

3rd Eruption [Galvani's Law]
Gas and steam become sub-atomic from the solar furnace forming the substance and atmoshere of all the planets.
Plate 3
ICE MOUNTAINS
KUIPER BELT
PLUTO
NEPTUNE
URANUS
SATURN
JUPITER
MARS
EARTH
VENUS
MERCURY
BORIUM GAS
MATTER ELEMENTS, ATOMS, PARTICLES
ETC PROTONS, NUETRONS & ELECTRONS
HYDROGEN GAS
HELIUM
NITROGEN GAS
LITHIUM GAS
CARBON DIOXIDE GAS
OXYGEN
METHANE GAS
Galvano/Metallic 'Splash' Explosion
Ejecting Nutrons and Electrons.
Sub Atomic Particles of Melted Ice
Explode into Space, caused from Water
& Molten Material Making Contact
BERNOULLI'S LAW
CARBON SHELL OF THE SUN
Solar Winds
Solar Winds
©

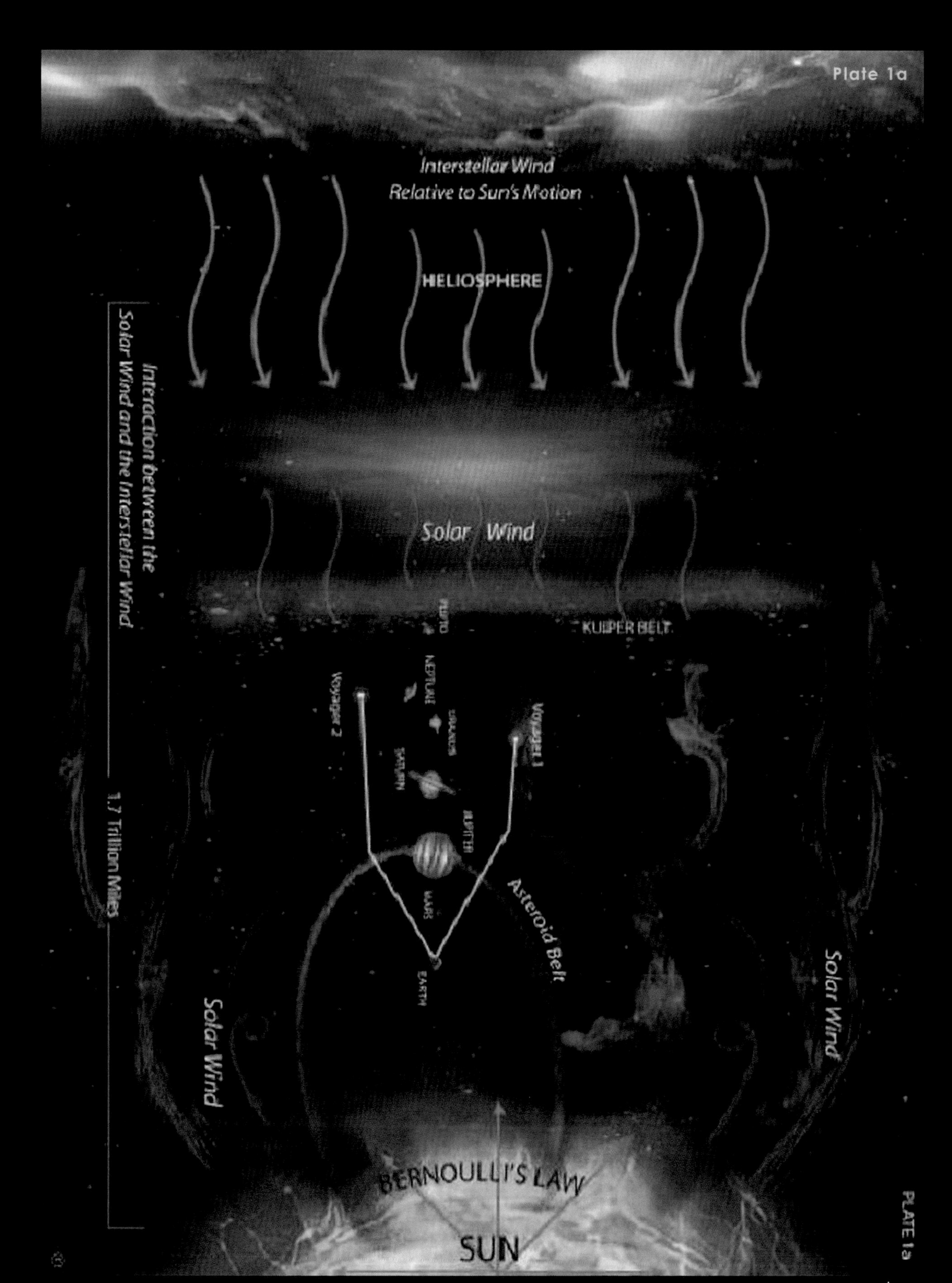
Plate 1a
Interstellar Wind
Relative to Sun's Motion
HELIOSPHERE
Interaction between the
Solar Wind and the Interstellar Wind
Solar Wind
KUIPER BELT
1.7 Trillion Miles
Asteroid Belt
Solar Wind
Solar Wind
BERNOULLI'S LAW
SUN
PLATE 1a

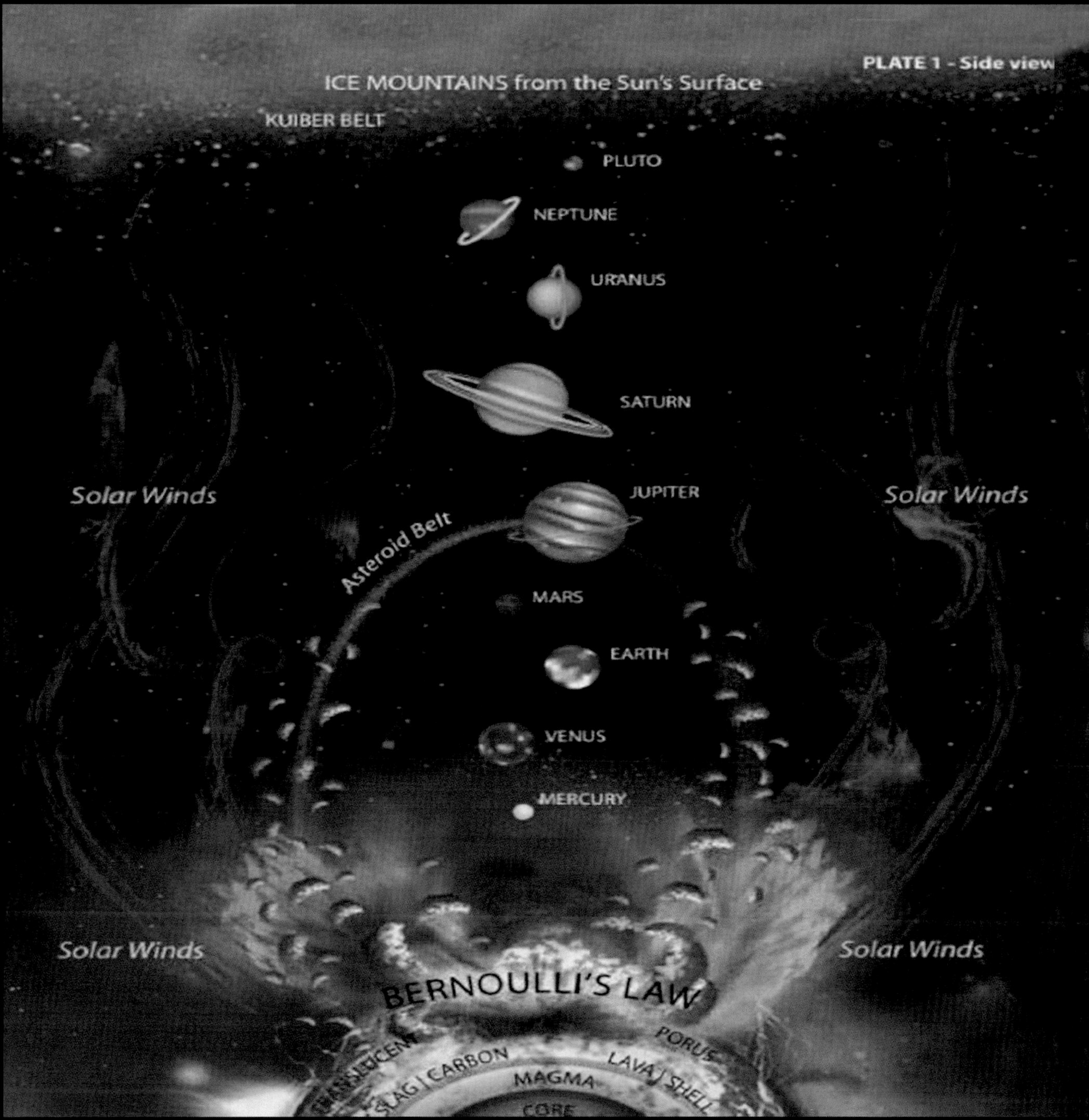

PLATE 1 - Side view

ICE MOUNTAINS from the Sun's Surface

KUIBER BELT

PLUTO

NEPTUNE

URANUS

SATURN

Solar Winds

JUPITER

Solar Winds

Asteroid Belt

MARS

EARTH

VENUS

MERCURY

Solar Winds

Solar Winds

BERNOULLI'S LAW

PORUS

CARBON

LAVA

MAGMA

SHELL

SLAG

CORE

FRONT VIEW

Plate 4

ICE MOUNTAINS FROM THE SUN'S SURFACE

1st Eruption of the Sun's Mantle

KUIPER BELT

2nd Eruption of the Sun's Mantle

PLUTO

NEPTUNE

URANUS

3rd Gas Matter Eruption of the Sun's Mantle

SATURN

4th Eruption of the Sun's Mantle

JUPITER

Jupiter ejection from the sun created the Asteroid Belt

METEOR/ASTEROID BELT

METEOR/ASTEROID BELT

MARS

EARTH

VENUS

MERCURY

5th Eruption of the Sun's Mantle

BERNOULLI'S LAW

Mantle of the Sun

1st to 5th eruption, caused by Water/Molten "Splash" & GALVANO Explosion

Solar Winds

Solar Winds

Slag Shell

SUN

Carbon Mantle

THE EARTH'S PATH DURING THE FOUR SEASONS

Plate 2

The Earth Travels from South 23.5° to North 23.5° Changing Climate Every Three Months

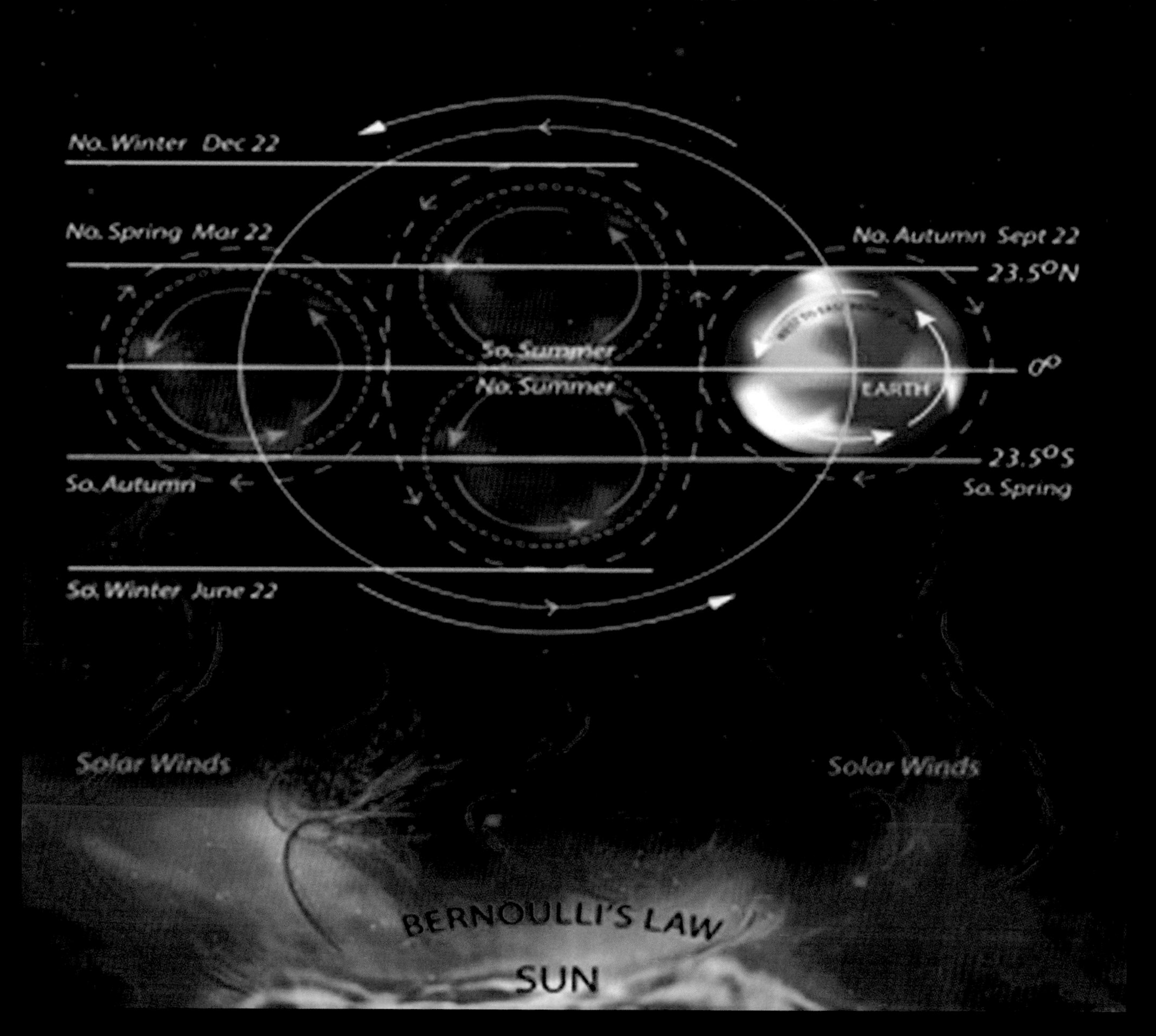

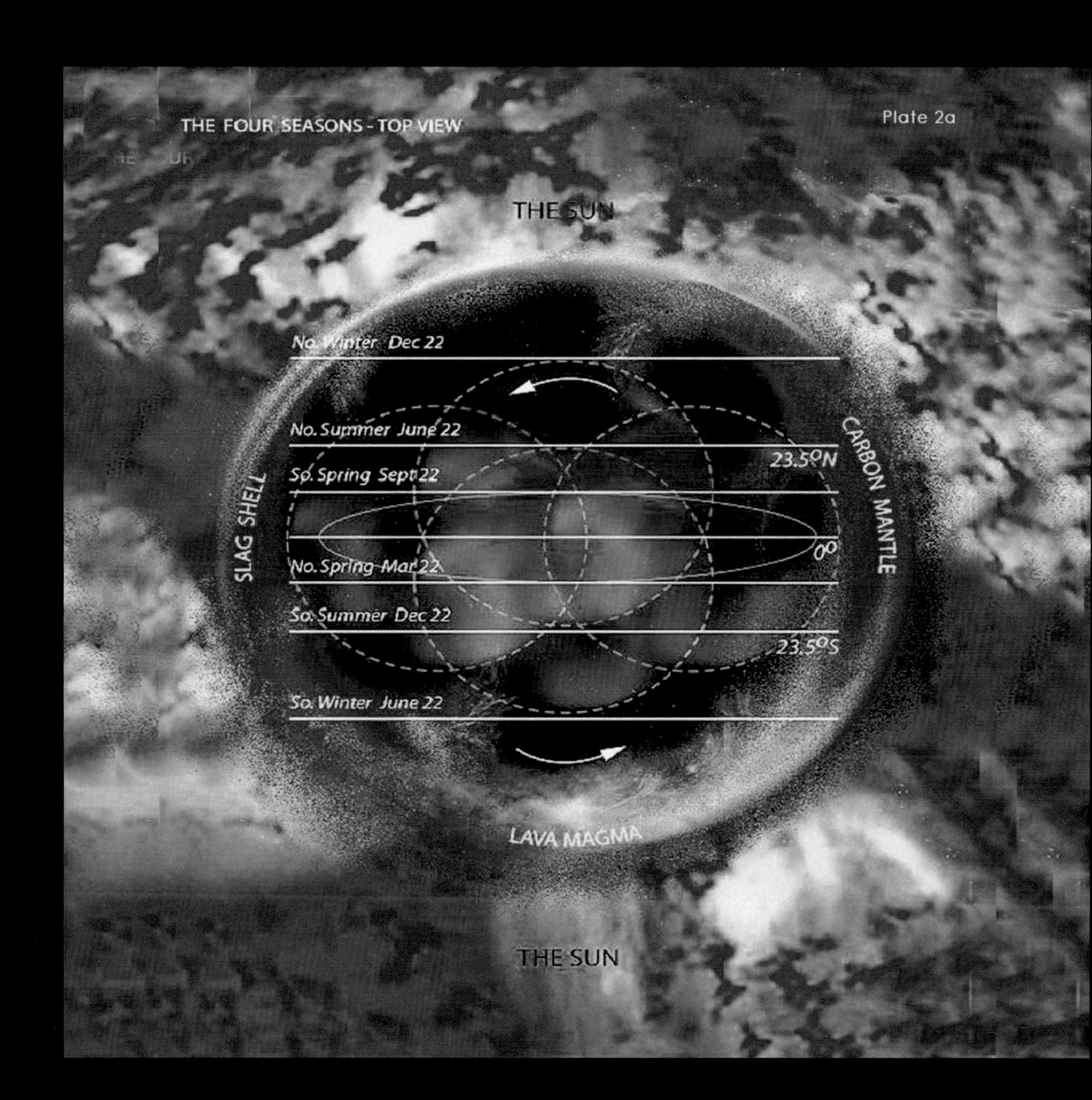
THE FOUR SEASONS - TOP VIEW
Plate 2a
THE SUN
No. Winter Dec 22
No. Summer June 22
So. Spring Sept 22
No. Spring Mar 22
So. Summer Dec 22
So. Winter June 22
23.5°N
0°
23.5°S
SLAG SHELL
CARBON MANTLE
LAVA MAGMA
THE SUN

Plate 5

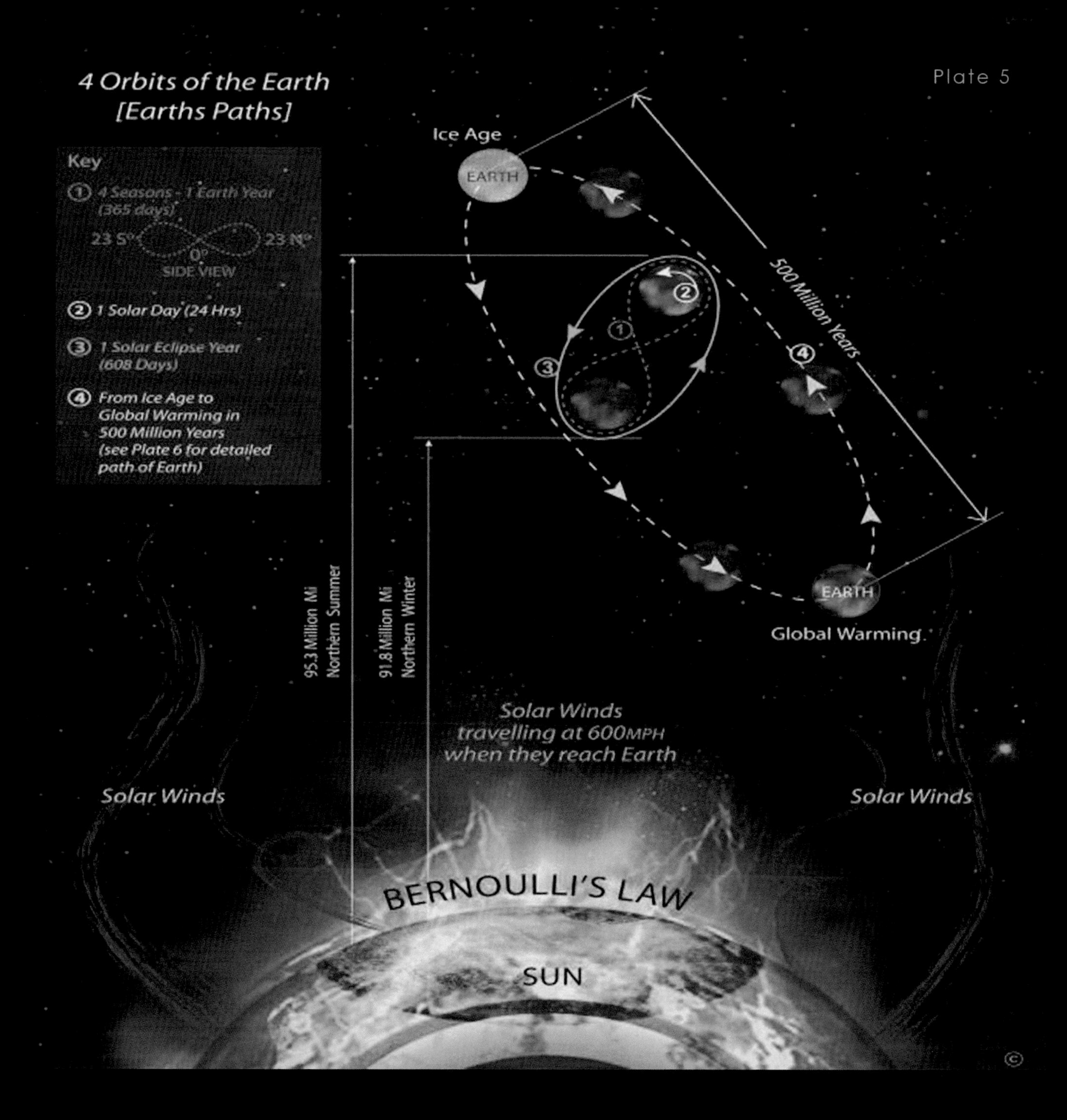

4 Orbits of the Earth
[Earths Paths]
Key
① 4 Seasons - 1 Earth Year (365 days)
23 S°
23 N°
0°
SIDE VIEW
② 1 Solar Day (24 Hrs)
③ 1 Solar Eclipse Year (608 Days)
④ From Ice Age to Global Warming in 500 Million Years (see Plate 6 for detailed path of Earth)
Ice Age
EARTH
500 Million Years
EARTH
Global Warming
95.3 Million Mi Northern Summer
91.8 Million Mi Northern Winter
Solar Winds travelling at 600MPH when they reach Earth
Solar Winds
Solar Winds
BERNOULLI'S LAW
SUN
©

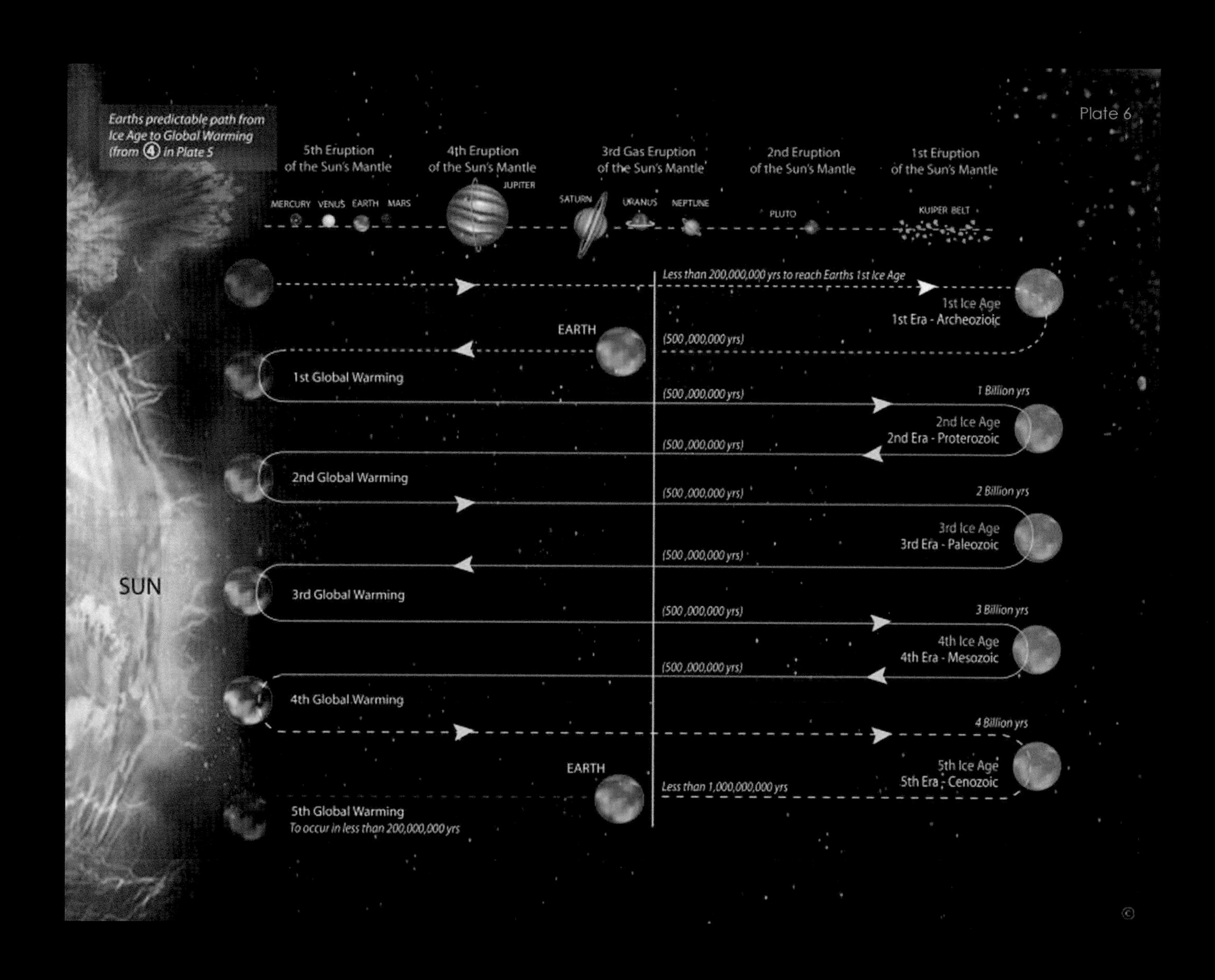

Earths predictable path from Ice Age to Global Warming (from 4 in Plate 5)
Plate 6
5th Eruption of the Sun's Mantle
4th Eruption of the Sun's Mantle
3rd Gas Eruption of the Sun's Mantle
2nd Eruption of the Sun's Mantle
1st Eruption of the Sun's Mantle
MERCURY
VENUS
EARTH
MARS
JUPITER
SATURN
URANUS
NEPTUNE
PLUTO
KUIPER BELT
Less than 200,000,000 yrs to reach Earths 1st Ice Age
1st Ice Age
1st Era - Archeozioic
EARTH
(500,000,000 yrs)
1st Global Warming
(500,000,000 yrs)
1 Billion yrs
2nd Ice Age
2nd Era - Proterozoic
(500,000,000 yrs)
2nd Global Warming
(500,000,000 yrs)
2 Billion yrs
3rd Ice Age
3rd Era - Paleozoic
(500,000,000 yrs)
SUN
3rd Global Warming
(500,000,000 yrs)
3 Billion yrs
4th Ice Age
4th Era - Mesozoic
(500,000,000 yrs)
4th Global Warming
4 Billion yrs
5th Ice Age
5th Era - Cenozoic
EARTH
Less than 1,000,000,000 yrs
5th Global Warming
To occur in less than 200,000,000 yrs
©

CHAPTER 5

TWENTY-FIRST-CENTURY ASTRONOMY

The Formation of Water and Our Solar System

The universe and the naturally occurring laws of physics, thermodynamics , gravity, mass and matter (solid, liquid, gas, and plasma) had all been well established 4.5 billion years ago. The cosmos and everything in it is comprised entirely of solar material or other 'star stuff', and the universe has existed for more than 14 billion years. It was calculated by going back 13.4 billion years at the speed of light for the earliest discovered starlight—that we know of—to reach Earth. This prehistoric galactic light was captured by the Hubble telescope using infrared lenses.

The four universal laws of thermodynamics:

- The general and overall law of thermodynamics: If two systems are in thermal equilibrium with a third system, they must be in thermal equilibrium with each other. This law helps define the notion of temperature.

- The first law of thermodynamics: Due to the fact that energy is conserved, the internal energy of a system changes as heat flows in or out of that system. Equivalently, machines that violate the first law (perpetual-motion machines) are impossible in nature or by experiment. Heat is transferred with the flow of thermal energy from one object to another; this rule applies both in nature and through experimentation.

- The second law of thermodynamics: The measure of any closed or isolated system cannot decrease. Such systems spontaneously evolve towards thermodynamic equilibrium—the state of the maximum measured amount of the system. Equivalently, machines that do not abide by the second law (perpetual-motion machines) are impossible and cannot exist.

- The third law of thermodynamics: The total amount of any pure substance in thermodynamic equilibrium approaches zero as the temperature approaches zero. The total amount of a system at absolute zero is typically zero and in all cases is determined only by the number of different ground states it has.

There have been suggestions of additional laws, but none of them achieve the generality of the four accepted laws, and they are not mentioned in standard scientific textbooks. The laws of thermodynamics are important fundamental laws in physics, and they are also applicable in all other natural sciences . Also read Maxwell's equations at your local library for proof and further insight.

Kepler's three laws of planetary motion, deduced by prolonged and tedious consideration of the observed position of Mars, are: (1) the planets move in ellipses with the sun at one focus; (2) the areas swept out by the radius vector in equal time intervals are equal; and (3) The cubes of the mean distances (half the major axis of the orbit) are proportional to the squares of the periodic times. Hot air in the presence of cold air results in inertia. These laws are sufficient to determine the position of a planet at any later time if its position is known at one time, and the dimensions and orientation of the orbit are known. The AP Theory obeys these laws.

Plate 3468

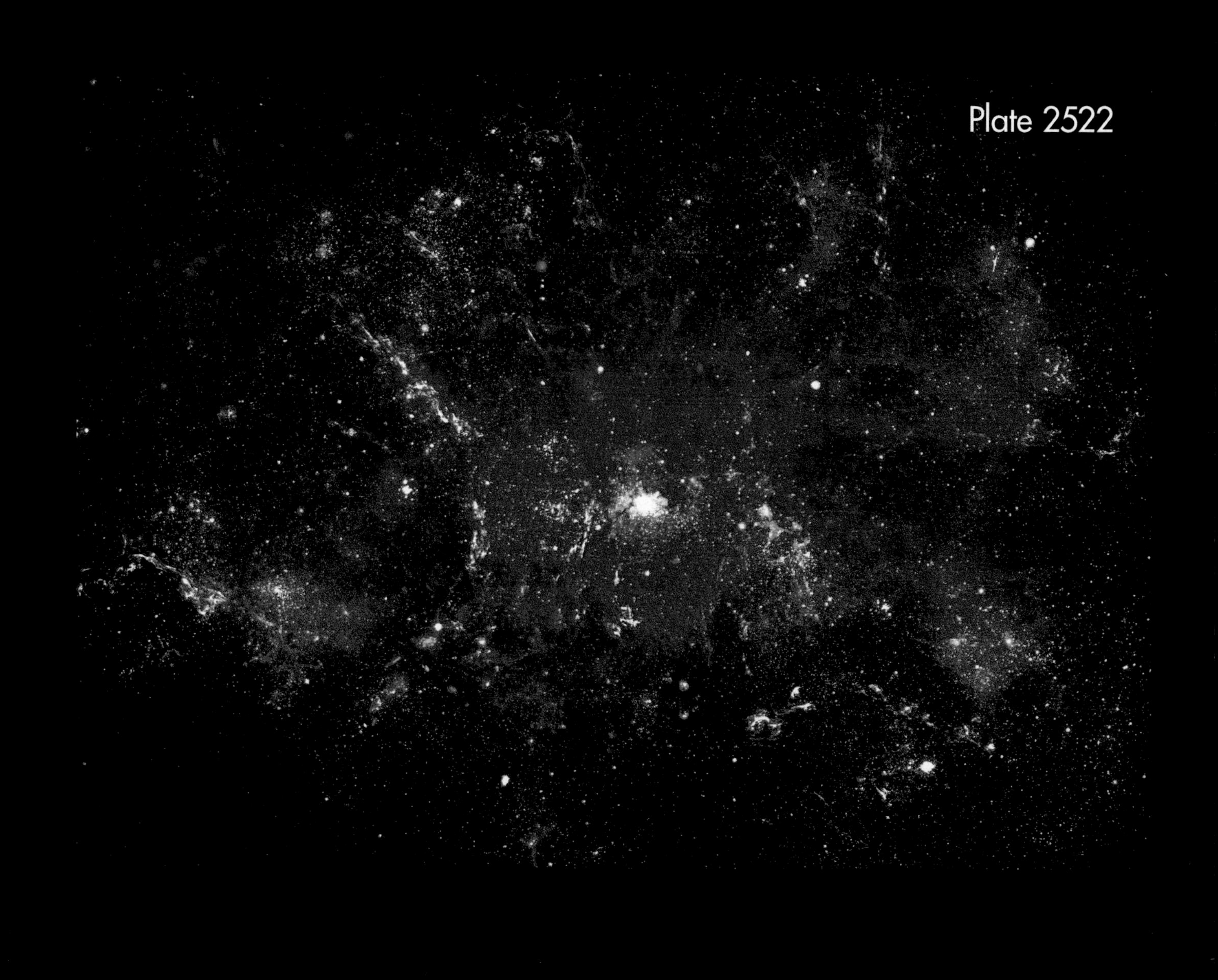
Plate 2522

Plate 0856

Plate 0597

Plate 8806

The Formation of Our Solar System

About 5 billion years ago, in our Milky Way galaxy, a vast molecular gas cloud of interlaced molecules of unstable nebula comprised mostly of Atomic Hydrogen (Plates 3468, 2522, 0856)combined with other neighbouring clouds of Noble gases (Plate 1149). Kinetic energy, inertia, and galactic winds drifted the intergalactic cloud of mixed gases into a 'dusty' area of the galaxy. This area was comprised of a large amount of electrically charged intrinsic fissile, benign atoms of sub-atomic particles (cosmic rays) and minerals left over from exploded supernovas. The mixed cloud of gases and galactic nebula continued to drift around the galaxy in a predestined path, propelled by the galactic winds and inertia. The drifting cycle eventually brought the cloud through an absolute-zero (–273.15 degrees Celsius) region of the galaxy.

This caused the mixed integrated gases and charged particles, which were collected on its travels, caused from shock waves from exploding stars around the galaxy, to condense and solidify into gigantic solid mountains of a solidified mixture of gases and ionized fissile atoms of captured particles and galactic nebula. The condensed mixture of galactic elements and gases were comprised of all the naturally occurring elements in our solar system (Plate 1234). Embedded in the ice-bound, frozen protostar were some extremely hot, gas-consuming ions of fissile elements. A fusion process began when friction from the electrically charged atoms ignited, and this was fed by the newly forming flammable elements of the condensed cloud. The ignited gases fused and formed a hot spot within the newly forming cloud, and our star was born (Plate 4619).

Our Sun is 125 times greater than the diameter of Earth and inhales its Hydrogen fuel from the galactic gases by heaving in Hydrogen and exhaling or "heaving" out Helium every six minutes as the mature hot yellow dwarf star it is today. Our Sun exhales helium, which forms the Heliosphere, and is expected to continue to respire in this manner for the next 5 billion years.

The fused Hydrogen formed a hot spot, which then rendered molten elements from the impurities in the molecules of the galactic elements. The impurities from the molten elements in the hot spot produced a slag cauldron within our newly forming star. The extremely hot cauldron then began to thaw our partially frozen young star 4.9 billion years ago.

The mountains of condensed gases thawed and breached the Sun's shell and then its iron core of liquefied mixed gases and matter. When the charged liquids made contact with the molten elements, it caused an internal cataclysmic galvanic explosion (Plate 8741). This event most likely caused a gas fission explosion to occur within the hot spot of the cauldron in accordance with Galvani's law. The explosion caused the superhot neutrons to lose their gluon and melt. In that hostile environment, the extreme heat then stripped the positively charged and negatively charged atoms bare to a state of passive, unstable ionized and un-ionized electrons (–) and protons (+). Fusion caused the Hydrogen fuel in the Sun's core to convert into Helium. This event was the first solar nuclear reaction from nuclear energy produced by our Sun. The Sun is said to be a nuclear fusion reactor in perfect balance.

Wolfgang Pauli's law of conservation of energy predicted there must still be an undiscovered particle. That particle turned out to be a Neutrino. A neutrino is electrically neutral, is the smallest particle and cannot be broken down. Pauli's exclusion principle states, 'Two electrons in an atom cannot simultaneously occupy the same quantum or energy state.' His law of conservation of energy explains how the fission process could have begun from the extreme heat and pressure within the cauldron. The process then continued and caused the bare nuclei to collide and produce enough energy to separate the superheated passive atoms. The collision of the stripped positively charged and negatively charged atoms in that extreme environment caused them to split into a subatomic state of protons, neutrons (discovered by Chadwick in 1932), electrons (discovered by Thompson in 1897), and photons. The separated atoms then split again into their sub molecular state of quarks, neutrinos, and leptons and then into the Higgs Boson Exitrons (Plate 4668), recently discovered by the Higgs Boson experiments at CERN (European Centre for Nuclear research) in Switzerland's Large Hadren Collider. The combination of the ionic energy from the fusion and the internal gas steam pressure from the fission caused an explosion with the force of an exploding star. This cataclysmic explosion (Plate 8741) caused the first nuclear internal solar Coronal Mass Ejection (CME) of unstable subatomic particles and pressurized impregnated subatomic steam from gas molecules of Hydrogen, Oxygen and others.

Plate 1234

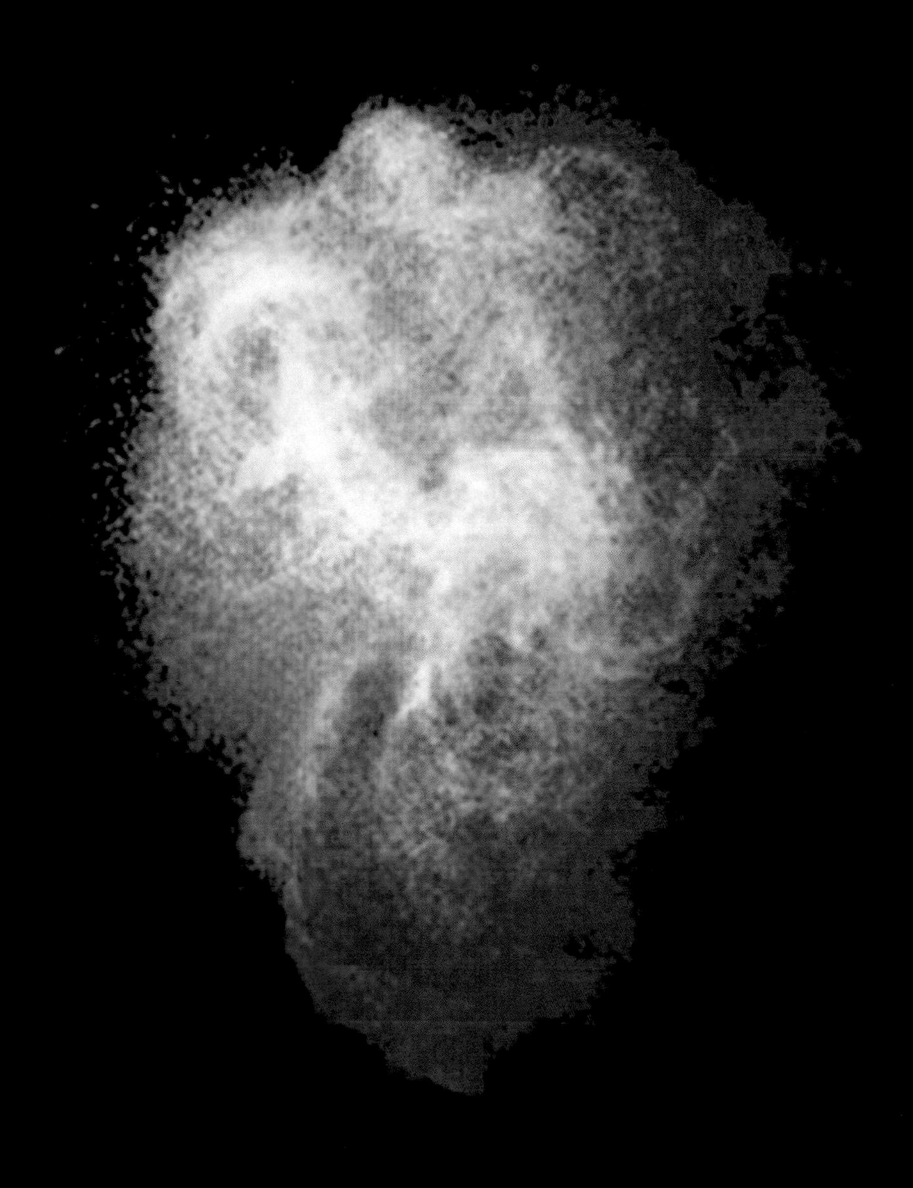

Plate 4619

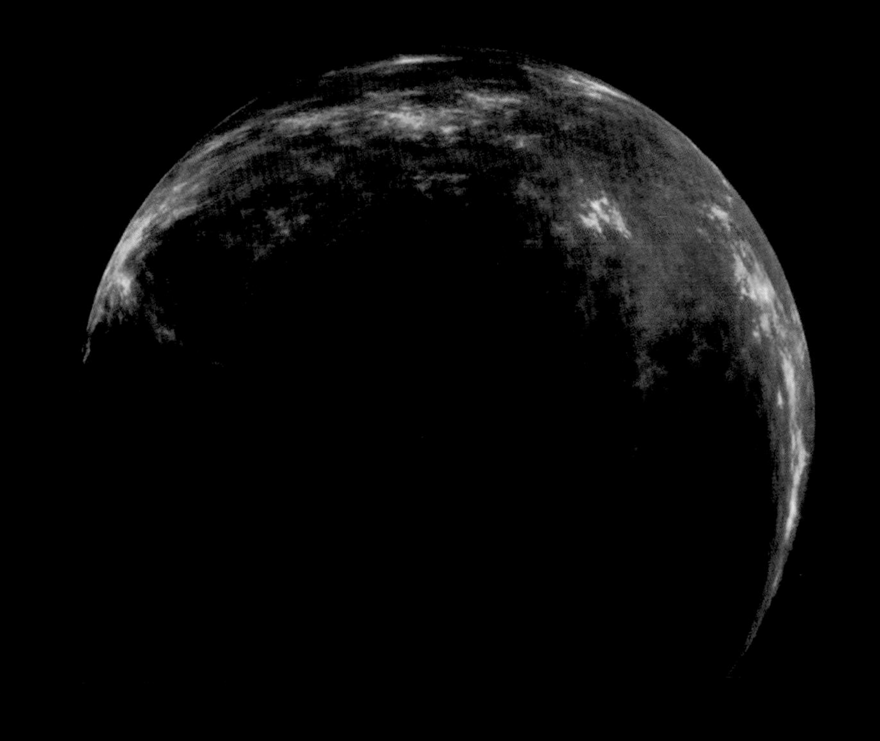

The great variation in temperature and newly produced chemical reactions caused a tremendous explosion (CME) within the cloud with a force equal to a small exploding star forming the Oort Cloud and the Kuiper belt. The tremendous internal explosive pressure pulverized and (CME) ejected the mountains of thawing unstable subatomic photosphere and solidified gases into each other. The extreme explosion (Plate 0917) ejected the hot molecules of gases and matter upwards with tremendous force from within the confined area of the cocooned cauldron of our newly forming Sun. The tremendous gas steam pressure from the first (CME) internal atomic explosion blasted the passively separated lighter in atomic weight atoms of Atomic Hydrogen into the atoms of oxygen which were heavier in atomic weight. This combination of extraordinary events caused the particles to become an unstable compound of H2O.

The molecules of the other subatomic noble gases that are light in atomic weight were blasted from our protostar into the other subatomic particals (cosmic rays) of gases that are heavier in atomic weight. This internal explosion impregnated with the hydrogen atoms the oxygen atoms at a ratio of H2O , compounding them into one molecule of unstable water. Nuclear fission caused another powerful (CME) internal explosion, resulting in the compounding of the Hydrogen and Oxygen from the pressurized split molecules. The explosion continued blasting the subatomic particles of Hydrogen and Oxygen into one another until they became a compound of H2O and unstable water as we know it.

When the compounded Hydrogen and Oxygen atoms were exploded into the absolute-zero temperature (–273.15 degrees Celsius) of the cosmos they fused (Plate 8555). The severe, solidifying cold and the tremendous high-pressure ejection from the hot spot of the Sun into the absolute-zero temperature of the cosmos compounded them further. The extremely low intergalactic air pressure gradient and temperature froze and fused the split molecules of water vapour into a cloud of frozen compounded gases called The Thiea Cloud. The explosion then compounded and fused the molecules into H2O, CO2, N2 + O2, and so on forever and ever and thrust those outwards into interplanetary space.

Plate 0856

CHAPTER 6

The first solar eruption (CME) blasted off all of its shell and simultaneously established the Kuiper belt discovered in 1992 and extends 3 billion miles from Earth. The first eruption also produced the Oort (Heliospheric) cloud, discovered by NASA's Voyager 1 in 1977 which encompasses our entire solar system. The Oort cloud surrounds the solar system and contains an infinite number of carbon shelled comets. Its outermost boundary is deemed to be the source of our comets. The Heliopause separates the interplanetary region of Oxygen and Nitrogen from the intergalactic medium of Hydrogen and Helium-3 and extends about 11.3 trillion miles out from the Sun.

The galactic gases of mostly Atomic Hydrogen are surrounding the Oort cloud which is made up of interstellar medium and the Sun's fractured mantle which are the comets of today. Helium gas is not as light in atomic weight as the surrounding Hydrogen which it mixes with because of close atomic numbers. The noble Heliospheric gases of Helium which surround and compress the Earth's atmosphere are keeping it from rising above the Exosphere. Gravity does not hold down our atmosphere, lighter gases are. These low-pressure Heliospheric gas atoms, which are light in atomic weight and of a smaller size are heavier as a whole and are primarily comprised of Helium. The decreased air pressure of Heliospheric interplanetary gases extends about 11.3 trillion miles out from their solar origin and is bordered by the Heliopause, where they finally slow and then stall. The Heliopause is an invisible border separating smaller Hydrogen intergalactic gases that are lighter in atomic weight from the heavier Heliospheric sun-produced Helium gases. When the Heliopause was established, it separated the Heliospheric gases from the intergalactic gases 4.8 billion years ago. Earth's mesosphere is about 65 miles above sea level at the Kármán line , which is the doorway to the Heliosphere. It's the border separating the larger atoms of Earth's atmospheric gases of Oxygen and Nitrogen, which are heavier in atomic weight and have high air pressure, from the sun-produced Heliospheric gases of Helium, which have lower air pressure, lighter atomic weight, and smaller atoms than our atmospheric gases. The heliosphere extends to the helioshock.

The stratosphere or ozone (O3) layer separates the Earth's expanding atmospheric gases are pushing against the Heliospheric gases with an air pressure at 14.7 pounds per square inch from the decreased air pressure (1.5 pounds per square inch) of the compressing mesospheric gases of Hydrogen, Helium, etc., which are lighter in atomic weight. These low-pressure sun-produced Heliospheric gases are mostly Helium, which surround and compress Earth's atmospheric gases of Oxygen and Nitrogen. Our atmosphere is surrounded with a higher air pressure of 14.7 pounds per square inch (760mm Hg) than the Heliosphere with 1.3 Lb. psi.. Astronomers try to tell us without an explanation of how gravity is selective and why it is only attracting and suspending Oxygen and Nitrogen but not attracting the other gases like Hydrogen A.W.1, Helium A.W. 4, CO, Methane, Radon A.W. 86 discovered by Fredrick Earns Dorn in 1900, Xenon A.W. 54, Krypton A.W. 36 and Neon were discovered by Morris W. Travers. Argon was discovered by John Strutt in 1894 and has an A.W.18 Carbon Dioxide (CO2) and N2O etc. This evidence proves it is not gravity holding down Earth's atmosphere but rather Heliospheric gases, which have smaller, lighter atoms and form the natural Helium ceiling that prevents the atmosphere from expanding any further.

Outer space is not a perfect vacuum but is made up of tenuous plasma awash with charged photons and particles from exploded stars, electromagnetic fields, and the occasional star. As gases expand they filter the impurities out and become lighter (purer) as they rise and become cosmic. The AP theory states the Heliopause, discovered in 1986 is an invisible border separating the low-pressure interstellar gases from the Sun's atmosphere. Hydrogen has an atomic weight of 1 atomic mass unit or more, from the intergalactic gases of the heliosphere. Atomic Hydrogen, which have smaller atoms, atomic weight and an air pressure that is lesser than that of Heliospheric gases of Helium. The universal gases have a predicted atomic weight of .0998 or less and are yet to be discovered.

A *second* internal, high-pressure galvanic eruption (CME) propelled more subatomic particles of fractured, condensed photosphere upwards into space, establishing Pluto, Eros, and the Kuiper belt 4.7 billion years ago. The Kuiper Belt was discovered in 1992. The solar debris of rockey bodies exploded from the Sun's photosphere and still comprises a large area of our solar system. Most of the comets in our solar system can still be found in the outer reaches of the Oort cloud.

The Sun's ruptured lava shell allowed the melting condensed mountains of previously formed water and subatomic particles of mixed gases and matter to flood into the solar cauldron. This mixture of liquid and magma coming into contact with each other caused a powerful galvanic explosion. The tremendous internal explosion caused the gigantic cloud of newly formed water vapour to be ejected through the solar shell into the absolute-zero environments of the cosmos and it drifted into interplanetary space where newly formed and ejected Earth collided with it this event is known as the Theia collision (Great Impact). The melted remainder of the water flooded into the photosphere's gaping volcano-like hole, which was blasted open and exposed by the first and second eruptions.

A *third* hydrostatic eruption (CME) of extremely high pressure from the subatomic gas steam molecules and subatomic elements of water vapour manifested from fission, then blasted into the absolute-zero temperature of interstellar space. The Sun's photosphere and noble gases with the lowest atomic weight (e.g. $H2$ + He) travelled farthest, where they solidified and formed the gaseous planets Neptune and Uranus. The third solar explosion propelled Saturn 840 million miles from its planetary exit point from the Sun (Plate 1). The remaining split atoms of elements and gases with heavier atomic weights (e.g. $CO2$) came to rest closer to the Sun, forming the atmospheres of some regions of the terrestrial planets Mercury, Venus, Mars, and Jupiter (plate 3).

Earth's ecosystem formed its atmosphere of Oxygen gas from living plants and its nitrogen gas from its amino acidic and alkaline of disintegrated vegetation and elements (ammonium nitrate, etc.), making contact with each other in the presence of water. These disintegrated elements were produced from Earth's natural and constant vibration. All the naturally occurring elements on Earth have atomic weights of between 1 (hydrogen) and 94 (plutonium) atomic mass units, and they all originated from and were produced by our Sun.

JUPITER

When Jupiter exploded from within the Sun's extremely hot cauldron during the Sun's fourth eruption (CME), it fractured more than 10 per cent of the Sun's outer shell. Jupiter's violent exit formed the asteroid belt from ejected, pulverized photosphere and solar mantle material. The asteroid belt is located in an arc from the Sun's surface between Mars and Jupiter (plate 4).

Every twelve years, Jupiter travels directly behind the Sun and cannot be seen from Earth for only twelve hours. Considering this, it takes 5.10 years of the 12-year cycle for Jupiter to reach 30 million miles above the plane of the Sun, and after another 5.10 years, it returns to a point directly behind the Sun, as seen from Earth, without crossing the Sun's equator. Jupiter is in a position in the sky to be seen every day of the year and is travelling in a 'great circle'. These facts imply that Jupiter is not circling around the equator of the Sun but that Jupiter and all the other planets are circling above the celestial poles of the Sun at slower speeds and over shorter distances than are presently supposed. In order for A (Jupiter) to orbit B (the Sun), A must cross B's equator twice before it can be described as an orbit. Jupiter does not cross the equator of the Sun twice in its 10.8-year journey above the celestial poles of the Sun.

For example, as seen from the Earth, the Sun is 0.5 degrees across and the height is 0.25 degrees. At the distance (from Earth) to Jupiter when it is at the other side of the Sun, the size of the region blocked is 0.25 degrees—that is, sin (0.25 degrees × 577 + 93) million miles. This works out to a bit less than 3 million miles. The distance of Jupiter above the plane is 577 × sin (3 degrees) or 30 million miles. The 3-degree inclination is more than enough to see Jupiter over the top of the Sun.

The remainder of the terrestrial planet's substance and matter was ejected during the Sun's fifth internal eruption (CME), forming the terrestrial planets entirely from solar material and following directly after Jupiter's ejection (Plate 9533).

The solar wind eventually established the anticlockwise elliptic and predictable paths above the Sun's celestial poles of all the planets. Each and every one of the planets is affected by the solar wind. This 1.8-million-mile-per-hour solar force causes the planets to spin on their axis and rotate in their orbits. The powerful solar wind and atmospheric pressure are also the sources of their gravity, auroras, and magnetism. The solar force also perpetuates the planets in their orbital and elliptical paths. Solar wind power is the force which makes the planets circle above the Sun's celestial poles and its 100,000-cubic-mile vast, gaping planetary exit point from the Sun. The solar wind will continue to animate the solar system millions of miles above the Sun's celestial poles for another 5 billion years as it has for the past 4.7 billion years in accordance with Bernoulli's law (plate 1A). If the Sun were the face of a clock, the exit point of the planets and solar wind would be between 2 and 3 o'clock.

Jupiter's violent eruption exposed the Sun's hydrogen fission molten cauldron and unleashed the powerful solar wind at a speed of 400 kilometres per second or 9.6 x 107 (Plate 1 A) kilopascals or 1.8 million miles per hour. The solar wind (plate 4560) carries exhausted Helium gas out 100,000,000 astronomical units from the Sun, reaching 12 trillion miles out to the Heliopause, where it slows down, loses inertia, and stalls (Plate 1A). The expelling solar gases established the Oort cloud (aka the Heliosphere) 4.7 billion years ago, and the interplanetary boundaries are defined by the Heliopause, which borders the Oort cloud and separates it from the intergalactic gases. The ozone (O3) layer is the border that separates the Earth-produced Oxygen and Nitrogen gases from the solar-produced interplanetary gases of Helium, Lithium, etc. **and are animated by the solar wind,.** These interplanetary gases of Helium etc. are lighter in atomic weight then the interstellar low-pressure Hydrogen gases which extend to the heliopause and are presumed to separate the heliospheric Helium gases from the galactic gases of Hydrogen. Hydrogen gases are lighter in atomic weight and have a predicted lower air pressure than the interplanetary Helium gases.

As our new solar system cooled, the fission process continued to consume and further heat up our newly forming stellar oven. The Sun is 10,000 times denser than Earth and is 1.3 million times greater in volume . The fission process converted 0.15 per cent of a benign gas-and-cosmic-dust cloud, which was once frozen and then thawed, into our solar system and the 8,000-degree-Celsius star we know today as our Sun. The fission process continued, consuming the cloud for approximately the next 500 million years while forming our solar system and increasing its temperature. The Sun inhales galactic Atomic Hydrogen as fuel and exhales (solar wind) Helium a distance 12 trillion miles out from the Sun to form the Heliosphere. The result of the fusing of the remainder of the cloud was a 27-million-degree-Celsius yellow dwarf star that evolved into our Sun only 4.3 billion years ago (Plate 0924). The Sun is estimated to be made up of 90% Hydrogen and will continue burning for at least another 4 billion years.

Plate 4560

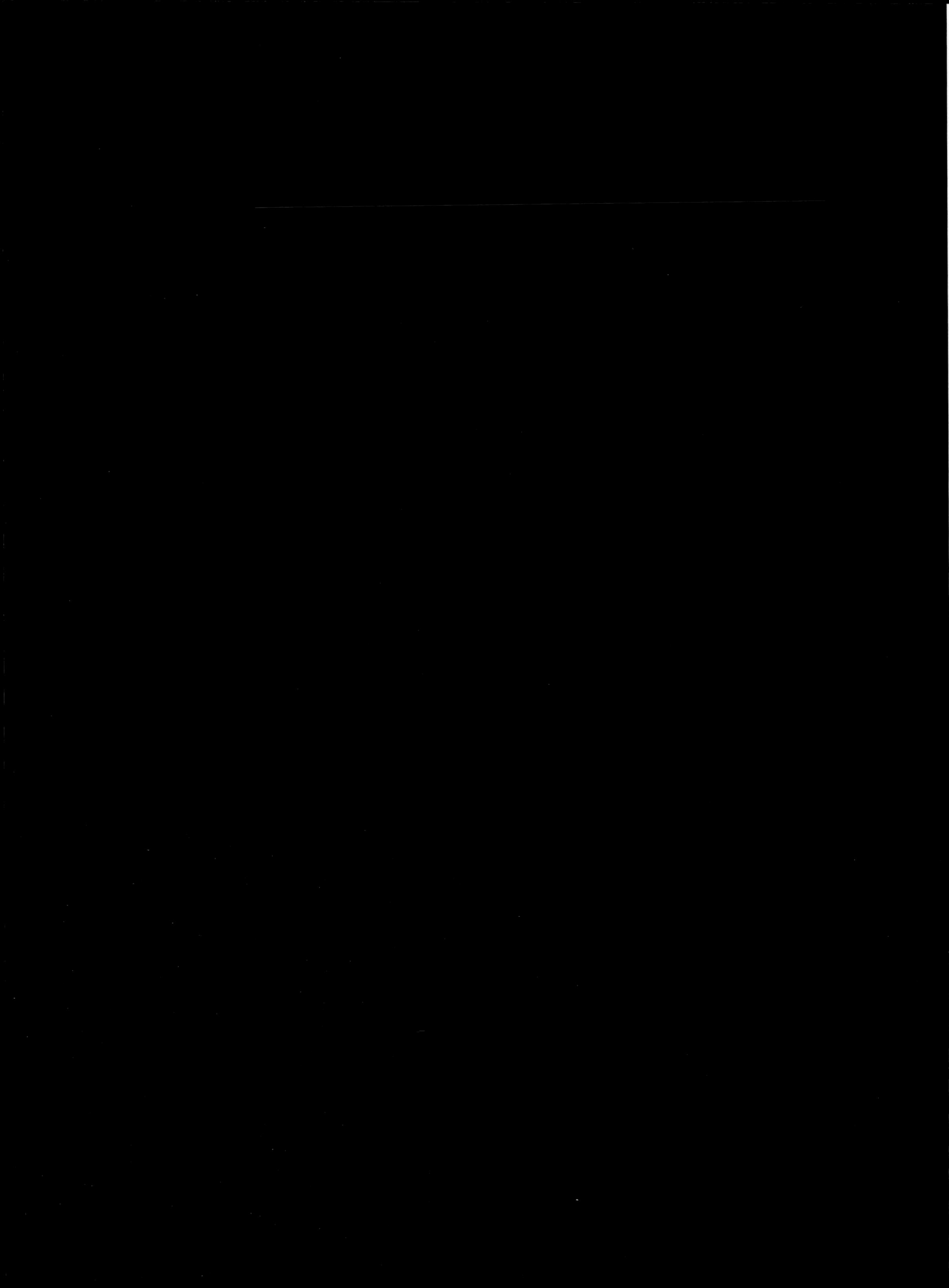

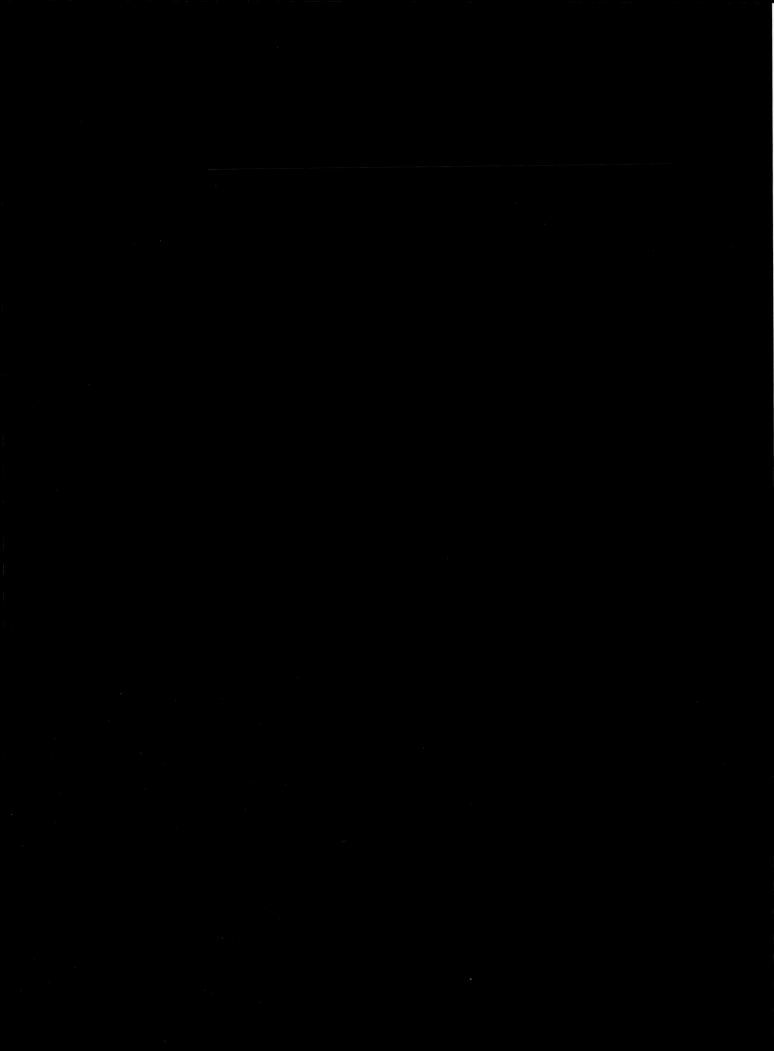

Plate 0924

Question 1: Angle measure of arc of eclipsing Jupiter .

the answer is simply the angular measure of the sun look at the diagram below

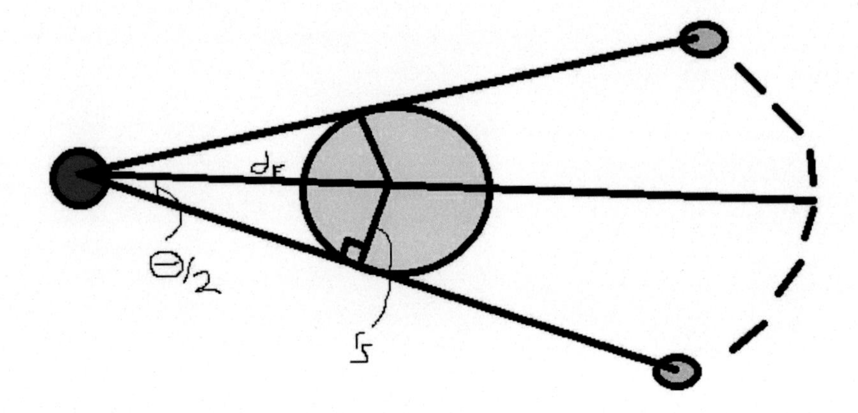

r_s- radius of the sun $(6.955 \times 10^5 \text{ km})$ and d_e- distance of the earth to the sun (1 AU $\approx 1.50 \times 10^8$km)

thus that arc measure is going to be $\text{Sin}\left[\frac{\theta}{2}\right] = \frac{r_s}{d_E} \rightarrow \theta \approx 0.53\,° = 31.8"$ (Minutes of angle)

Question 2:

How long will this be observed for?

This is quite complicated the solution requires you to factor in the orbital velocity of Jupiter as well as the Earth, after some extremely lengthy calculations you can not get the solution in closed form (that means you can't get an exact solution) but the final answer you get is $\sim$ 16.7 hrs given orbital velocities $v_{\text{earth}} = 30 \text{ km}/s$ $v_{\text{jupiter}} = 13 \text{ km}/s$ and distances from the sun of $d_E = 1$ AU and $d_J = 5$ AU

this is easier to estimate if you just assume Jupiter is fixed (which is pretty good since earth is moving twice as fast) in that case you can figure out when there will be a clear line of sight (LOS) from earth to Jupiter when the vector between the earth and Jupiter is tangent to the sun, it just lay down the simplest vector algebra but will not solve it entirely as there is no closed form as said above.

Fig 1. where Jupiter and the earth is at time t = 0

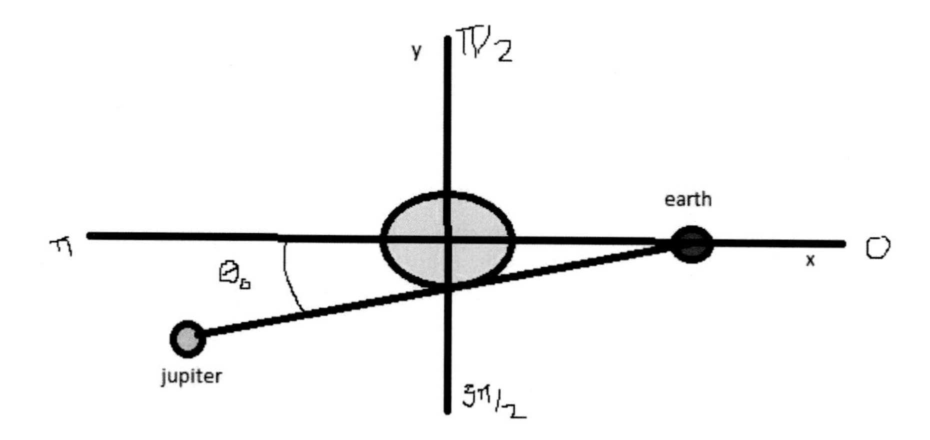

OK so lets assume Jupiter is frozen at a fixed point given in Cartesian coordinates as $J = (d_J \text{Cos}[\pi + \theta_0], d_J \text{Sin}[\pi + \theta_0])$ where $\theta_0 \approx \theta/2 = 0.27°$ this is the original pt where we have a line of sight to Jupiter from earth obviously if it were at π we wouldn't be able to see it since wed have to look right through the sun (refer to fig 1) so we need that original θ_0 off set we calculated before.

Now some time later the earth will have moved to the other side so we will have a new line of sight fig 2. we see the earth has moved to the new Position where we will see Jupiter again but on the other side of the sun

Fig 2. at time t = t'

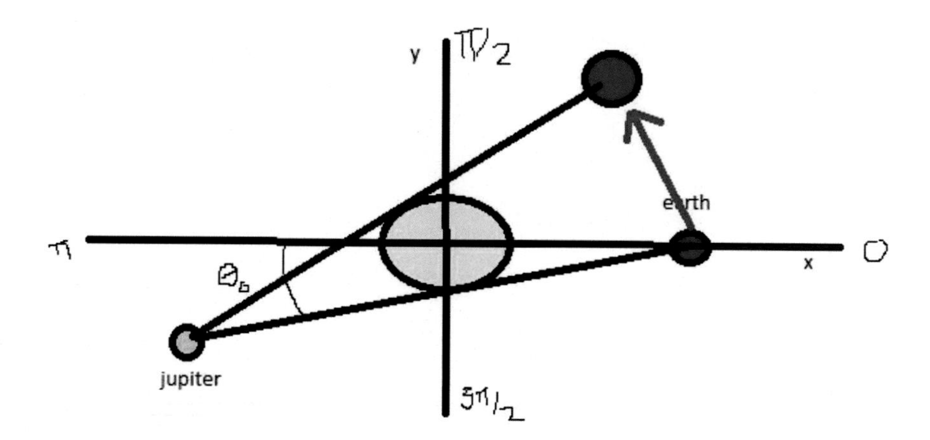

What is this time t'?

we describe the location of the earth in Cartesian coordinates as $E = (d_E \operatorname{Cos}[\omega_E t], d_E \operatorname{Sin}[\omega_E t])$ where the location depends on time as it should. Now the condition that we have a LOS to Jupiter must be that the vector connecting Jupiter and earth given as $R = E - J$ for a vector pointing from Jupiter to earth. must be tangent to the surface of the sun. Well any tangent unit vector of the sun is given as $T = (-\operatorname{Sin}[\theta], \operatorname{Cos}[\theta])$ where θ is generically some pt on the sun using the same coord system as we have above to clarify, suppose you want to know the tangent vector to the sun at the pt $\theta = 0$ thats the right most surface of the sun in the picture this gives $T = (0, 1)$ which is a vector point upwards as you'd expect now suppose you wanted to know the tangent vector at the "north pole" of the sun (upper most point of the sun in the picture) that corresponds to the angle $\theta = \frac{\pi}{2}$ giving $T = (-1, 0)$ which is a vector pointing to the left as you'd expect.

So if vectors R and T line along the same line then $R = \lambda T$ where λ is some real constant. we have to solve when this is true

looking at each component of the equation

$$\frac{d_E \operatorname{Cos}[\omega_E t] - d_J \operatorname{Cos}[\pi+\theta_0]}{\left(d_E^2+d_J^2-2 d_E d_J \operatorname{Cos}[(\omega_E-\omega_J) t-\theta_0]\right)^{1/2}} = -\lambda \operatorname{Sin}[\theta], \quad \frac{d_E \operatorname{Sin}[\omega_E t] - d_J \operatorname{Sin}[\pi+\theta_0]}{\left(d_E^2+d_J^2-2 d_E d_J \operatorname{Cos}[(\omega_E-\omega_J) t-\theta_0]\right)^{1/2}} = \lambda \operatorname{Cos}[\theta]$$

Notice I divided by $\left(d_E^2 + d_J^2 - 2 d_E d_J \operatorname{Cos}[(\omega_E - \omega_J) t - \theta_0]\right)^{1/2}$ this makes things considerably easier as R is now a unit vector so T and R must have the same magnitude ie 1, from the picture its easy to see that at time $t = 0$ it must be $r = T$ where r denotes the normalized vector R and then we must have at time $t = t'$, $r = -T$

This leads us to

$$\frac{d_E \operatorname{Cos}[\omega_E t'] - d_J \operatorname{Cos}[\pi+\theta_0]}{\left(d_E^2+d_J^2-2 d_E d_J \operatorname{Cos}[(\omega_E-\omega_J) t-\theta_0]\right)^{1/2}} = \operatorname{Sin}[\theta], \quad \frac{d_E \operatorname{Sin}[\omega_E t'] - d_J \operatorname{Sin}[\pi+\theta_0]}{\left(d_E^2+d_J^2-2 d_E d_J \operatorname{Cos}[(\omega_E-\omega_J) t-\theta_0]\right)^{1/2}} = -\operatorname{Cos}[\theta]$$

this can be put in the form

$$\frac{d_E \operatorname{Cos}[\omega_E t'] - d_J \operatorname{Cos}[\pi+\theta_0]}{d_E \operatorname{Sin}[\omega_E t'] - d_J \operatorname{Sin}[\pi+\theta_0]} = -\operatorname{Tan}[\theta]$$

from a geometric argument shown in fig 3., $\theta = \frac{\pi}{2} - \theta_0 + \omega_E t'$

fig 3.

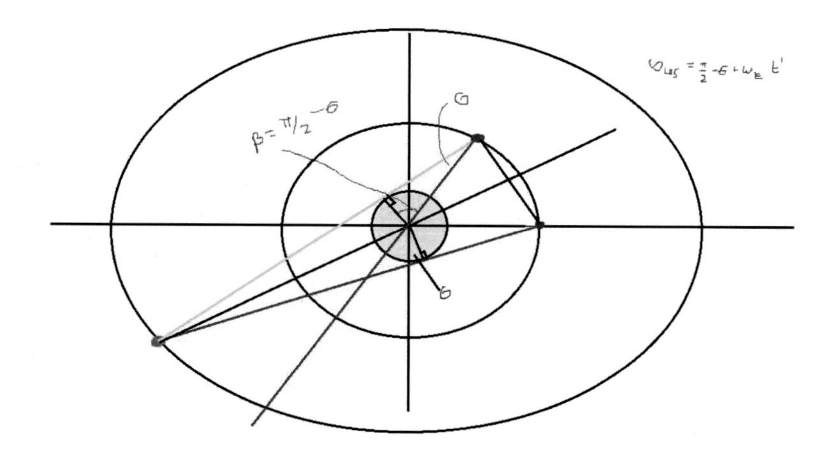

we arrive at

$$\frac{d_E \operatorname{Cos}[\omega_E t'] - d_J \operatorname{Cos}[\pi+\theta_0]}{d_E \operatorname{Sin}[\omega_E t'] - d_J \operatorname{Sin}[\pi+\theta_0]} + \operatorname{Tan}\left[\frac{\pi}{2} - \theta_0 + \omega_E t'\right] = 0$$

thus with the following $\omega_E = 2.\times 10^{-7}\ \text{rad}/s$ $\left(\omega_E = \frac{v_{\text{earth}}}{d_E}\right)$, and the distances used before $t' \approx$ 15 hrs adding the effect of a moving Jupiter extends the time to $t' \approx 16.7$ hrs. Since the orbital inclination of Jupiter is ~1° this effect won't always be observable for any given year (Jupiter may pass above or below the sun or through the sun but not through its center as we assumed here). We would have incorporate the 1° inclination to be able to solve the transit time completely but this gives you a maximum time so it will definitely be around ~10 hrs depending on how close Jupiter comes to passing through the light of sight that goes through the sun.

CHAPTER 7

Earth's comfortable distance from the Sun (Goldilocks zone) and its temperature and atmosphere, which exclusively sustain the only life in our solar system, are evidences of Earth being the vortex of the moisture in our solar system. The other planets are without Oxygen or Nitrogen and the temperatures make life unsustainable. When the newly formed molten ball of iron that we now call Earth exploded out from the Sun, it passed through a frozen cloud of ice and water vapour we call the Theia Cloud. The extremely hot sphere collected the thawing water, spreading it over the entire surface of the Earth as it travelled through the frozen cloud of ice water. This single event known as the Theia Collision is credited with the first presence of water on Earth (Plate 0856).

The Earth's atmosphere and its ecosystem began when naturally occurring acidic, lipids, alkaline minerals, amino acids and other bases of life came into contact with each other in the presence of water. When the acid and alkali are joined in water Nitrogen gas is produced and released. Decaying vegetation is also the source of Nitrogen in our atmosphere. In the presence of rainwater lightning, Nitrogen Oxide gas is produced. Under the proper conditions, photosynthesis from sunlight caused this combination to inevitably form life, for instance fungi, amoebas, bacteria, and algae.

These microorganisms were the first life that formed on Earth. If the process were projected into the future, these evolving new life forms could have eventually contracted an herbivore parasite. If projected further, the herbivore parasite could have then just as likely contracted a carnivorous parasite. Thus life as we know it today could have formed and evolved from this scenario.

There seems to be a mass extinction about every 25 million years coinciding with Earth's Ice Ages.

The AP theory predicts that the Earth will be discovered to be circling 57 million miles per year above the Sun's celestial pole. It also predicts that all the planets will be discovered to be orbiting above the Sun's 100,000-cubic-miles planetary and solar-wind exit point. The Earth's predicted orbital speed when calculated will be about 6,500 miles per hour—not 67,000 miles per hour, as is commonly supposed from the now disproven Theory of Relativity and because of being bereft of reason or any new ideas it's being taught detrimentally through misinformation. If that were true, then the Moon and the international space station which is not affected by gravity would also have to travel at 67,000 miles per hour to keep up with the Earth, which we now know does not happen. The Earth is not orbiting 586 million miles per year around the Sun's equator. The AP theory predicts that this will be discovered with advanced instruments that the Earth is travelling through space at much slower speeds and over much shorter distances than supposed.

The Earth was formed 4.5 billion years ago, it cooled 3.8 billion years ago, and it cannot be attracted to the Moon because the Moon and Earth have one positive and one negative gravitational system, which keeps them repelled from each other. Earth's gravity is caused by its 1040 mph spin, its atmospheric pressure of 14.7 lb. per sq. in. and the 90 degree tilt from north to south over a 12 month period. Gravity has never been proven to have any influence over the Moon's movements. The Moon has its own gravity and moves away from Earth at a rate of 1.5" per year further proving the moon's gravitational independence from Earth. A rocket must travel at 7 miles per second to escape Earth's gravity; Earth's magnetic field is created by its iron core. Magnetic fields exist in the cosmos and are caused by shock waves. Photons attract and stick to individual atoms which then form molecules which than form everything else. Attraction does not work at the cosmic level.

The Shoemaker–Levy asteroid was gravitationally attracted to Jupiter from a greater distance than Earth is from the Moon. This is further evidence that the positive and negative electromagnetic fields of the planets keep them equidistant from each other and not from an unexplainable force of cosmic gravity. If in fact gravity existed in the cosmos (contrary to Newton's law) we would be able to observe solar debris gravitationally attracted to each other to form large clumps instead of the individual objects of the asteroid belt that we can observe today. In 1974 Sky lab and Apollo proved that there was zero gravity (micro) in space.

CHAPTER 8

The Earth also has four separate paths of motion:

1. The Earth travels a total of 90 degrees from North to South and back to North, lasting 365 days, in a 23.45-degree elliptic orbital inclination. This North–South movement is the cause of the four equinoxes in the northern and southern hemispheres (Plates 2 and 2A). Earth is farthest from the Sun's plane during the northern summer and southern winter.

2. Earth and all the other planets spin anticlockwise as viewed from above the North Pole, at a speed of 1,040 miles per hour. One 360-degree rotation is equal to one sidereal day of 23 hours, 56 minutes, and 4.09 seconds. Earth is being propelled by the solar wind, which has the power and force of 1.8 million miles per hour. This compelling evidence and previous discoveries now prove that the solar wind can spin the Earth, thus causing centrifugal force and attracting solids to its centre.

Earth's gravity is similar to the amusement park spinning ride named the Rotor. The Rotor's high-speed spin and 90 degree angle (like Earth's 0.37% per day progression) pins the riders to the walls of a spinning rubber-lined drum. By using the principles of centrifugal force, the riders are 'stuck' to the rubber walls. Earth's corkscrew silhouette (when viewed waterless), its rapid 1,040-mile-per-hour rotation, its 90 degree tilt, and its 14.7 lb. per. Sq. in. atmospheric pressure, are in this author's opinion, the forces responsible for the source of Earth's gravity and the four daily tides on Earth. The 1040 mph spin also causes Earth's magnetic field.

The magnetic poles, characterized by the auroras and the solar wind and their vast distance from each other, keep Earth and the other planets from colliding but are not the source of their gravity. The solar wind is spinning the planets, and this can be proven by using Bernoulli's principle of incompressible fluids, which can be obtained by integrating the Nervier –Stokes equation. This proves that the spinning process induced by the solar wind and its atmospheric pressure are what cause the planets to have gravity. Auroras occur on all the planets in our solar system, and there caused by charged particles carried by the solar wind from the Sun. The heliosphere protects Earth from cosmic rays.

3. A solar-eclipse year lasts 608 days and occurs about six times every ten years and casts a 100-mile-wide shadow in a set path over Earth. Eclipses, Jupiter's failure to cross the Sun's equator twice during its 11.8-year cycle, and the fact that Jupiter is in a position in the sky that can be seen every day of the year from Earth (except for a 12-hour period every 11.8 years). These pieces of indisputable evidence. Accurate observations over the years strongly suggest that Earth and Jupiter are circling above the Sun's celestial poles and not around its equator (Plate 5).

4. **Earth** has a theorized 500 million year regularly recurring cycle, part of the 26,000-year-precession of the equinoxes. This cycle is taking the Earth from perihelion, causing global warming, to aphelion, resulting in an ice age (plate 6). Since the rotation axis is processing in space, the orientation of the celestial equator also processes during the same period. This means that the position of the equinoxes is changing slowly with respect to the background stars. This precession of the equinoxes means that the right ascension and declination of objects changes very slowly over a 26,000-year period. This effect is negligibly small for casual observing but is an important correction for precise observations. Transient stars are regularly recurring and predict astronomical occurrences.

The Earth changes direction every 500 million years or so causing Ice Ages. It has experienced three Eons: Proterozoic, Achaean and Phanerozoic, 10 Eras and about 22 Periods. The Earth has just entered the Cainozoic Era of the Phanerozoic Eon and is warming after its fifth ice age. The fifth ice age wiped out 70% of all life on Earth as did the previous one's. It was NOT an asteroid shower that killed the dinosaurs, it was the last Ice Age which started more than one hundred years ago and ended about 15,000 yrs. ago. Earth's warming cycle has approximately 300 million years remaining until it climaxes with an average temperature of 95 degrees Fahrenheit (32 degrees Celsius) and just before it begins its cooling cycle all over again. Ice age temperatures are estimated to fall to 32 degrees Fahrenheit (0 degrees Celsius), but when 20 degrees Fahrenheit is added for the blanketing ozone layer, the temperature of the Earth would be 52 degrees Fahrenheit during an ice age (Plate 6). The temperature of Earth today is 39 degrees Fahrenheit and warming. When the 20-degree-Fahrenheit atmospherically blanketed trapped heat (ozone layer) is added to the Earth's temperature of today, it raises the Earth's true temperature to 59 degrees Fahrenheit.

When the Earth started to warm up about 15,000 years ago after the last Ice Age it warmed where the angle of the Sun shone for the longest period. The parts of Earth to warm first after the ice age ended about 25,000 years ago simultaneously were Saudi Arabia and Australia now are.

Earth's iron core and its spin are what causes Earth's magnetic field which protects us from cosmic rays (subatomic particles) and causes magnetism on Earth. The solar wind protects the solar system which extends 12 trillion miles out from the Sun, from interacting with galactic magnetism.

CHAPTER 9

HISTORIC EVENTS AND LATEST SCIENTIFIC DISCOVERIES

The Bible credited God with creating the universe and orbiting it around the Earth. Galileo said, "In questions of science, the authority of a thousand is not worth the humble reasoning of a single individual". For Galileo Galilee, this thought meant everything. He went against everyone and believed what he thought was true by stating, the Earth is NOT the centre of the universe. History has shown him to be mostly correct.

Galileo started applying mathematics to physics. This helped him form theorems about the centre of gravity of solids. Heavy bodies were supposed to fall at speeds of 32 feet per second squared seeking the centre of the Earth.

Due to Galileo, we have excelled in our science. He developed many theories on motion behaviour and discovered many exhilarating and worthwhile ideas. He made the astronomy telescope which advanced science. His greatest achievement was discovering Jupiter's moons and satellites.

Galileo's *heliocentric* theory detrimentally replaced the word God *(geocentric)* with the word circumnavigation of our Sun. Solar wind energy is the source responsible for the orbiting of our planets all in the same direction above the Sun's celestial plane and not gravity. As a result of his reinvision, Galileo found himself in trouble with the church, which feared their authorities' opinion would now be open to question and they would loose their position of power as "owner of the game".

Galileo recanted and withdrew his theory under great duress in 1633, claiming it was a 'violation of the laws of physics'. Although the theory was flawed, an abundance of evidence proved the observations of Galileo as well as Copernicus and Giordano Bruno, the rebel Dominican monk, were all correct by stating that the planets were not circumnavigating the

Earth but rather the Sun. It appears the scientific community today is sadly travelling the same path as the church did 400 years ago by promoting an already discredited Accretion Theory.

Galileo's observations were accepted by the Astronomical Society , unseating the church as the astronomical authority and placing it on the scientific community, where it remains to this day. This was an important case where the church authorities were proven to be just plain wrong.

CHAPTER 10

The earliest light travelling at 186,000 miles per second was discovered coming from the farthest galaxy so far which was 13.8 billion years away. Although it is the oldest light source discovered, it isn't the first light from the beginning of time. Many believe the universe has an alleged 80-billion-year cycle and posit the creation of the universe from kinetic (dark) energy.

The Big Bang Theory written by Edgar Allen Poe in 1848 (cosmic microwave background) alleged that the origin of matter, space, and time began with a tremendous explosion about 13.8 billion years ago. It also posits that it produced all the atoms of all the elements in the universe (Plate 4668).

It took 13.4 billion of those years for the light from the oldest known exploded supernova to reach Earth. The Hubble microwave telescope showed light from the ancient stars as they reached Earth from a part of the universe which was expanding and accelerating in all directions at the same time some 13.8 billion years ago, leaving many unanswered questions.

Hubble's breakthrough discovery of an expanding universe in the 1930s and Arno Penzias and Robert Wilson proving the existence of cosmic background radiation in the 1950s allegedly from the Big Bang explosion positively and conclusively made null and void the long-supposed theory of general relativity, quantum gravity theory, string theory, static state theory and nebular hypothesis (accretion theory). These discoveries were the final evidence needed to place the final nail in the coffin of the formation theories based on cosmic gravity.

These findings shattered long-held beliefs, leaving the scientific community in chaos. The worst-case scenario was the unacceptable idea that the scientific community could no longer 'own the game' as the church did before it. Bereft of a new formation idea, the VIPs and decision-makers uncritically accepted a flawed alternative with blind trust. They reverted back to Pierre Laplace's now-infamous, accretion theory even though it was improvable, flawed, and wrong. It wasn't long before they realized they were backing the wrong theory. This takes us to the re-envisioned Twenty-First-Century Astronomy where only logical deduction and provable probability are considered.

CHAPTER 11

Viktor Safronov's (1917–99) lifelong research of Pierre Laplace's presently accepted *accretion theory (nebular hypothesis),* alleging disks of gas and dust were coalesced by gravity, collided, formed proto-planets, 'stuck' together, and reformed into our solar system. Accretion was later discovered to be basically flawed and improvable. The main objection was Isaac Newton's formula proving gravity cannot attract, hold, or condense gases, nor does it exist in the cosmos. There are no examples of Earth's gravity attracting gas anywhere on Earth.

According to famed chemical engineer Robert Ulanowicz in his 1986 book *Growth and Development,* Laplace's theory met its end with early nineteenth-century developments of the concepts of irreversibility, more advanced experimental knowledge, and the second law of thermodynamics. In other words, Laplace's problem was based on the premise of reversibility and classical mechanics; however, Ulanowicz points out that many thermodynamic processes are irreversible so that if thermodynamic quantities are taken to be purely physical, then no such objection is possible as one could not reconstruct past positions and moments from the current state. Computer simulations are now available to prove this point to be true. The highest measurement of thermodynamics takes a very different view, considering thermodynamic variables to have a statistical basis which can be kept separate from atomic physics.

Due to its conforming to the accepted rule, Laplace's theory was incompatible with mainstream interpretations of quantum mechanics that stipulate uncertainty, and computer simulations now confirm it. Indeterminacy is the majority position amongst physicists. The interpretation of quantum mechanics is still open for debate, and there are many who take opposing views, such as Bohme's and many others' interpretations of the same event. Isaac Newton's math proving that there is no gravity in the cosmos and our gravity readings on our way to the moon with the Apollo mission, were the final nail in Accretions' and Einstein's Relativity coffin.

The nebular hypothesis (accretion theory) was never proven and is now being exposed as improbable because of the many unanswerable questions and the long history of unproven experiments. Our most sophisticated computer simulations could not duplicate the coalescence of Uranus or Neptune. Nor could it duplicate or explain the specifics of how water originated or formed in any model tried. After many more failed experiments, the scientific community downgraded their opinion and decided to their detriment to leave the original gravity source and the formation of water completely unexplained.

In his book *Principia: Math ematical Principles*, Newton's laws of physics state, 'gravity does not exist in the cosmos now confirmed by the LIGO.' Webster's dictionary disprove the *accretion and other gravity-based theories* by defining gas as a fluid (as air) that has independent shape or volume which tends to expand indefinitely due to its molecular structure and cannot be attracted or held by gravity. Gas expands indefinitely (escape velocity) and settles in layers according to its atomic weight and atmospheric air pressure. Gravity had very little or nothing to do with the formation of our solar system.

CHAPTER 12

The small-atom interplanetary Helium gases, which are heavier in atomic weight and have greater air pressure, cannot pass through the small-atom intergalactic Hydrogen gases, which are lower in atomic weight and have decreased air pressure. Galactic gases cannot expand through universal gases. This fact complies with the laws of physics.

Heliospheric gases, like lithium, expand to helium and can expand no further, helium expands to hydrogen and can expand no further, and hydrogen expands to the next lightest universal gas, which is predicted to be discovered beyond the Heliopause with an expected lower atomic weight of less than 0.989 and is predicted to be located in an environment with a decreased air pressure. Albert Michelson and Edward Morley called this invisible wave aluminiferous ether. It was later proved that they were right in believing all space is comprised of gas with atoms of different sizes and atomic weights. Today we call this light Heliospheric gas helium-3 with the utmost confidence that a lighter universal gas will be discovered beyond the galaxy.

The AP theory also proposes that universal gases compress galactic gases of Atomic Hydrogen and have smaller atoms and lower atomic weight and a decreased air pressure than Heliospheric Helium gases. Heliospheric gases comprised of smaller Helium atoms have an air pressure of 1.5 pounds per square inch. Heliospheric air pressure is much, less than the compression of Earth's atmospheric air pressure of 14.7 pounds per square inch (1 ton per square foot). Universal gases are predicted to have a lower atomic weight than 0.1 (atomic mass unit) compared to the Heliospheric Helium gases with larger-sized atoms than Atomic Hydrogen and an atomic weight of 4.003 atomic mass unit. These intergalactic gases (Hydrogen etc.) are beyond the Heliopause with a decreased air pressure. Universal gases have a lower atomic weight of about 0.989 atomic mass unit or less and smaller-sized atoms than Hydrogen atoms and are found in-between the Galaxies.

The solar wind slows and then appears to stop altogether at about 10.3 trillion miles out from the Sun at the Heliopause because its inertia caused from the solar wind isn't strong enough to travel into and against the resistance of the more-compressed intergalactic Hydrogen gases which, in turn are compressed by universal gases. These findings are based on information allegedly received from Voyager 1 as it travelled beyond the Heliopause and conveyed to us by Robert Decker, a space scientist at the Johns Hopkins University Applied Physics Laboratory in Laurel, Maryland .

CHAPTER 13

NEWTON

Isaac Newton born in 1643 was one of the most famous English physicists, mathematicians, and astronomers of his day.

In a world of ignorance and superstition, Newton published Principia Mathematica showing us how gravity attracts solids to Earth but is absent in the cosmos. He thought if something could be projected beyond the boundaries of Earth, it would be free of gravity and would float freely because he rightfully believed the cosmos had no gravity. Newton's 'cannonball' experiments and observations proved gravity to be an attracting force keeping all solids earthbound and yet having no effect on gases.

Newton discovered the relationship between external force and motion, and he also composed the famous principle of equal and opposite reactions, which are still relevant today.

The Sir Isaac Newton's laws were used to disprove the nebular hypothesis and the conservation of momentum of Earth and the celestial bodies. It was later discovered that the solar wind animated the solar system, not gravity.

The incident of Newton seeing an apple fall straight down from a tree is known to us all. Newton deducted that the gravitational force was originating from the centre of the Earth and that the same gravitational force was present across and around the entire Earth.

Newton's accurate calculations of the orbital period of the Moon and of the planet's predictable motions were just some of his important discoveries.

Isaac Newton's discoveries in the fields of gravitational behaviour, mathematics, and science were second to none and are still being practiced today. Newton is referred to as the father of calculus, and his math skills allowed him to derive a new algebraic formula for pi. The AP theory is based on many of Newton's findings and compliments them in many of its gravitational conclusions.

Einstein tried to take Newton's work to another level (Relativity and Special Relativity) but failed.

CHAPTER 14

Universal laws of motion are three physical laws that together laid the foundation for classical mechanics. They describe the relationship between a body and the forces acting upon it, and its motion in response to said forces. They have been expressed in several different ways over nearly three centuries [without ever varying] and can be summarized as follows:

First Universal law of motion: When viewed as an inertial reference frame, an object either remains at rest or continues to move at a constant velocity, unless acted upon by an external force which then changes its motion.

Second Universal law of motion: F = ma. The vector sum of the forces **F** on an object is equal to the mass m of that object multiplied by the acceleration vector a of the object.

Third Universal law of motion: When one body exerts a force on a second body, the second body simultaneously exerts a force equal in force and opposite in direction on the first body.

The dictionary and Sir Isaac Newton defined gravity as 'the force of attraction by which terrestrial bodies tend to fall towards the centre of the Earth'. Gravity is one of the four natural forces. The other three natural forces are strong and weak, electromagnetic, and lastly, nuclear. We now know that atmospheric pressure and the Earth's spin are the forces causing gravity. Newton's equation is the only gravitational theory that was ever mathematically proven.

The law states the following:

The law of universal gravitation states that any two bodies in the universe attract each other with a force that is directly proportional to the product of their masses and inversely proportional to the square of the distance between them. [Separately it was shown that large spherically symmetrical masses attract and are attracted as if all their mass were concentrated at their centres.] This is a general physical law derived from empirical observations by what Newton called induction [and has never changed in any way].

If SI are units, F is measured in Newton's (N), m1 and m2 in kilograms (kg), r in meters (m), and the constant G is approximately equal to 6.674 × 10−11 N m2 kg−2. The amount of the constant G was first accurately determined from the results of the Cavendish experiment conducted by the Henry Cavendish in 1798. Cavendish did not himself calculate a numerical value for G [because Newton did not use it].

This cornerstone experiment was the first test of Newton's theory of gravitation between masses in the laboratory. It took place 111 years after the publication of Newton's Principia and 71 years after Newton's death, which meant none of Newton's calculations could use the value of G; therefore he could only calculate a force relative to another force [with successful results].

We now know that G (strength of gravity's attraction) can be expressed as: G= (6.67191+/- 0.00099) X 10" Cubic Metres per Kg. per square second it has a 0.015% uncertainty.

The strength of the escape velocity of gas is ½mv2, where m is the particle mass and v is its velocity. Maxwell and Boltzmann deduced that the mean kinetic energy is proportional to T. This statement is usually written as:

1/2 mv2 = 3/2 kT

In this equation, k is a fundamental constant called the Boltzmann constant, which has the tiny value of 1.38 × 10-23 Joules per Kelvin. We can see the main features of this equation easily. The temperature is proportional to the square of the average velocity and it is proportional to the mass of the particle.

This proves that the escape velocity of gas is greater than Earth's gravity.

Newton's groundbreaking law of gravitation resembles Coulomb's law of electrical forces, which is used to calculate the magnitude of electrical force between two charged bodies. Both are inverse-square laws, in which force is inversely proportional to the square of the distance between the bodies. Coulomb's law has the product of two charges in place of the product of the masses, and the electrostatic constant in place of the gravitational constant.

CHAPTER 15

The Shoemaker–Levy asteroid was gravitationally attracted to Jupiter from a greater distance than the Moon is from Earth. This evidence further proves that Jupiter is not made up entirely of gas. The asteroid also proved that the electromagnetic fields of the planets are the ones keeping them equidistant from each other and not the forces of an erroneous cosmic gravity which is the reason being offered without any explanation or knowledge of its origins.

All planetary objects are not orbiting at the Sun's equator from the influence of gravity nor are they being propelled by gravity. The planets are being propelled by the solar wind. The solar wind's existence was mathematically proven by Prof. Eugene Parker of the University of Chicago in 1958. *Mariner 2* , a Venus mission launched on 27 August 1962, was the first to confirm the Heliosphere and solar wind's existence with its particle detector. Mariner's discovery was further evidence towards proving that the solar wind and not cosmic gravity is the energy source of our solar system's animation.

The famous astronomer Nicklaus Copernicus, whose name was originally Nicklaus Kopernik , is known as the father of geometry. His father became a civic leader in Torun as a magistrate and married Barbara Watzenrode, who came from a prominent family from Torun in about 1463. Nicklaus and Barbara Kopernik had four children, two sons and two daughters, of whom Nicklaus Copernicus was the youngest. He published a second book, De Revolutionibus Orbium Coelestium, which was banned by the church not long after his death in 1543.

Copernicus's naked-eye observations were flawed due to the lack of a telescope (not discovered yet) and his ignorance of the Sun and solar system's size, its distance from the Earth, and the existence of anything beyond Saturn. This lack of information made it impossible to produce an accurate model portraying the planets circumnavigating the Sun at its equator. Despite these shortcomings, Copernicus's observations and geometry skills confirmed that the Sun and not the Earth is the centre of our solar system. Copernicus finally concluded that the entire solar system (minus the Sun itself) is self-contained and orbiting the Sun.

CHAPTER 16

Albert Einstein was born in 1879 in Germany to a family of non-observant Jews. He had a brilliant mind from the start and always curious about the way the world around him worked. He was a pacifist and atheist and moved to Switzerland to avoid the war. As a theoretical physicist, he had many failings.

In 1903, Einstein married his first wife, Mileva Maric; they had two sons, and then separated in 1914. They divorced in 1919, and then he married Elsa Lowenthal, a woman who was his first cousin on one side and second cousin on the other. In 1933 they immigrated to the United States, and by the end of 1936, Elsa had died of heart problems.

In 1939 Einstein spoke out in favour of the allies building an atomic bomb, which he completed as a member of the Manhattan Project in Chicago and it was used in 1944 against Japan. He said that he only did so because he felt that there was significant risk that the Germans would resort to such devices. He later expressed regret at having done so. In 1940 he became an American citizen and took a professorship at Princeton. Learn more at your local library.

Albert Einstein, a victim of depression, died on 17 April of 1955 of a ruptured abdominal aortic aneurysm, which caused internal bleeding.

$$R_{\mu\nu} - \tfrac{1}{2}g_{\mu\nu}R = 8\pi T_{\mu\nu} - \Lambda g_{\mu\nu}$$

EINSTEIN'S FIELD EQUATION may look simple, but enter at your own risk. The equation is the basis of general relativity.

Einstein admitted that his Relativity and Special relativity theory equation (above) was 'fudged' and referred to his formula allegedly proving his equation of planetary circumnavigation is allegedly caused by gravity as his 'biggest blunder' and a 'deliberate untruth'. He cheated because he wanted to take Newton's place in history. Einstein confessed to deceptive scientific practices and breaking all the established rules of scientific integrity by introducing a cosmological constant into his Special Relativity formula, which conclusively proved gravity, could not do what Einstein had claimed. Einstein's confession of uttering 'deliberate untruths' conclusively confirmed that the Quantum gravity, Relativity and Special Relativity theories were deeply flawed and truly his 'biggest blunder'. He thought that the force of gravity was an illusion. It did not end well for Albert Einstein. Eventually a combination of quantum mechanics, Fred Hoyle, and Edwin Hubble exposed his ominous actions and disproved his theory.

Georges Lemaître , a respected astrophysicist, and Russian meteorologist and mathematician Alexander Friedmann (1888–1925) found the answer but needed proof. Friedmann then put forward a mathematical model of an expanding universe. It was presented on the possibility of a world with constant negative curvature of space. Friedmann published in 1922 and again in 1924. Lemaître stressed the possibility of an expanding universe as Hubble finally proved, but Einstein rejected it out of hand, telling Lemaître, 'Your math is correct, but your physics are abominable!' It was later found to be a deliberate untruth.

Einstein had earlier admitted to Willem de Sitter that his special relativity theory was mathematically 'fudged' because a cosmological constant was introduced to the formula though there was no evidence for it. Lemaître, Hoyle, Friedmann, Humason, and Hubble all get credit for producing enough evidence to finally prove Einstein wrong. They all concluded that Earth's gravity could not rotate the Earth at 1,040 miles per hour, cause planetary circumnavigation of the Sun, cause four tides per day on Earth, keep the planets equidistant from each other, or hold down our atmosphere. Isaac Newton proved gravity could not have formed our solar system in the cosmos. He also showed our solar system could not form from the power of gravity and nor could the solar system form from an unknown gravity source. This is what we're now being falsely informed, by the flawed teaching of the nebular hypothesis. Einstein's works were contrary to Newton's findings and publications, and he (Einstein) was finally discredited.

The discredited standard model equation which he worked from the answer foreword and then introduced a "bridge" (cosmological constant) allegedly proving that quantum gravity could do all the above and the inability to prove or explain quantum gravity, suggests that a re-imagined and more fundamental theory exists. In 1931 Einstein visited Mount Wilson to thank Hubble personally for his discovery of an expanding universe even though it and quantum mechanics had proved him (Einstein) wrong because "c" cannot be squared. Einstein's E = mc2 (where E is the energy in ergs, m the mass of the matter in grams (in the beginning there was no mass), and c the cosmic speed of light in centimetres per second) was a conversion of matter to energy (c2 = 9 x 1020), not energy into matter. Einstein was further embarrassed when it was found that the universe was not unchanging and had not existed in its present state forever as he predicted.

Einstein never proved matter could be taken from energy because energy is an essence without mass. Nor could he prove the fabric of space and time can bend from gravity. The AP theory is a reinterpretation with a more logical, monistic process that describes the forming of our solar system from galvanic explosions from our Sun and should be introduced to replace the now disproven Accretion theory. His quantum mechanics where light was a stream of particles and a wave both at the same time. Einstein's interpretation was also disproven by John Bell (1928-1990) and the '60s scientific (hippy) community who found Niels Bohr to be right by claiming photons were real and random until they became a particle. Einstein thought it was predestined, he was wrong. His black holes (Sagittarius A-Stat) and worm holes are also improvable. The jury is still out. The astronauts don't weigh anything on board of the International Space Station (ISS) because there's no gravity in the cosmos. They pre-measure and weigh everything on Earth.

CHAPTER 17

English scientist and physicist Michael Faraday (1791–1867) is known for his brilliant discoveries of electromagnetic induction, electromagnetic rotations, the magneto-optical effect, diamagnetism, field theory, and many more. Many famous historians regard him as the most influential and exemplary experimentalist in the history of science. The incredible scope of Faraday's work spanned over a 60 year period. He is considered one of the top figures of the nineteenth century for his remarkable contributions to the field of electricity and gas behaviour under freezing conditions.

The *Experimental Researches in Electricity and the Chemical History of the Candle* were both written by Faraday. These works should be in every scientific library as invaluable reference books. Michael Faraday earned a doctorate in civil law and was elected as a foreign member of the Royal Swedish Academy of Sciences. He was also recognized by the French Academy of Sciences for his work in gas behaviour.

James Clerk Maxwell (Maxwell's laws) discovered that light, electricity, and magnetism all travel at the same speed of 186,000 miles per second. The speed of light is the fastest known speed, ,and nothing can exceed that speed as proven by the Large Hadron Collider where it was found that as an atom reached the speed of light it actually slowed down. This conclusively proved that the speed of light cannot be squared as Einstein suggested in his infamous *E = mc2* formula making it null and void.

The AP theory is based on Michael Faraday's experiments, which were the forerunners of the successful experiments on electric motors, transformers, ultraviolet light prisms (Plate 0597) and frozen gas. Faraday also discovered a static field and the unity of magnetism, electricity, and light. James Maxwell mathematically translated Faraday's static field into waves of electric force. Faraday also proved that Noble gases such as, Oxygen, Hydrogen; Lithium , etc. can be frozen to a liquid and/or solid state. The intergalactic medium temperature is at least –270 degrees Celsius. Hydrogen freezes at –260 degrees Celsius and oxygen at –160 degrees Celsius. These temperatures prove that mixed gases can be frozen and thawed in space by a natural chain of events. Free-floating clouds of frozen oxygen and other noble gases are commonly observed and photographed in outer space. Carl Sagan turned Voyager 1 toward Earth to photograph the planets. That photo helped prove the AP Theory when he oversaw the successful photographing of "a portrait of the planets" showing the planets orbiting above the celestial plane of the Sun. He called the Earth "a little blue dot".

CHAPTER 18

Global warming, the *greenhouse gas effect, and climate* change are euphemisms for global pollution caused by mankind. Here's an example in nature proving that gravity removes the carbon from CO and CO_2, eliminating the long-term threat of irreversible air pollution. The eruption of a Greenland volcano polluted the environment for thousands of square miles around. The pollution was so bad that air traffic had to be diverted from the area for weeks. Prior to the volcanic eruption the air quality in the vicinity was excellent. A year after the volcanic eruption the air quality of Greenland was again excellent, proving that carbon, a solid, cannot stay in the atmosphere. The amount of CO_2 in our atmosphere today is 0.038 per cent , just enough to feed the plant life on Earth. This is also true of how a collapsing thunderstorm cloud becomes a dust storm. Dust storms are a yearly event in the western US.

Dust does not stay in the atmosphere because like carbon, dust is a solid and falls back to Earth.

Carbon dioxide is a combination of superheated carbon (a solid) mixed with oxygen (a gas). As the superheated, unstable carbon atoms cool, the solid carbon separates from the oxygen and returns to Earth. The Oxygen in mixtures of CO and CO_2 returns to the atmosphere and the carbon is gravitationally attracted to Earth. This process allows the cooling oxygen gas to escape into the atmosphere, where it is unaffected by gravity. The total amount of CO_2 in our atmosphere today is 0.038%.

The purveyors of misinformation who are desperately trying to own the game by contradicting the overwhelming evidence and obvious findings or any of the proven facts that do not suit their agenda have been proven to be just plain wrong. These astronomers are the purveyors of fear. Einstein's 'deliberate untruths' are being overwhelmed by contradicting and convincing examples in nature and of many successful laboratory experiments proving his alleged conclusions wrong. This tactic is a ploy and leads us to believe this fear mongering is purely economic and self gratifying. Despite many futile, pointless, and unsuccessful attempts, it has never successfully been proven that gas can be attracted, held, or influenced by Earth's gravity. Nor has anyone ever offered one successful experiment or example in nature to prove our atmosphere is being held to Earth by gravity or even shown to be possible that it can be.

Examples of Gas Behaviour without Gravity's Effect

The best proof of gravity not attracting gas is in nature, where we know for a fact that lighter and smaller gas atoms of helium completely surround our atmosphere, (Plate 2481) offering a natural ceiling of Helium; this is what keeps our atmosphere from expanding, not gravity which has nothing to do with attracting or holding gas. In physics, it can be proven just by knowing that large atoms (Plates 0204, 3596, 6279, 5506) (Oxygen) cannot expand through the smaller (Hydrogen) atoms. If gravity attracted gas, the smaller Hydrogen atoms would pass through the heavier and larger Oxygen and Nitrogen atoms and rest on Earth.

It would be like pouring rocks through sand; it can't be done. In industry, Nitrogen blanketing is used to keep hydrocarbons from evaporating and citrus oils from oxidizing. Gravity has nothing to do with these processes. Also the Pike River coal mine in New Zealand used nitrogen gas to successfully purge the mine of poison gases.

The Earth's average present temperature is 39 degrees Fahrenheit (4 degrees Celsius). The ozone layer adds 20 degrees Fahrenheit, making the Earth's average temperature 59 degrees Fahrenheit (15 degrees Celsius). If all of Antarctica thawed, it would not produce enough thawed water to fill all the below-sea-level and desert areas on Earth. The Earth's atmosphere contains 79 per cent nitrogen, 19 per cent oxygen, and 2 per cent other gases including 0.038 per cent of carbon dioxide (0.055 at sea level). The level of carbon dioxide in our atmosphere today is the lowest in Earth's history of 4.5 billion yrs. carbon dioxide is not the bad guy. Carbon dioxide is an essential building block to sustain all life on Earth and in our seas and represents only 0.038 per cent of our atmosphere.

The origin of life, our ecosystem, and our atmosphere on Earth began in tidal zones when newly formed water from exploded subatomic solar gas molecules of hydrogen and oxygen, which formed in the Sun were explosively ejected. The absolute-zero environments caused the frozen gases to form The Theia cloud of fused and frozen H2O. Water (Plate 8555) accumulated on Earth after its ejection from the Sun and when it experienced a collision with a frozen cloud as previously explained after (water formation) H and O fused into H_2O. The extremely hot planet Earth comprised of molten materials collided with the frozen cloud this event is known as the Theia collision. When the hot planet entered the frozen cloud, the thawing process covered the Earth with water.

Plate 2481

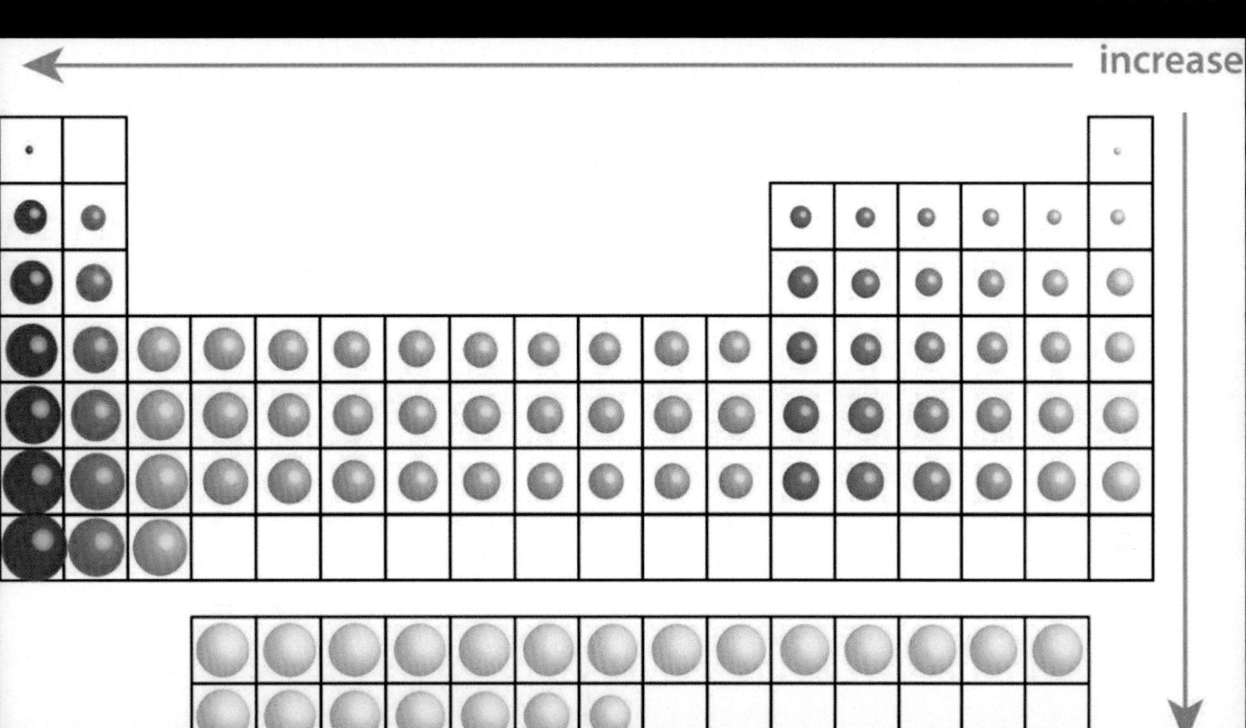
increase

Plate 3596

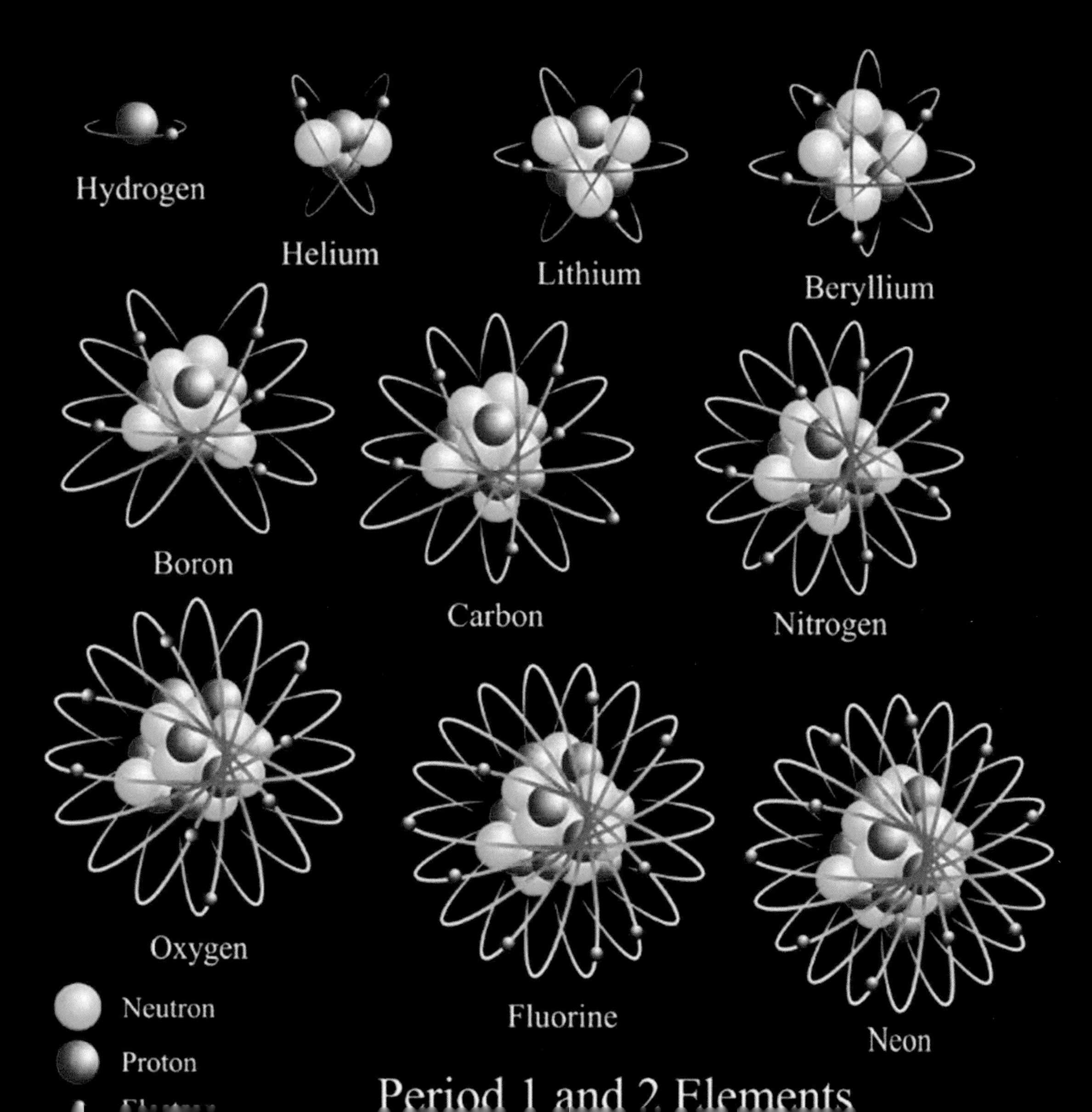
Hydrogen
Helium
Lithium
Beryllium
Boron
Carbon
Nitrogen
Oxygen
Fluorine
Neon
Neutron
Proton
Period 1 and 2 Elements

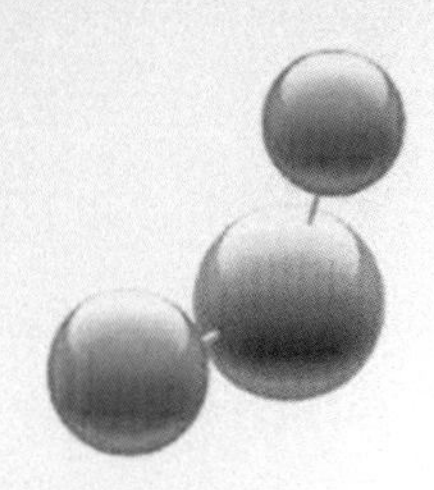

CO_2
Carbon dioxide

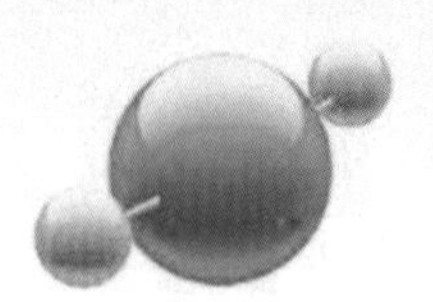

H_2O
Water

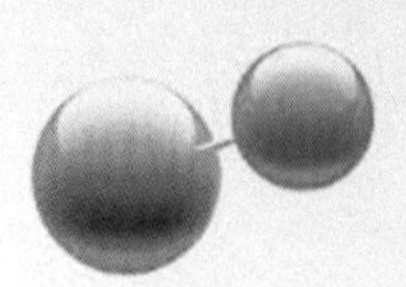

CO
Carbon monoxide

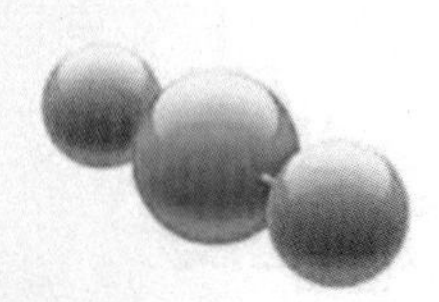

N_2O
Nitrogen monoxide

CH_4
Methane

SO_2
Sulfur dioxide

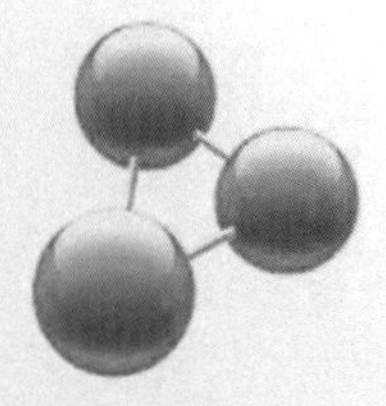

O_3
Ozone

O_2
Oxygen

N_2
Nitrogen

atomic radii [pm]

H 31																	He 28
Li 128	Be 96											B 84	C 76	N 71	O 66	F 57	Ne 58
Na 166	Mg 141											Al 121	Si 111	P 107	S 105	Cl 102	Ar 106
K 203	Ca 176	Sc 170	Ti 160	V 153	Cr 139	Mn 139	Fe 132	Co 126	Ni 124	Cu 132	Zn 122	Ga 122	Ge 120	As 119	Se 120	Br 120	Kr 116
Rb 220	Sr 195	Y 190	Zr 175	Nb 164	Mo 154	Tc 147	Ru 146	Rh 142	Pd 139	Ag 145	Cd 144	In 142	Sn 139	Sb 139	Te 138	J 139	Xe 140
Cs 244	Ba 215	La 207	Hf 175	Ta 170	W 162	Re 151	Os 144	Ir 141	Pt 136	Au 136	Hg 132	Tl 145	Pb 146	Bi 148	Po 140	At 150	Rn 150
Fr 260	Ra 221	Ac 215															

Ce 204	Pr 203	Nd 201	Pm 199	Sm 198	Eu 198	Gd 196	Tb 194	Dy 192	Ho 192	Er 189	Tm 190	Yb 187	Lu 197
Th 206	Pa 200	U 196	Np 190	Pu 187	Am 180	Cm 160							

WATER - H_2O

Chemical Reactions

$$2H_2 + O_2 \rightarrow 2H_2O$$

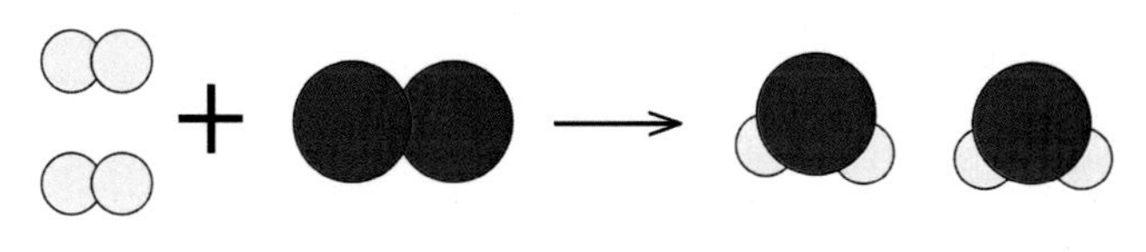

2 hydrogen molekules + 1oxygen molekule yelds 2 water molekules

2×(2.02 amu) + 32,00 amu yelds 2×(18,02 amu)
4,04 amu + 32,00 amu yelds 36,04 amu

36,04 amu reactans

Water Molecule

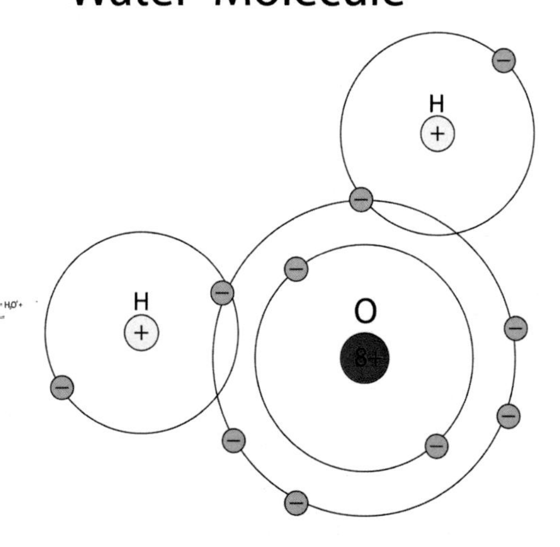

4 step formation of a molecule of water

1.
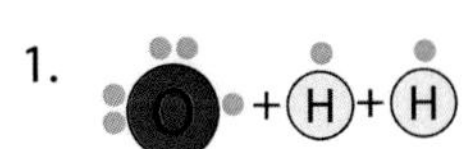

2.
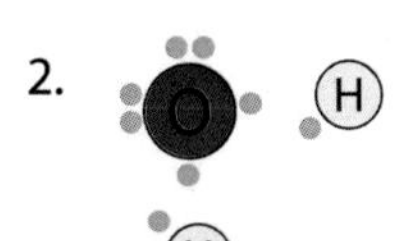

3.
O H H

4.
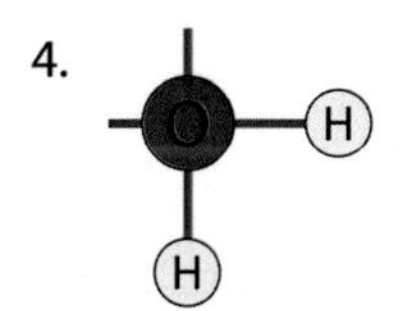

Hydrogen peroxide is thermodynamically unstable and decomposes to form water and oxygen

$$2H_2O_2 \rightarrow 2H_2O + O_2\uparrow$$

Under the influence of the very high temperatures or electric current water decomposes into molecular oxygen and molecular hydrogen:

$$t\uparrow \quad 2H_2O \rightarrow 2H_2\uparrow + O_2\uparrow$$

As Earth cooled to about 30–40 degrees and formed rainwater pools. Lightning travelling at 90,000 mps or one half the speed of light can produce nitrogen oxide, and sunshine (for photosynthesis) became the vehicles of life. Earth's vibration caused amino acids and alkaline minerals to pulverize and come together to form Nitrogen gas in the atmosphere. When a mixture of an acid and an alkali is in the presence of rainwater, their pH changes, and nitrogen gas is released. The freed Nitrogen gas became 79 per cent of Earth's atmosphere and formed one of the building blocks of life on Earth in the process. When Nitrogen Oxide gas is introduced to rainwater in the presence of lightning, air, great variations of temperature, and sunlight (photosynthesis), it forms amoebas, fungi, and algae. These Oxygen-producing microbes combined with Earth's abundance of carbon were among the first forms of carbon based life on Earth. The first life forms were vegetation and were the first organisms to supply Oxygen, which makes up 19 per cent of the Earth's atmosphere.

This scenario can be carried one step further by presuming that the algae, after many years, acquired a vegetarian parasite. The vegetarian parasite then contracted a carnivore parasite, and with time, the rest is evolutionarily possible. Oil comes from decaying life forms and Earth's vibration. Earth's gravity and vibration moves the natural oils from decaying animals and plant life deep into the sub layers of Earth. Prehistoric oil becomes coal after many eras of subterranean pressures.

The dinosaurs did not perish because of an asteroid shower; they died out with the other 70 per cent of Earth's life form during the last great ice age, which lasted for approximately 500,000 years. Our Earth has been warming for about 50,000 years and still has many millions of more years to go before it starts its cooling cycle again. Think about it. Don't blindly accept unreasonable explanations. Ask Questions.

CHAPTER 19

EXPERIMENTS TO PROVE THE AP THEORY IS TESTED HYPOTHESIS, NOT SPECULATION

The AP theory posits that the Earth's atmosphere is NOT being held down to Earth by the force of gravity as is presently supposed. Earth's produced atmospheric gases of Oxygen and Nitrogen exert an air pressure of 14.7 pounds per square inch or 2116.10 pounds per square feet at sea level. Earth's atmospheric gases of Nitrogen and Oxygen are larger, have an atomic weight and air pressure greater than their surrounding interplanetary gases of Helium and Hydrogen. Low-pressure interplanetary gases and smaller atoms of Hydrogen, Lithium, and Helium are surrounding and compressing Earth's atmospheric gases of nitrogen and oxygen. These low-pressure interplanetary gases of Helium—which are sun-produced completely surround our atmosphere. Heliospheric gases of Helium with smaller atoms, lighter atomic weights, and lower air pressure—are forming a natural ceiling and keeping the Earth's atmospheric gases from rising through them and from rising above the mesosphere.

One of the proven laws of nature is that a gas with high air pressure and heavier atomic weight cannot rise (expand) through a gas with lighter atomic weight and air pressure . Interplanetary gases of mostly Helium exert lesser air pressure than the greater air pressure exerted from Earth's atmospheric gases of Oxygen and Nitrogen. This is further evidence in nature proving that Earth's atmospheric air pressure cannot rise past the lighter Heliospheric air pressure and its gases of Helium, Lithium, etc. These hemispheric altitude pressure changes can be felt when we fly or scuba-dive, which always causes us to 'clear our ears' to compensate for the greater or lesser outside pressure.

The air pressure in between the mesosphere and the ionosphere is 1.5 pounds per square inch, and the higher we go, the lesser the air pressure becomes. Yet the difference in gravitational pull at 50,000 feet is nominally different than it is at sea level.

When NASA's astronaut parachuted from the fringes of space in 2013, the force of gravity was close to the same as it is at sea level. When no traces of Earth's atmospheric gases were found beyond 70 miles above sea level, NASA's experiment proved Earth's heavy atmospheric gases are being restricted from rising by the Heliospheric gases (Helium, and etc.), which are lighter in atomic weight and low in air pressure.

These Heliospheric gases were calculated in 2012 from The Voyager 1 space craft data to have a diameter of 11.25 trillion miles extending out from the Sun to the Heliopause.

Above 120 miles, atomic oxygen is the principal atmospheric constituent for several hundred miles. However, Helium is even lighter than atomic oxygen, and has smaller-sized atoms, so its concentration falls less rapidly with altitude, and it finally replaces the atomic oxygen as the principal atmospheric constituent above a certain altitude, which varies with the sunspot cycles between 400 and 900 miles.

> And at still higher altitudes atomic hydrogen finally displaces helium as the principal constituent. The hydrogen extends many earth radii out into space and constitutes the telluric hydrogen corona, or geocorona. The temperature of the upper atmosphere, and hence its density, varies with the intensity of solar ultraviolet radiation and this, in turn, varies with solar activity in general. The solar radio noise flux is a convenient index of solar activity since it can be monitored at the earth's surface. The minimum night-time temperature of the upper atmosphere above 300 kilometres has been expressed in terms of the 27-day average of the solar radio-noise flux at 8-centimeter wavelength. This varies from about 600 K near the minimum of the sunspot cycle to about 1,400K near the maximum of the cycle. The maximum daytime temperature is about 33% larger than the nighttimes' minimum. (Written and published by: Van Nostrand's Scientific Encyclopaedia)

Heliospheric gases and our atmospheric gases are not being held down by Earth's gravity; they're kept from rising by the galactic ceiling of lighter, smaller Hydrogen gases, which are in turn held down by even lighter universal gases yet to be discovered.

Experiment 1

To prove gases settle in layers according to their atomic weight, take a tank, add three kinds of gases of different atomic weights, attach a GC/MS to the outlets, and allow the gas to escape slowly. The gases with the smallest atoms and lowest atomic weights will exit the tank first.

Take two 1-litre flasks, each of which can be closed by a stopcock but also opened so that gases can flow from one flask to the other (while the two flasks are sealed off from the atmosphere). Fill one flask with hydrogen and close the stopcock. There is air in the other open flask; close the stopcock to that flask also. Connect the flasks and open them to each other. Set the flasks one above the other so that if you have narrow-neck round-bottom flasks, the apparatus will look like a vertical dumbbell. Leave it that way for a while. Then close the stopcocks and disconnect the flasks.

Say we do this under normal ambient conditions—for example, 20 degrees Celsius, 1 standard atmosphere (pressure). The ideal gas law works reasonably well under these conditions. Using the ideal gas law, 1 mole of gas (20 degrees Celsius, 1 standard atmosphere) is about 24.1 litres, so 1 litre is about 0.0415 moles or 41.5 mill moles. Hydrogen gas is 2 grams/mole or 2 milligrams/mill mole. The weighted average of dry air (78 per cent nitrogen (28 grams/mole), 21 per cent oxygen (32 grams/mole), and 1 per cent argon (40 grams/mole)) is 29 grams/mole.

Now weigh each flask and subtract from each the weight of the flask itself (the tare). If the theory is correct, one flask will be filled with hydrogen weighing 83 milligrams (41.5 mill moles × 2). The other flask will be filled with air weighing about 1200 milligrams (41.5 × 29 mill moles). The difference would be easily measurable on most laboratory balances.

Experiment 2

An experiment to prove air pressure exists is to take a 1-gallon screw-top can and add 100 millilitres of water to it. Place the can on a stove burner and turn on the heat. When steam comes out of the screw-top opening, screw the cap on to the can and remove it from the heat. As the can cools, the outside air pressure will crush the can, proving our atmosphere is surrounded by 14.7 pounds per square inch of crushing and compressing Heliospheric force.

Venus's atmospheric pressure of interplanetary gases that have light atomic weights has an air pressure force 100 times greater than Earth's. Venus's atmospheric air pressure crushed two Russian spaceships and is a hundred times more extreme than Earth's. A law of gas behaviour in nature is that an atmospheric gas that has a heavy atomic weight, such as oxygen, cannot rise beyond and pass through a gas that has a lighter atomic weight, such as helium.

Japanese lanterns work on the same principle. As the Oxygen in the inner shell of the lantern is consumed, the nitrogen gas raises the lantern upward into space. Scuba gear and space suits are pressurized to keep the occupants from being crushed underwater or exploding in space.

Nitrogen blanketing is a term used in the chemical industry all around the world. It's a process where nitrogen gas is introduced into the headspace of a partly full container of hydrocarbons (with heavy atomic weights) or other expensive ingredients. Nitrogen blanketing is administered in order to keep out the Oxygen from our atmosphere and prevent oxidization and evaporation of the valuable hydrocarbons (with higher atomic weights), spirits, and other expensive ingredients.

A recent example of 'blanketing' was used in the Pike River (New Zealand) coal mine disaster. Nitrogen gas was pumped into the mine to push out the heavy poisonous gases. These natural and industrial examples prove our atmosphere is not being held to Earth by gravity but by gases of lighter atomic weights which are surrounding the heavier gases.

These examples are played out in industries every day and in nature; the examples are many and can be easily proven. Every time we take a deep breath and fill our lungs with air until we cannot add any more to our lungs, we are proving that gas has volume, weight and mass but it also has an escape velocity. When we fill our SCUBA tanks, we prove air has weight.

Here's another example in nature that proves that gravity removes the solid carbon from CO and CO_2, eliminating the long-term threat of irreversible air pollution. The recent eruption of a Greenland volcano polluted the environment for thousands of square miles. The pollution situation was so bad that air traffic had to be diverted from the area for weeks. Prior to the volcanic eruption the surrounding air quality in the vicinity was excellent. A few months after the volcanic eruption the air quality of Greenland was again excellent as it was before the volcano. The cooling solid carbon in the mixtures of CO and CO2 caused by the volcano was gravitationally attracted to Earth. This process allowed the expanding Oxygen gas to escape into the atmosphere when cooled.

Those who continue to try to own the game by contradicting this or any other proven fact that doesn't suit their agenda are being overwhelmed by convincing examples in nature and successful laboratory experiments which are proving them wrong. Despite their many futile and unsuccessful attempts, no one has ever successfully proven that gas can be attracted, held, or compressed by Earth's gravity. Nor has anyone offered one successful experiment or example in nature to prove that our atmosphere is being held to Earth by its gravity.

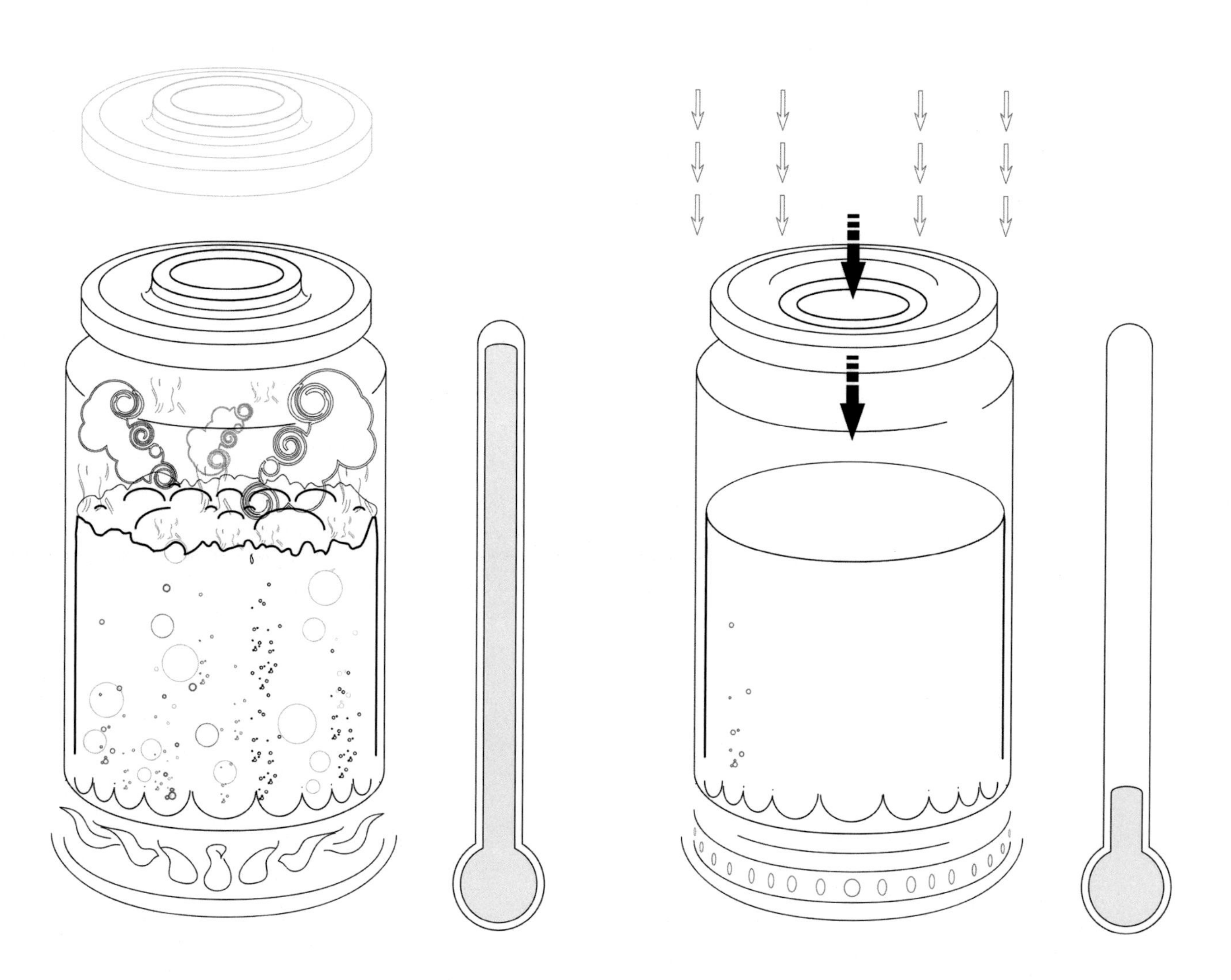

Experiment 3

The AP theory claims thawed gases on the surface of our newly forming Sun melted and found their way to its extremely hot cauldron within the core of our newly forming Sun.

To prove our entire solar system could have been formed from molten splashes one only has to read Luigi Galvani's work on the gallium process. Here's an example to prove Galvani's work. Get a high heat–tolerant container, add oil or metal, and bring it to a boil. When the mix begins to boil, add water or frozen gas to the container. When the water makes contact with the molten metal or oil, a cataclysmic galvanic explosion will occur, splashing the boiling contents outwards with extreme force. Care must be taken when conducting this experiment.

Experiment 1 is similar to the galvanic explosion that happened inside of our newly forming Sun when thawed gases reached its fusion-heated molten cauldron. We now know that 0.15 per cent of the material from the Sun's cauldron formed the entire solar system 5 billion years ago.

Experiment 4

Here is an easily proven experiment to show that the solar wind is the force spinning our planets and holding them in place using Bernoulli's principle.

Take a vacuum cleaner wand, set the vacuum on blow, and place a beach ball over the wand. The 'blow' of wind from the vacuum wand will surround, spin, and support the ball and keep it from falling. The spin of the ball, which is produced by the vacuum's 'blow' wind, is similar to the spin the 1.8 million-mile-per-hour blowing power of the solar wind exerts on the planets.

The tremendous power of the solar wind causes the planets to spin millions of miles above the celestial poles of the Sun. Contrary to public belief; gravity plays no part in the spin of the planets or their paths around the Sun. The solar winds are solely responsible for their spin and their path as the vacuum is solely responsible for the spin and their path of the ball.

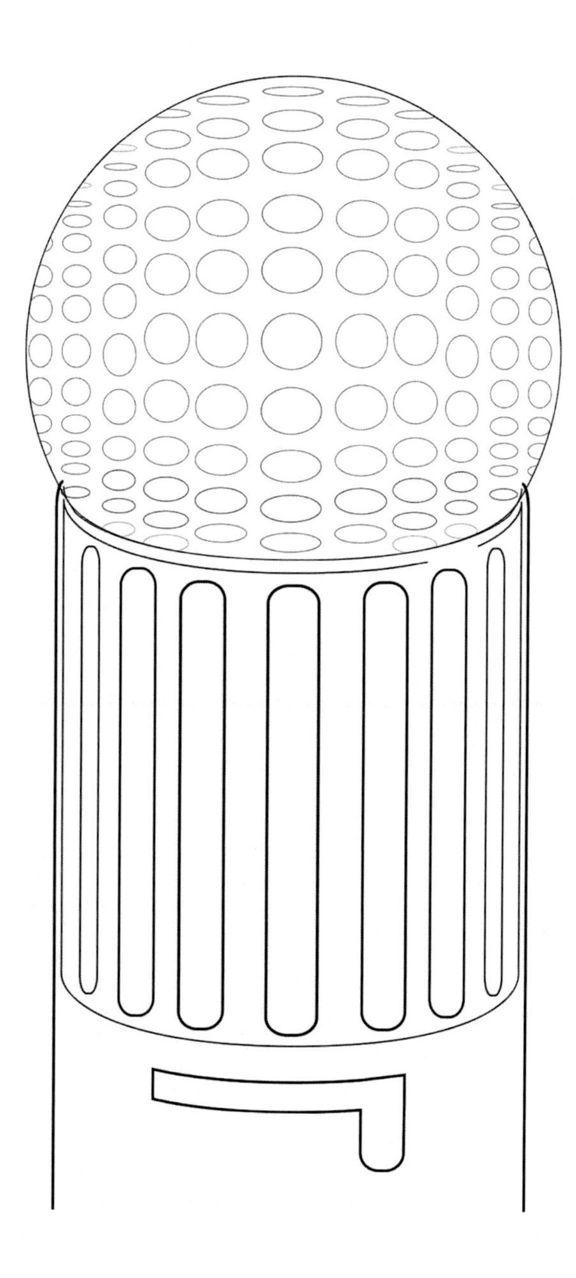

Experiment 5

Carbon dioxide and carbon monoxide are mixtures of a solid and a gas fused in extreme heat. They are poisonous when they are mixed in high concentrations in a confined area.

When the hot CO2 mixture of a solid and gas rise to cool, Earth's gravity pulls the solid carbon atom away from the Oxygen gas atoms, leaving two benign elements. The gravitationally attracted solid carbon falls to Earth and benefits the plant life, while the oxygen gas rises into the atmosphere and benefits animal life. Carbon dioxide is not something to fear; after all, plants thrive on it to survive, and it's in every carbonated beverage. The amount of atmospheric Argon is 0.93%, Carbon Dioxide in our atmosphere today is 0.038 per cent, Neon 0.0018%, Helium 0.0005 and N2O is 0.00003. Air pollution is being treated as a red herring. These percentages are hardly enough to worry about. Air cleanses itself in nature from gravity, rain and wind.

Water pollution is the thing to worry about. Once water is polluted, it must be mechanically or chemically treated at great costs. This is placing mankind in a precarious position of dependency on the same group who polluted the water in the first place. Water cannot be created or destroyed but it can be permanently polluted air cannot be permanently polluted.

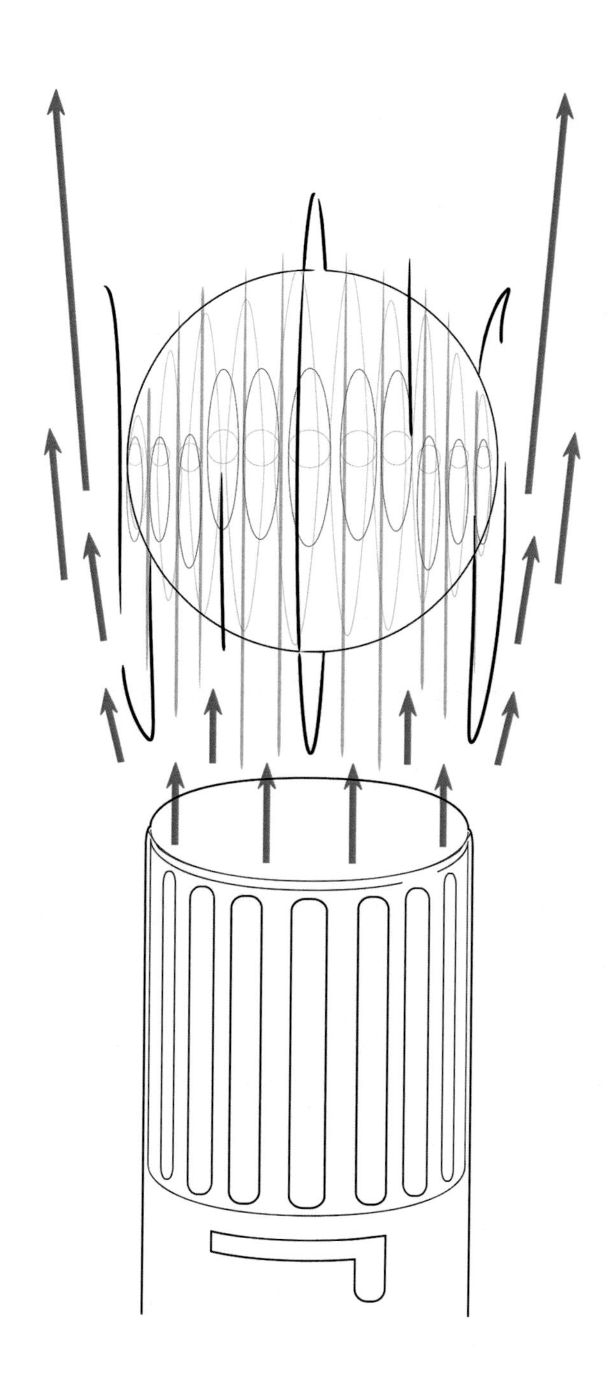

Experiment 6

Mix together 95% of the ingredients of life which are Carbon, Nitrogen, Hydrogen, Oxygen, Phosphorus and Sulphur and try to duplicate Mother Nature's results. Life on Earth started when rainwater accumulated on Earth as it was cooling. When Earth cooled to about 30–35 degrees or so, rainwater and sunshine became the vehicles of life. When Earth's amino acids and alkaline minerals come together in the presence of rainwater, and their pH changes, and nitrogen gas is released. When nitrogen gas is introduced to rainwater in the presence of lightning, air, and 35-degree sunlight, a photosynthesis process forms solar protons that reach Earth and forms fungi and algae. These were the first forms of life on Earth. This evolutionary scenario can be carried one step further by presuming that the algae, after many years, acquired a vegetarian parasite. The vegetarian parasite then contracted a carnivore parasite, and with more time, the rest is evolutionarily possible.

Oil is a renewable, natural resource produced from rotting vegetation and animal life.

As all living organisms perish and their hydrocarbon remains dissolve and gravitate deep below the surface of the Earth they become our energy source of today—oil and coal.

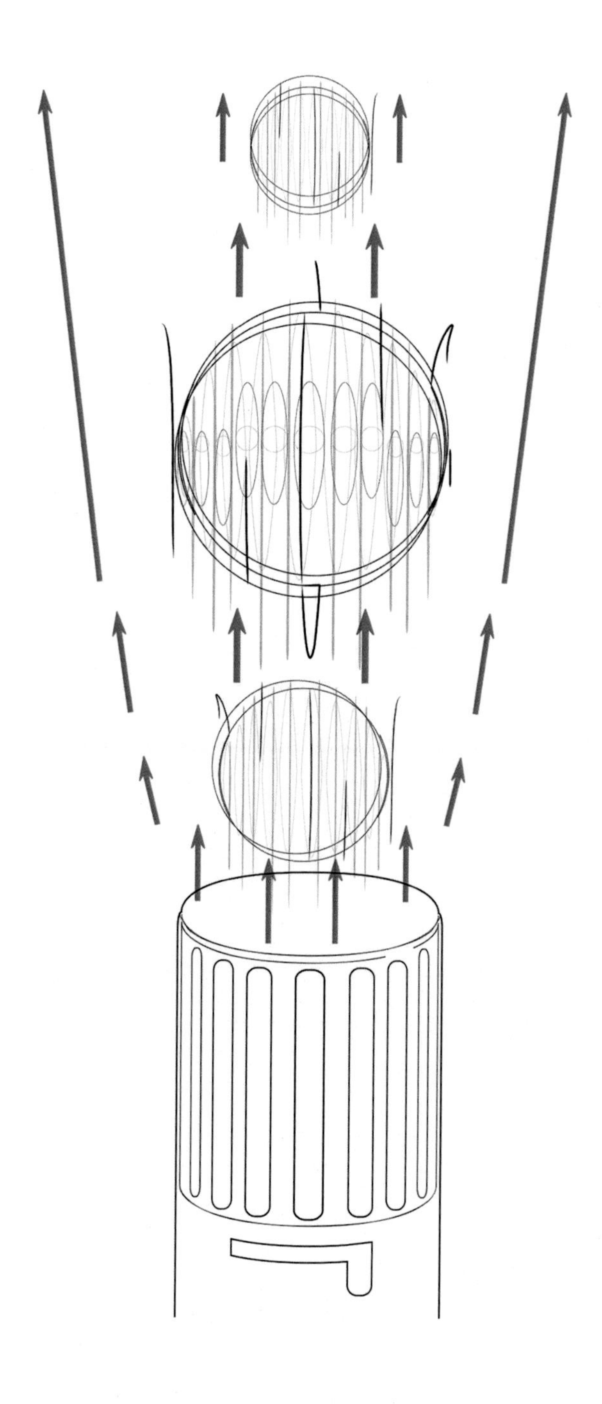

BIBLIOGRAPHY

Kerrod, Robin , The Star Guide by Robin Kerrod (Viking Press and Penguin Books).

Van Nostrand's Scientific Encyclopaedia .

Man, John, Illustrated Encyclopaedia of Astronomy (Chancellor Press).

Levy, David H., and John O'Byrne, Sky watching: (R. D. Press and the Nature Co.).

Beazley, Mitchell, Exploring the Universe (California Institute of Technology and Carnegie Institution of Washington).

National Geographic.

Harvard Museum of Natural History.

Yenne, Bill, The Illustrated Guide to Astronomy (Bison Books: London).

Air Liquid and Edward Elsevier, Gas Encyclopaedia: The Book.

The World Almanac and Book of Facts.

Harvard–Smithsonian Centre for Astrophysics.

Hubble Telescope Mission.

Massachusetts Institute of Technology.

Whitaker's Almanac.

Webster's Encyclopaedic Unabridged Dictionary.

Isaacson, Walter, Einstein: His Life and Universe.

Building a Model Solar System Vol. 34 & 35 (Eagle moss Publications Ltd.: London).

Chemical Principles (W. B. Saunders Co.: Philadelphia and London).

Astronomy/Sky Magazines February 2007 etc .

California Institute of Technology

'Metallurgical and Materials Transactions', ProQuest Science Journals (April 2004), 35B, 2.

Hal Levinson, PhD , University of Arizona and the Arizona State Observatory.

Surface-Oriented Melt/Substrate Heat Transfer Model in Aluminium Strip Casting .

Nelson, Lloyd, Paul Brooks, Ricardo Bonozza, and Corradini, Triggered Steam Explosions.

Journal of Nuclear Science and Technology, 40/10 (October 2003), pp. 783–795, and 37/12 (December 2000), pp.1049–55.

Hong, Kim, and Shin , Insights from the Recent Steam Explosion Experiments in TROI.

Journal of Loss Prevention in the Process Industries.

'Impact, Recoil and Splashing of Molten Metal Droplets', International Journal of Heat and Mass Transfer.

Special Thanks to: www.biography.com and those unnamed sources who contributed helpful and important biographical information.

5087 TRIVIA Questions and Answers By: Marsha Kraves, Fred Worth & Steve Tamerius

Edited By: Michael Driscoll - Black Dog & Leventhal Publishing

NASA Ames Research Centre

www.askanastronomer.com.

www.sciencedirect.com/science.

Time Almanac, 2004, p. 430

'Steam Explosion', Wikipedia.

Bara, Mike, superearth.com.

'The Exploded Planet Hypothesis', <metaresearch.org>

Copywriter photos: Shutter stock

Original art work: Angelo Pettolino

ACKNOWLEDGMENTS

Melie Art: Amelia Batchelor, re-imagined from original art by Angelo Pettolino.
Computer Scientists: Kevin and Brenda Hayne.
For comments: aptheory@aptheory.info.   Web site:www.aptheory.info

Periodic Ta

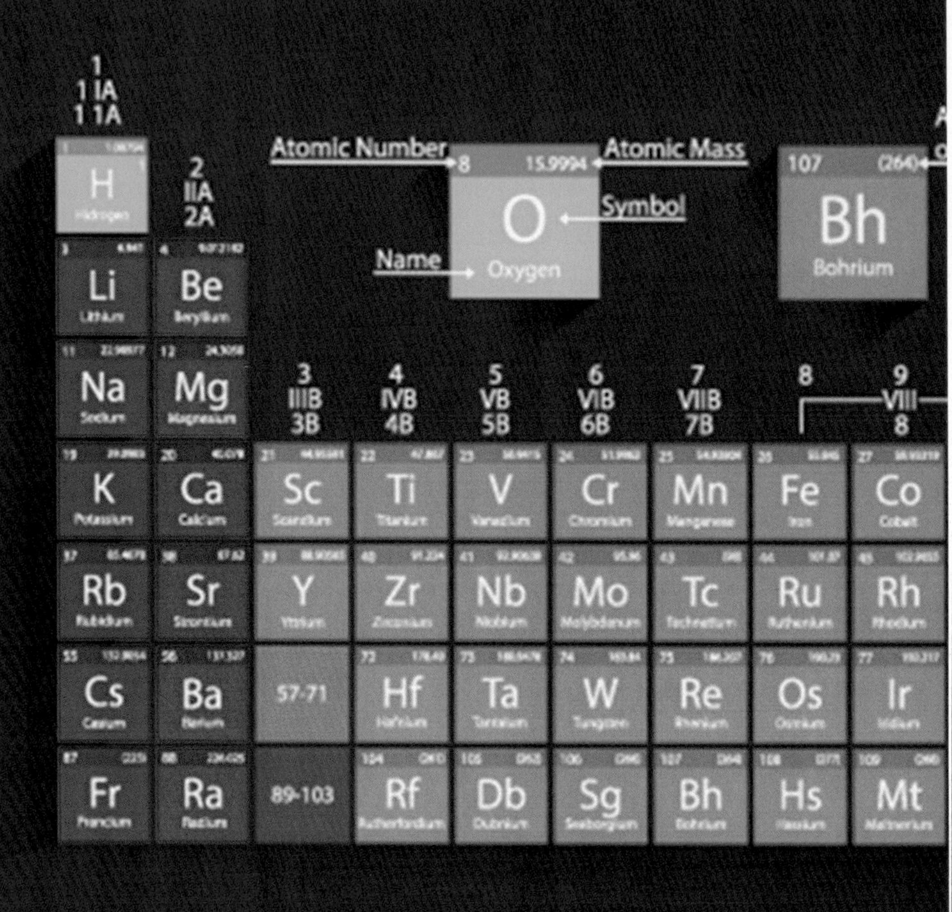

Alkali Metal | Alkaline Earth | Transition Metal | Basic Metal | Semimetals

Plate 9694

le Of Elements

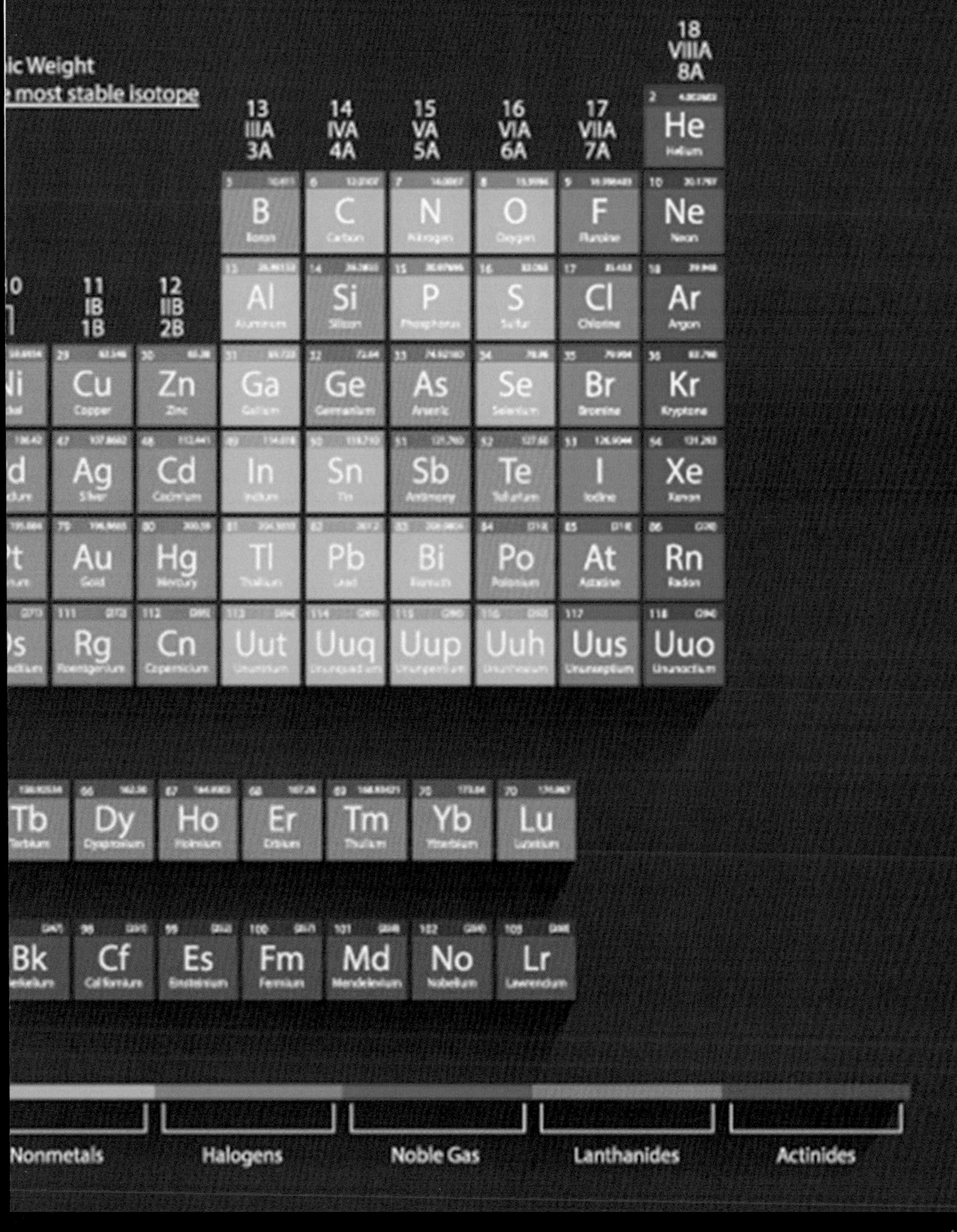

Plate 1149

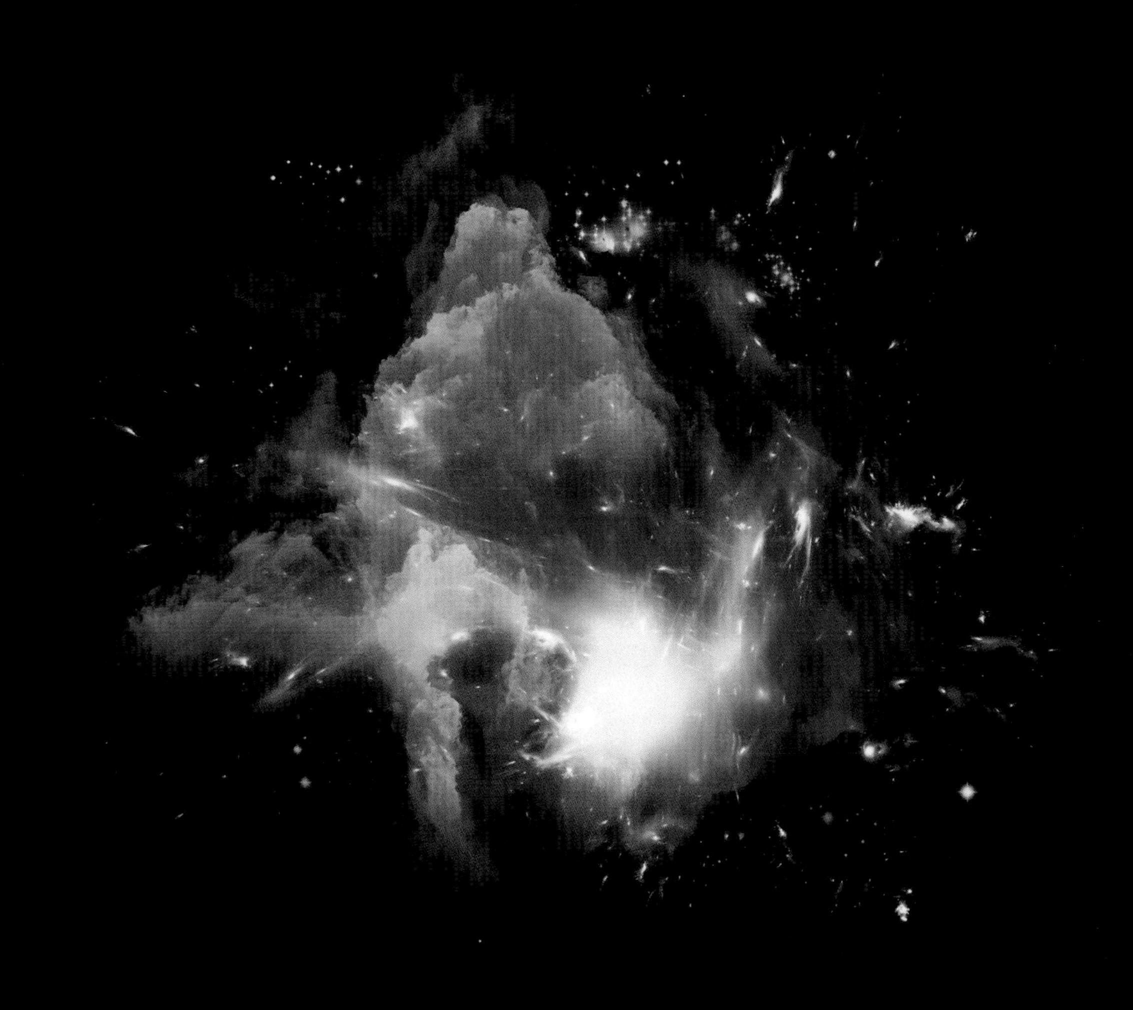

Plate 5506

atomic radii [pm]

H 31																	He 28
Li 128	Be 96											B 84	C 76	N 71	O 66	F 57	Ne 58
Na 166	Mg 141											Al 121	Si 111	P 107	S 105	Cl 102	Ar 106
K 203	Ca 176	Sc 170	Ti 160	V 153	Cr 139	Mn 139	Fe 132	Co 126	Ni 124	Cu 132	Zn 122	Ga 122	Ge 120	As 119	Se 120	Br 120	Kr 116
Rb 220	Sr 195	Y 190	Zr 175	Nb 164	Mo 154	Tc 147	Ru 146	Rh 142	Pd 139	Ag 145	Cd 144	In 142	Sn 139	Sb 139	Te 138	J 139	Xe 140
Cs 244	Ba 215	La 207	Hf 175	Ta 170	W 162	Re 151	Os 144	Ir 141	Pt 136	Au 136	Hg 132	Tl 145	Pb 146	Bi 148	Po 140	At 150	Rn 150
Fr 260	Ra 221	Ac 215															

Ce 204	Pr 203	Nd 201	Pm 199	Sm 198	Eu 198	Gd 196	Tb 194	Dy 192	Ho 192	Er 189	Tm 190	Yb 187	Lu 197
Th 206	Pa 200	U 196	Np 190	Pu 187	Am 180	Cm 160							

ABOUT THE AUTHOR

Angelo Pettolino has been a researcher for 37 years with a long-standing interest in astronomy . He studied cosmology at Adler Planetarium in Chicago and is a passionate, ambitious, modest collector of solar system formation evidence and information found in this book. Through his persistence and tenacity he has found the logical answers to some of the many unanswered questions for the formation of our solar system. This cutting edge, one of a kind knowledgeable book "21st. CENTURY ASTRONOMY" describes up to the moment discoveries and presents them in an unbroken chronological chain of collected information and provable events. The results of this collected evidence has "filled in the blanks" with a more acceptable and demonstrable explanation for the formation of water and our solar system.

As with all new ideas, the author is no stranger to unfounded criticism. Through it all, the AP theory has stood the test of time from its first printing in 1974. By presenting logical evidence and proving most of its statements, the AP theory has stood up to its detractors with vested interests and at the same time encouraged constructive criticism. For the past 3 years a $25,000 reward is being offered to anyone who can disprove The AP Theory. As of this writing there have been no successful explanations.

Plate 8806

ABOUT THE BOOK

First Revision of: *The Formation of Water and Our Solar System from a Fission Process with an Improved Heliocentric Model* . Xlibris book #500581 – Nonfiction and informative this text book is suitable for teaching all college levels of Astronomy.

This first Revision is an expounded version, with more graphics and is titled: "21st. CENTURY ASTRONOMY". The AP Theory: ISBN 9781456869359 Xlibris classroom text book # 698058 is a definitive, controversial, insightful, nonfiction book embraces The AP Theory and expounds and answers the unanswered questions asked about The AP Theory Xlibris Book #500581, further helping to clarify and prove it. This cutting edge revision includes 55 new images and features up to the minute astronomical facts and latest discoveries with original and copywriter art work by the author and never seen before color plates to better help to explain and clarify the author's vision. This book is the only one to point out the errors of the past and logically explain the new discoveries.

Since the introduction of indisputable proof by the scientific world confirming an expanding universe/solar system, many inconsistencies in the Accretion Theory, which are all now discredited, have come to light. Since its demise some Astronomers have been feverishly trying to explain the formation of our solar system and water with theories like Special Relativity, General Relativity, String, Steady State, Nebula and Super symmetry Hypothesis and The gravity formation Accretion Theory all without success.

This accurate yet controversial book titled "21st. CENTURY ASTRONOMY" completes the puzzle, answers every question and ticks every box of doubtful questions with provable and logical explanations and experiments to prove the AP Theories' point. The "21st. Century Astronomy" text book tries to explain what's right not who's right and should be the benchmark in learning for all classrooms and can be used exclusively in college class rooms from first year through to graduation.

A re-imagined theory to the now-disproved Accretion theory is central to what is called Twenty-First-Century Astronomy—or the AP Theory by: A. Pettolino. This 'bold truth' book based on the past 100 years of proven facts and the latest, up-to-the-minute discoveries taking us one step closer to the logical conclusions. The book attempts to answer the unanswered questions and dispel previous misinformation and misconceptions in a chronological manner.

This contradictory, insightful new book offers information and gives a logical explanation for the formation of our solar system and water, which has been a mystery up until now. The AP theory also unlocks the riddle of how our solar system formed only 4.8 billion years ago. The theory chronologically describes the unbroken chain of events explaining how fusion and fission reacted—within a 100,000-cubic-mile area in our infant, partially frozen (Sun) cloud—produced and provided all the materials to form water and our entire solar system.

This comprehensive description completes the puzzle of how it all began with hot air coming into contact with cold air, causing wind which started inertia which started the process of formation. Our solar system formed from exploded star nebula and frozen gases as does everything in the cosmos.

Index

Symbols

A

B

C

D

E

Printed in the United States
By Bookmasters